PROGRESSIVE ENLIGHTENMENT

TRANSFORMATIONS: STUDIES IN THE HISTORY OF SCIENCE AND TECHNOLOGY

JED Z. BUCHWALD, GENERAL EDITOR

Myles W. Jackson, *Spectrum of Belief: Joseph von Fraunhofer and the Craft of Precision Optics*

Paul R. Josephson, *Lenin's Laureate: Zhores Alferov's Life in Communist Science*

Mi Gyung Kim, *Affinity, That Elusive Dream: A Genealogy of the Chemical Revolution*

Ursula Klein and Wolfgang Lefèvre, *Materials in Eighteenth-Century Science: A Historical Ontology*

John Krige, *American Hegemony and the Postwar Reconstruction of Science in Europe*

Janis Langins, *Conserving the Enlightenment: French Military Engineering from Vauban to the Revolution*

Wolfgang Lefèvre, editor, *Picturing Machines 1400–1700*

Staffan Müller-Wille and Hans-Jörg Rheinberger, editors, *Heredity Produced: At the Crossroads of Biology, Politics, and Culture, 1500–1870*

William R. Newman and Anthony Grafton, editors, *Secrets of Nature: Astrology and Alchemy in Early Modern Europe*

Gianna Pomata and Nancy G. Siraisi, editors, *Historia: Empiricism and Erudition in Early Modern Europe*

Alan J. Rocke, *Nationalizing Science: Adolphe Wurtz and the Battle for French Chemistry*

George Saliba, *Islamic Science and the Making of the European Renaissance*

Suman Seth, *Crafting the Quantum: Arnold Sommerfeld and the Practice of Theory, 1890–1926.*

Leslie Tomory, *Progressive Enlightenment: The Origins of the Gaslight Industry, 1780–1820*

Nicolás Wey Gómez, *The Tropics of Empire: Why Columbus Sailed South to the Indies*

PROGRESSIVE ENLIGHTENMENT

THE ORIGINS OF THE GASLIGHT INDUSTRY, 1780–1820

LESLIE TOMORY

THE MIT PRESS

CAMBRIDGE, MASSACHUSETTS

LONDON, ENGLAND

© 2012 Massachusetts Institute of Technology

MIT Press books may be purchased at special quantity discounts for business or sales promotional use. For information, please email special_sales@mitpress.mit.edu or write to Special Sales Department, The MIT Press, 55 Hayward Street, Cambridge, MA 02142.

This book was set in Engravers Gothic and Bembo by Toppan Best-set Premedia Limited. Printed and bound in the United States of America.

Library of Congress Cataloging-in-Publication Data

Tomory, Leslie, 1974–
Progressive enlightenment : the origins of the gaslight industry, 1780–1820 / Leslie Tomory.
    p.   cm. — (Transformations: studies in the history of science and technology)
Includes bibliographical references and index.
ISBN 978-0-262-01675-9 (hardcover : alk. paper)
1. Gas-lighting—Great Britain—History. 2. Gas light fixtures industry—Great Britain—History. 3. James Watt and Company (Birmingham, England) 4. Distillation—Refeach—Europe—History 5. Industrial revolution—Europe. I. Title.
TP733.G6.T66   2012
338.4'79621324094109033—dc23
                                                                            2011026389

10   9   8   7   6   5   4   3   2   1

to my family

CONTENTS

PREFACE IX

ABBREVIATIONS XI

INTRODUCTION I

I THE ROOTS OF GAS LIGHTING

1 GAS LIGHTING AND PNEUMATIC CHEMISTRY 13

2 INDUSTRIAL DISTILLATION 37

II A QUESTION OF SCALE

3 BOULTON & WATT 67

III BUILDING A NETWORK

4 FREDERICK WINSOR AND THE NATIONAL LIGHT AND HEAT
COMPANY 121

5 THE GAS LIGHT AND COKE COMPANY 169

CONCLUSION 239

NOTES 245
BIBLIOGRAPHY 319
INDEX 341

PREFACE

Many years ago, I read Jules Verne's *Around the World in Eighty Days*. One episode in the story tweaked my curiosity. Some time after Phileas Fogg and Jean Passepartout leave Fogg's London home on their journey, Passepartout recalls that he has left a gas light burning in his rooms. He states to inspector Fix that the lamp is burning at his expense, and that he will have to pay the entire bill when he and Fogg return from their round-the-world journey. Once back in London, Fogg shares with Passepartout some of the money he has won from his bet with his colleagues at the Reform Club, but deducts the amount owed for the gas bill. I wondered: What was this nineteenth-century gas light? How could it be that the lamp could remain lit unattended for eighty days? How could it be metered, like electricity? What was the company that sold the gas? What was the source of the gas? I never answered those questions at the time, but when I took a seminar course in the history of chemistry at the University of Toronto, I decided to address those questions. This book is the result.

I have accumulated many debts in preparing the book. I would like to thank the people who read the manuscript or parts of it and provided me with valuable comments: Marco Beretta, Victor Boantza, Bert Hall, Chris Hamlin, Janis Langins, Trevor Levere, Bill Luckin, and Ian West. In addition, many referees have contributed important insights and suggestions that have greatly improved the book and the articles it was based on. I have also received support from the Institute for the History and Philosophy of Science and Technology and the library at the University of Toronto. I would like to thank Marguerite Avery, Jed Buchwald, and Trevor Pinch for helping this book come out with the MIT Press. I'd also like to thank Paul Bethge, Katie Persons, and the others at the press for their tremendous work. Peter Morris, Marco Beretta, Jim Voelkel, Ian West, and David Hale provided help with obtaining images. The reproduction of images has been supported by a grant from the Society for the History of Alchemy and Chemistry. Most images have been kindly supplied by the Birmingham Central Library, the Chemical Heritage Foundation, and the Science Museum in London.

Portions of this book have appeared in modified form in "The origins of gaslight technology in eighteenth-century pneumatic chemistry" in *Annals of Science*, in "Building the first gas network, 1812–1820" in *Technology and Culture*, in "Fostering a new industry in the Industrial Revolution: Boulton & Watt and gaslight 1800–1812" in the *British Journal for the History of Science*, and in "Gaslight, distillation, and the Industrial Revolution" in *History of Science*.

ABBREVIATIONS

BWA   Boulton & Watt Archives

cwt   hundredweight (100 or 112 pounds)

DNB   *Dictionary of National Biography*

GAL   George Augustus Lee

GLCC   Gas Light and Coke Company

GW   Gregory Watt

JWjr   James Watt Jr.

JWP   James Watt Papers

LMA   London Metropolitan Archives

MCD   Minutes of the Court of Directors

MCW   Minutes of the Committee of Works

MII   Muirhead II collection in Boulton & Watt archives

MIV   Muirhead IV collection in Boulton & Watt archives

MS   manuscript

MRB   Matthew Robinson Boulton

Some of the most prominent features of the skyline of nineteenth-century London, including the dome of St. Paul's Cathedral, the Parliament buildings, and Tower Bridge, continue to be visually notable in that city. Some of the largest edifices of nineteenth-century London, however, are no longer present, and survive only as a few skeletal frames. Those edifices never had any aesthetic value, and they long ago passed into desuetude, later to be torn down, cut up, and sold as scrap. These landmarks were the gasholders of the manufactured gas industry, built to store a day's supply of gas for the light-hungry metropolis. By the end of the nineteenth century, some of these iron containers were bigger than St. Paul's Cathedral. These huge gasholders were the most visible components of a technological system whose physical infrastructure included the gasworks where tons of coal were gasified daily and a network of mains originating there, with ramifications spreading out under city streets and feeding gas lamps in streets, home, shops, parks, and public buildings. The physical size and visual presence of the gas infrastructure were but one way in which this technology was integrated into urban life in the nineteenth century. Gas was part of the reshaping of urban infrastructure in the nineteenth century that created the modern city through the construction of water, sewage, rail, road, electricity, telegraph, and telephone infrastructures. Between 1812 and 1820, London's gas infrastructure grew from nothing to a point where gas mains reached most areas of the city.

As important as gas lighting was in the nineteenth century, the origins of the industry are still poorly understood, having received little attention from historians of technology since the 1960s.[1] The industry, however, occupies an important position in the history of technology at the turn of the nineteenth century. Gas was among the first of a new wave of technologies of the Industrial Revolution, a wave in which technologies and the industries based on them had some features that distinguished them from some of the classic technologies of the eighteenth century. The characteristic attributes of the first-wave technologies were that they were invented and deployed by individuals or small partnerships,[2] that they required relatively little capital, and that they were largely or exclusively dependent on artisanal skills for their invention and development, owing little to the content of contemporary science or natural philosophy in an immediate way. The classic textile machines of the Industrial Revolution, including James Hargreaves' spinning jenny (1765), Richard Arkwright's water frame (1769), and Samuel Crompton's mule (1779), all possessed these traits, as did Abraham Darby's invention of the coke smelting of iron (1709), Henry Cort's puddling-and-rolling techniques for iron production (1784), and Josiah Wedgwood's various pottery inventions. All these technologies were invented and developed by people who, although they may have relied on methodologies and mentalities learned from a scientific worldview, drew largely from their artisanal skills for their inventive work. The financial resources needed for the invention and deployment of these technologies were within the scope of what individuals or partnerships could muster. Though some textile and iron firms grew quite large in the late eighteenth century and the early nineteenth century, even these were owned by a small number of wealthy individuals. These characteristics are, of course, approximate; exceptions can be found. Thomas Newcomen's steam engine (1712), for example, was based on the understanding that the atmosphere had a weight, which had been developed in the seventeenth century by Evangelista Torricelli and Otto von Guericke. Likewise, the development of chlorine bleaching depended in part on Carl Wilhelm Scheele's discovery of chlorine in 1774, and in part on Claude-Louis Berthollet's subsequent investigation of its bleaching properties. The Leblanc process for soda production was based on a reaction discovered in a scientific context, although the nature of the reaction was not understood at the time. In all these cases, however, a great deal of innovation was needed to find practical applications for ideas, products, and processes discovered in a largely scientific context.

In contrast to these first-wave technologies, the nineteenth century saw the emergence of technologies and industries that were more immediately dependent on contemporary science and formal organized research, required more capital, and therefore were usually built and run by larger companies or institutions. These new technologies signaled the beginning of a new phase in the industrialization of the West. The railways are a salient example of this, especially in the latter two points. Enormously expensive to build, they were far beyond the reach of individuals and partnerships, and thus they were built by large companies. Some industries of the later nineteenth century, such as the chemical industry and the electrical industry, showed these characteristics even more clearly. The development of much of the electric network, for example, depended on scientific research into the nature of electricity and magnetism, was deployed by large companies, and required significant capital investment.

I argue in this book that gas lighting was one of the first technologies to possess all the characteristic attributes of the second wave. Its invention owed much to contemporary scientific research, not only in terms of the discovery and description of gases, but also in regard to the laboratory instruments of pneumatic chemists. If chlorine bleaching and the Leblanc process owed their existence respectively to a single chemical and a process discovered in a scientific context, the contribution of science to the development of gas lighting was far more extensive, more akin to nineteenth-century industries such as electricity, telegraphy, and chemical dyes. The first inventors of gas lighting used knowledge, techniques, and instruments that were in most cases created in the chemical laboratories of men of science. In addition, like the railways, urban gas networks required substantial capital to deploy. In contrast to other technologies developed in the eighteenth century, and in common with some other important infrastructures constructed during the Industrial Revolution, including canals and ports, gas networks had capital requirements that were beyond the capacities of individuals or small firms, and assumed the legal form of a limited-liability joint-stock corporation that facilitated the pooling of capital. Gas, then, represented a merger of two important contemporary trends: the construction of large infrastructures and the development of new technologies. Beginning in the 1830s, the railways would repeat this merger, but gas led the way by twenty years.

The nexus between science and technology (or, in more contemporary language, natural philosophy and the arts) operative in the creation of the gas industry points to another major theme of this book: the role of the Enlightenment in the Industrial Revolution, recently most extensively explored by Joel Mokyr.[3] One reason why gas lighting was the first of the second-wave technologies is that it was the earliest case in which the Enlightenment dream of science at the service of industry actually worked in an important way.[4] Historians have described how the Enlightenment adopted the Baconian ideal of natural philosophy serving society. Mokyr, in particular, has claimed that the reason the Industrial Revolution did not die out, as many other bursts of technological creativity did, is that this ideal eventually created a wide and accessible base of useful knowledge that fed technological dynamism in Europe in the eighteenth century and the early nineteenth century.[5] Despite the earnest hopes of Enlightenment intellectuals, however, the knowledge that proved useful in this period was not the content of formal natural philosophy. As Mokyr observes, "before 1850, the contribution of formal science to technology remained modest."[6] Rather, innovators were becoming ever more adept at creating practical applications from the growing, sometimes codified but usually oral mass of useful knowledge generated during the eighteenth century.[7] The dream of science contributing to industry was largely realized later. But gas lighting was a major first step in the fulfillment of that dream. It was an entirely new technology and industry, derived largely from the content of formal contemporary science. Eighteenth-century men of science exploring pneumatic chemistry usually didn't have any specific applications in mind, but the inventors of gas lighting were able to make use of the knowledge they had created, and they built a new technology from it, using their own ingenuity and skills. This technology eventually served as the basis for an entirely new industry. Parts I and II of this book explore how this scientific knowledge was generated, how it was transmitted to the inventors who made use of it, and how they eventually created industrial forms of gaslight machinery, thereby showing continuities between science and the arts in the eighteenth century.

The history of gas lighting also shows the wide and deep roots of the technological dynamism of the Industrial Revolution. The long-term movements that contributed to the ultimate emergence of this technology originated in the seventeenth century, not only in the pneumatic investigations of chemists but also in broad economic forces. The dating and the causes of what has been called the "great divergence" between the West (Britain in particular) and the rest of the world in technological and economic development have been disputed. Kenneth Pomeranz and others argue for an eighteenth-century source springing primarily from Britain's coal reserves and its colonial empire.[8] Arguing against this recent eighteenth-century dating of the divergence are Joel Mokyr, Margaret Jacob, Christine MacLeod, and Robert Allen, all of whom date the roots of the West's technological dynamism as far back as the Scientific Revolution, and in some cases earlier. This study of the origins of gas lighting shows how this new technology was riding a wave of ideas and culture dating back to the seventeenth century in much of Europe.

Where the case of gas lighting differs somewhat from Mokyr's Industrial Enlightenment is in the question of why Britain occupied a special place in Europe's technological and industrial history. Mokyr, who argues that the Industrial Enlightenment was a pan-European phenomenon, with knowledge and macro-inventions coming from France, Germany, and other countries, builds on the work of Jacob and Stewart to argue that Britain in particular had an "enlightened economy," as it most effectively implemented, albeit imperfectly and in a piecemeal way, many aspects of the Enlightenment program.[9] It was in Britain that anti-mercantilist and free-market ideas were pushed farther and faster, and that science informed industrial culture most profoundly. However, numerous critics have rejected arguments that a culture oriented toward natural and experimental philosophy was a cause of the Industrial Revolution.[10] These sorts of opinions concord with earlier descriptions of the Industrial Revolution—most conspicuously with that of David Landes, who had no place for natural philosophy in his monumental work *The Unbound Prometheus*.[11] There are also historians who, while accepting the Industrial Enlightenment for Europe as a whole, reject it as an explanation for Britain's lead. Robert Allen has recently explained Britain's anomalous position by arguing that its high-wage-and-cheap-coal economy meant that technology paid off there, whereas in regions where wages were lower the economic incentive simply wasn't strong enough.[12] Because gas lighting was invented and tried almost simultaneously in various forms in a number of European countries, it makes for an excellent comparative case study in industrial development during this period. I argue that, as J. R. Harris and others have claimed, British mechanical skills were indeed important for the establishment of gas lighting there, as the role of the firm of Boulton & Watt in this story demonstrates.[13] In addition, other factors, most importantly Britain's coal economy, were decisive in helping the industry develop there. Indeed, I argue that, far from the Continental versions of gas lighting being stillborn, technological innovation continued on the Continent, but produced a different sort of distillation industry—one oriented to wood—while Britain was developing gas lighting.

The final major theme of this book is gas as a network technology. I argue that gas lighting was path-breaking in this regard as well. The construction of large integrated networks has been understood as an important feature of nineteenth-century technology. The railways, in particular, have been seen by many historians, most notably Alfred Chandler in *The Visible Hand*, as a turning point because they represented a new kind of network. They were integrated and technically demanding, requiring the implementation of more structured management and more careful means of coordination to prevent disasters and to allow efficient operation. These sorts of technically demanding networks are tightly coupled—that is, changes in one part of the network can influence other parts of it quickly and in important ways. Electricity is a good example of this. The first builders of the electrical network had to develop a host of technical measures and business practices to stabilize their network, and designed parts of it by considering carefully how it would affect the rest of the network. In contrast to these later networks, the infrastructure networks of the eighteenth century and the early nineteenth century, such as roads, canals, water supply, and sewers, were simply not large, integrated, and tightly coupled enough to require the careful coordination that the railways and electricity did. I argue here, however, that the gas infrastructure was in fact an integrated and tightly coupled network, like the railways that followed it. It was far from easy to manage and operate, and the first gas company spent its first years developing ways to stabilize it. These included various technical measures, business measures (e.g., new management structures), and social measures (including means of controlling its users). I argue that the gas network was an important step up in the complexity of large infrastructure networks. I also explore the role of users in shaping this network as their expectations of how the technology should be used sometimes clashed with the expectations of the Gas Light and Coke Company, leading that company to introduce new methods of stabilization and control. The network strategy for gas lighting was not, however, inevitable. On the contrary, the emergence of the network form for gas lighting was the result of deliberate strategic choices, political negotiations, and pragmatic accommodations.

Part I of the book describes the contribution of pneumatic chemistry (chapter 1) and industrial distillation (chapter 2) to the eventual development of gas lighting. It also explores the bifurcation between the Continental and British traditions in distillation technology after 1800. Part II (chapter 3) explores how Boulton & Watt established the industry by building gas plants large enough to light mills, and the difficulties associated with that transformation. Part III describes the deployment of the network strategy, beginning with Frederick Winsor's drive to incorporate the Gas Light and Coke Company (chapter 4). Chapter 5 deals with the implementation of the strategy between 1812 and 1820.

In November of 1801, Gregory Watt, the youngest son of James Watt, was spending some time in Paris. He had met a number of his father's contacts there, and one of them took him to see an interesting and novel public display at the Hôtel de Seignelay. As Watt approached the building, he was greeted by the sight of bright flames streaming from Argand lamps mounted on the facade. Unlike the candles and oil lamps normally used for lighting, these lamps were fueled by inflammable gas brought by pipes from inside the building. Entering the building, Watt saw that the gas was generated from wood heated by a fire in an enclosed oven. The heat caused the wood to give off gas, which, after being purified by passing through water in a vessel located near the stove, flowed through pipes to the lamps. Lamps and chandeliers were mounted throughout the building, and as Watt passed through the building to the rear he saw more lights in the gardens. Philippe Lebon, the French civil engineer who had invented, patented, and built the apparatus, spoke enthusiastically to his interested visitors about the possibilities of this new form of lighting, which, he claimed, could replace candles, oil lamps, and perhaps stoves. Lebon was seeking investors willing to fund him so that he could design and manufacture a practical version of his "thermolamp," as he called the device.

Gregory Watt was quite interested in what Lebon had to say, and not simply because of the novelty of the Frenchman's work. In fact, Watt was most surprised by the similarity of Lebon's ideas to those of William Murdoch, an employee of Boulton & Watt, the Birmingham steam engine manufacturing firm that Gregory Watt's father had co-founded. Several years earlier, Murdoch had built a device for generating inflammable gases from coal for the purposes of illumination. Though Murdoch had spent some time working on it since then, the project had fallen dormant. The similarity between the two engineers' ideas was so remarkable that when Gregory next wrote to his brother James Watt Jr. he asked him to "tell Murdock that a man here has not merely made a lamp with the gaz procured by heat from wood or coal but that he has lighted up his house and gardens with it and has it in contemplation to light up Paris."[1] When this letter reached Birmingham, it set in motion experimental work that would, in time, lead to the first sales of industrial-scale gas plants, and would eventually launch a whole new industry.

How did it happen that these two engineers, unknown to each other, one working in Britain and one in France, simultaneously came up with similar ideas and processes for using inflammable gases for lighting purposes? It wasn't purely coincidental. In fact, not only did Murdoch and Lebon have much in common in their backgrounds and interests that contributed to their work on gas lighting; those interests also were shared, to varying degrees, by many other people throughout Europe who, like Murdoch and Lebon, were working with and investigating processes and apparatus that produced inflammable gases. The work of these people drew from two distinct industrial and natural philosophical traditions. The first, and the one from which gas lighting emerged, was the destructive distillation of wood, turf, and coal as an industrial process to procure various derivative products, such as coke, charcoal, and tar. (Here 'distillation' refers to a process of heating materials in enclosed ovens to break them down into other substances.) The second tradition was the chemistry of air. Both of these traditions had been gaining in intensity and breadth in the second half of the eighteenth century, and as they grew in strength they created a ferment within which gas lighting emerged on a number of occasions with varying degrees of independence. The ultimately successful deployment of gas lighting in Britain, integrating as it did Murdoch's and Lebon's work, drew from both traditions. The gas industry was, in effect, born when inventors working in the industrial distillation tradition used knowledge and instruments coming from pneumatic chemistry to explore the possible uses of one of the derivative products of distillation: gases.

The investigation of the destructive distillation of wood and coal as an industrial process was the immediate context in which gas lighting as a commercial technology was created. Murdoch, Lebon, and other gaslight pioneers, including George Dixon and Jean-Baptiste Lanoix, were actively interested in the process of heating various substances in closed furnaces before they began to focus more narrowly on the lighting possibilities of inflammable gases. This distillation tradition contributed two important elements to the development of the gas industry. The first was commercial culture: all the people who came to gas lighting through industrial distillation had a commercial motivation or an entrepreneurial mindset, which they carried through to their work with gas lighting. Second, this commercial culture gave them a sensitivity to the importance of scaling up the process so it could be used in industrial contexts. This meant that they tried to transform their original workshop experiments to larger and industrially viable models, leading those who lacked the financial means to seek investors to help them make this transition. Though there were many people—apothecaries, chemists, instrument makers, public lecturers—who used inflammable gases for lighting in various ways at the same time as the industrial distillers did, in general they didn't have the business orientation or the desire for scale that the industrial distillers had.

Pneumatic chemistry, although it was not the tradition in which the gaslight industry directly emerged, provided much, in the way of concepts, materials, experience, and laboratory apparatus, that industrial distillers and later gas engineers used to establish gas lighting. Specifically, pneumatic chemistry made two crucial contributions to gas lighting. First, as eighteenth-century chemists worked with and explored the nature of gases, they developed and refined instruments for producing, manipulating, separating, purifying, isolating, and storing gases—instruments that served as the prototypes for components of later gasworks. Second, pneumatic chemistry led to a greater understanding of the properties of gases in general, and of inflammable gases in particular. This understanding proved useful for choosing the materials and processes that would be used for gas lighting, and eventually for producing a gas mixture with higher lighting potential, by eliminating impurities and favoring heavy over light inflammable air (in modern terms, ethylene and methane over hydrogen).

That gas lighting as a commercial technology drew from both of these traditions is effectively demonstrated by exploring the various versions of lighting with gases that came largely from the natural philosophical tradition. The most important of these versions were Alessandro Volta's inflammable air lighter (from 1776), Charles Diller's philosophical fireworks (from 1783), and Jan-Pieter Minckelers' gas lighting in Louvain (1784). Although all three used inflammable gases for lighting, none was transformed into a commercial technology; the reason for this is that the commercial impetus found among the industrial distillers was not present.

Understanding gas lighting as a refocusing of the distillation tradition through knowledge and techniques borrowed from pneumatic chemistry also explains the timing of the invention of gas lighting. Most internalist histories of gas lighting depict its invention as stemming essentially from the observation of the inflammability of coal gas or wood gas.[2] These histories give a long list of precursors to Lebon and Murdoch, going back to the seventeenth century, who made this inflammability observation; however, they don't make clear in what way they differed from Murdoch and Lebon. In effect, they reduce the invention of gas lighting to a "eureka" story of heroic invention in which the canonical inventors, while distilling wood or coal in a pot of some sort, notice the inflammability of the gas being produced, suddenly think of gas lighting, then rush off and develop the technology to implement their brilliant new idea. This history, however, raises the question of why gas lighting didn't develop earlier. If the observation of inflammability is at the root of gas lighting, why didn't gas lighting become an industry in the seventeenth century? A further question is why so many people tried lighting with gases in various ways, beginning in the 1770s. Because the inflammability-observation story can offer no answer to this, some historians have sought various causes to explain the timing of the invention. Morris Berman, for example, argued that it was "only a converging set of social needs that could render [gas lighting] popular or deem it necessary; and from 1809, owing to such needs, the gas industry emerged in England in full force."[3] The needs Berman identified as driving the development of gas lighting were the upper and middle classes' desire for security and social order when faced with seemingly rampant crime. Similarly, Archibald and Nan Clow suggested that gas lighting failed to emerge in the seventeenth century because work hours weren't yet long enough to require it, because the iron industry "would have found it still very difficult to supply the necessary equipment," and because there was prejudice against coal gas.[4] These explanations trivialize the technology, reducing it almost to an afterthought.

The two-traditions argument offers a different answer to the question of timing. Specifically, in the last quarter of the eighteenth century the historical evolution of pneumatic chemistry and industrial distillation reached a point at which they created conditions in which inventors and engineers were more likely to think of gas lighting and better able to pursue its development. By that point, there was more knowledge of gases and their properties, and many pneumatic instruments were available. Not only did the instruments exist; they were widely available throughout Europe, so that the engineers who eventually worked on gas lighting could obtain them easily.[5] Furthermore, economic conditions were creating interest in industrial distillation: the deforestation of Continental Europe had caused a run-up in wood prices, and the loss of Britain's American colonies had created a need for alternate sources of tar for maritime uses. Distillation was seen as a possible means to address these problems. The increasing strength of these two traditions in the late eighteenth century created propitious conditions for the invention of gas lighting.

The public nature of Enlightenment natural and experimental philosophy in the late eighteenth century was a crucial factor in the intersection of the two traditions, allowing knowledge to flow from pneumatic chemistry to the distillation industry. Awareness of the work of the few elite chemists who figure in the story of pneumatic chemistry (including Henry Cavendish, Alessandro Volta, and Antoine-Laurent Lavoisier) was not limited to a small select group of their peers. Rather, knowledge of their results spread far geographically and socially. The first chapter of this book is replete with the names of famous and elite chemists of the eighteenth century who worked in pneumatic chemistry; the second chapter is full of the names of little-known engineers and inventors who made use of this knowledge. The Enlightenment mechanisms that aimed to organize and spread knowledge, such as books, encyclopedias, journals, and public lectures and demonstrations, were successful in this case.[6] Indeed, cases of simultaneous invention, such as with gas lighting, are indicative of the easy flow of knowledge.[7]

The chapters in part I examine the two traditions that contributed to the various versions of gas lighting in the late eighteenth century. Chapter 1 briefly describes the nineteenth-century gas plant, then shows to what extent it was based on apparatus and concepts developed in eighteenth-century pneumatic chemistry. It goes on to explore how common the idea of lighting with inflammable gases had become by the 1790s. A discussion of forms of lighting with gases that drew mostly from pneumatic chemistry, and of how these also diffused knowledge, follows. Chapter 2 looks at why industrial distillation was becoming more important in the last quarter of the eighteenth century, and how this led to many cases of people investigating the process as a way to implement some form of gas lighting. The most important of these investigators was Philippe Lebon. Chapter 2 concludes by showing that Lebon's work, far from ending in failure, laid the foundations for an industry that produced charcoal, methanol, acetic acid, and other chemicals from wood.

1 GAS LIGHTING AND PNEUMATIC CHEMISTRY

1.1 INTRODUCTION

The degree to which the new technologies of the late eighteenth century and the early nineteenth century depended on contemporary natural and experimental philosophy has been much debated. Though it is generally accepted that technological development was at the very least helped by the use of methodologies learned from science, such as mathematics and systematic experimental investigation, there has been no consensus beyond these confines.[8] Historians have generally resisted the idea that entirely new industries emerged at that time—industries based on technologies derived from contemporary science in ways that became more prevalent in the second half of the nineteenth century, when the chemical and electrical industries, among others, drew heavily from knowledge produced in a scientific context. Such patterns, in which technologies can be shown to have been based on the content of natural philosophy, have proved elusive for the Industrial Revolution, with some exceptions (for example, the steam engine's link to the discovery of the weight of the atmosphere in the seventeenth century and chlorine bleaching's basis in the discovery of chlorine).[9]

Recent explorations of the subject, without entirely abandoning the claim that contemporary technologies were to some extent based on the content of natural philosophy, have located the dependence in a cultural and institutional context. It was the culture motivated by the spread of the Baconian and later Enlightenment ideal of natural and experimental philosophy being at the service of society that helped drive development in the arts. Enlightenment artifacts such as the *Encyclopédie* and the expansion of public science spread and strengthened a scientific mindset and scientific ideas, and helped to organize knowledge in such a way that useful information was easier to obtain than it had been.[10] In the eighteenth century, natural philosophy tended to be an open affair aimed at broad audiences. Learned academies and philosophical societies often had amateurs and industrialists as members, and there were many public lectures and demonstrations with scientific themes. This open science had various goals, including the education of pharmacists, the provision of entertainment, the promotion of the utility of science, and even moral improvement.[11] All this created the ferment within which the new technologies of the eighteenth century and the early nineteenth century were born.

Chemistry has occasionally been identified as presenting particular characteristics in the discussion on the links between science and technology because of the inherently productive nature of chemical operations. In addition, the practitioners of philosophical chemistry and their counterparts in practical chemistry were close because they often used the same instruments in their work, as well as following career paths that could overlap. The practical utility of philosophical chemistry, in other words, was less dependent on theoretical work than was the case with physics, for example, because the materials and techniques used in the context of these investigations were more likely to have some applications. Though the use and the interpretation of materials and instruments were not theory independent, in the context of eighteenth-century chemistry—with many new materials being discovered—the theoretical underpinnings mattered little if some newly discovered materials were found to have specific uses, such as chlorine for bleaching. In this vein, Ursula Klein has argued that, as the scope of chemistry expanded in the seventeenth and eighteenth centuries, and as more materials were produced with a growing range of instruments and techniques, some of these materials and the methods of making them proved to have practical value, particularly for pharmaceutical purposes.[12]

Gas lighting, more than any other technology of this period, depended on contemporary science in the many ways outlined above. People working in pneumatic chemistry did a great deal of work that proved useful in the foundation of the gas industry, not only by generating knowledge of materials—the gases themselves and the source materials—but also by designing the scientific instruments and techniques used to produce, purify, isolate, and store gases in the course of academic studies. The concepts, subject matter, experience, techniques, and laboratory apparatus of pneumatic chemistry provided the foundations for the technology of gas lighting.

In order for discoveries of new materials, techniques, and instruments to have any practical value, they had to be communicated to those who might use them. The Enlightenment ideal of open and public science, organized and presented in accessible ways, fostered such communication. In one of the first instances of a direct realization of the Enlightenment ideal, public science made knowledge of the pneumatic chemistry of gas lighting widely available and catalyzed the creation of a new technology and a new industry.

By looking at the development of pneumatic chemistry from the middle of the seventeenth century to the late eighteenth century, this chapter examines to what extent the technology of the gaslight industry was based on philosophical pneumatic chemistry. After describing an early-nineteenth-century gas plant (section 1.2), it explores how inflammable gases were investigated from 1650 to 1800 (section 1.3). Section 1.4 explores the development of pneumatic apparatus over the course of the eighteenth century. Section 1.5 describes some forms of lighting with gases that originated in pneumatic chemistry, and how some of these ideas were spread to a wide audience.

1.2   A GAS PLANT OF THE EARLY NINETEENTH CENTURY

In early-nineteenth-century gasworks (figure 1.1), gas was generated by distilling coal in retorts—enclosed ovens, usually made of cast iron, that were heated by furnaces. Coal heated in this way gives off coal gas, whose major constituents, in proportions determined by the temperature, are carburetted hydrogen (as methane and sometimes other hydrocarbon gases were commonly known as at this time) and hydrogen. Carburetted hydrogen burns with a brighter flame than hydrogen. By the 1820s, gas companies were trying to maximize the carburetted-hydrogen content of the coal gas by controlling the oven temperature, relying on the rule of thumb that the retorts should be cherry-red.[13] The minor constituents of coal gas are carbonic acid (carbon dioxide), carbonic oxide (carbon monoxide), olefiant gas (ethylene), and sulphur compounds. Ammoniacal liquor (ammonia compounds) and tars are also produced from the coal. All these other compounds were separated from the lighting gases by a three-stage purification process. The products were bubbled through water in a long cylindrical vessel, called the hydraulic main, where the tars and ammoniacal liquor condensed and remained in the water. Residual tars and other liquids were removed in a condenser, where the gases were passed through pipes immersed in cold water. The carbonic acid and sulphuretted hydrogen (hydrogen sulphide) were then removed by passing the remaining coal gas through lime water (aqueous calcium hydroxide), which reacts with those two gases. This step was essential, as sulphuretted hydrogen has a strong smell of rotten eggs and is highly toxic (even in small quantities). The purified gases were stored in a large container called a gasholder or gasometer (the latter being a misnomer, as it was not used for measurement, but reflecting the true

FIGURE 1.1

An early-nineteenth-century plant for the manufacture of gas. Left: gasometer and purifier. Right: gasification oven. Frederick Accum, *A Practical Treatise on Gas-light* (1816), frontispiece. Courtesy of Science Museum, London.

origin of the device as a scientific instrument).[14] From there, the gas was sent through pipes to burners throughout the city.

There were also other products of this process. The carbonized coal remaining in the retorts was coke (mostly elemental carbon), and was valued as a relatively clean fuel. The oils, tars, and ammonia compounds separated naturally according to their densities if they were allowed to sit and settle for some time. The purifiers produced lime water fouled with sulphur compounds.

I.3   PNEUMATIC CHEMISTRY, I650–I790

The roots of gas lighting go back to the seventeenth century. The years 1650–1750 witnessed increasing familiarity with inflammable vapors. Before 1650, inflammable exhalations coming from the earth were occasionally thought to be to the cause of comets, lightning, or the *ignis fatuus* sometimes observed in swamps.[15] From 1650 on, however, inflammable vapors were studied, and eventually artificial means of producing them were described in texts. In Britain especially, there was an interest in inflammable vapors from 1650 to 1700, particularly as coal mines went deeper and as inflammable and explosive exhalations in the mines became more familiar. Robert Boyle reported many cases of inflammable vapors being collected in coal mines or from springs.[16] In 1672 Boyle described how he generated inflammable "smoak" or "steam" from iron mixed with a "saline spirit" (hydrochloric acid).[17] After 1700, chemistry as a whole rose in social standing and acquired new theoretical bases and expanded ambitions, by, for example, being introduced into university medical curricula, being applied in geology and agriculture, and becoming a subject of popular lectures.[18] With this change, texts from many European regions increasingly included descriptions of natural inflammable exhalations and ways of producing them artificially, either by the dissolution of metals (typically iron, zinc, or copper) with vitriolic or muriatic acid (sulphuric or hydrochloric) or by the destructive distillation of coal, wood, and other animal and plant matter in closed retorts.[19]

After 1766, when Henry Cavendish published three papers in which he characterized inflammable air (hydrogen),[20] the first person to argue in print that there were different kinds of inflammable air was Alessandro Volta. Volta (1745–1827), born in Como, Italy, was raised in a devout Catholic family. Unlike many of his relatives, he was not attracted by an ecclesiastical career. He became interested in natural philosophy at the age of 14 after reading a book by Joseph Priestley on electricity. In 1772, at the age of 18, he wrote to Priestley about his interests and work in electricity, and Priestley responded amiably and encouragingly.[21] Volta pursued his interest in natural philosophy and became a teacher of physics at the local school in 1774. That same year, he improved the electrophorus, a device for storing electricity, and earned some renown. Volta continued his correspondence with Priestley, and in 1774 he read Priestley's published work on airs. It impressed him deeply and convinced him that a great future lay in store for pneumatic chemistry. He decided to pursue the subject in earnest.[22]

While conducting a series of experiments on inflammable air, Volta collected gas from rotting plant matter at the bottom of the stagnant marshes around Lake Como and from damp ditches. He described his discoveries and his subsequent analyses in a series of seven letters published in 1777.[23] Volta wasn't certain what his new marsh gas actually was, and his opinion evolved with time. He did not, however, think it was the same as coal gas.[24] News of his findings soon spread throughout Europe, and he was widely regarded as the discover of a new kind of inflammable air. Although it took a few years for the results to be firmly accepted, by 1790 French chemists had largely adopted the view that were several kinds of inflammable air, and in the next decade other chemists throughout Europe followed suit.[25]

Between 1765 and 1790 the knowledge of inflammable gases had progressed to a point where the basic properties of hydrogen and heavier inflammable gases were well known. This knowledge, when combined with various instruments and processes described in the next section, created the foundations for the new gas industry. Knowledge of other gases (especially carbon dioxide and hydrogen sulphide), and of how to separate them from inflammable gases, also proved important.

1.4   APPARATUS AND TECHNIQUES OF PNEUMATIC CHEMISTRY, 1700–1790

Pneumatic chemistry's primary contributions to gaslight technology were the instruments and processes used to produce, manipulate, purify, and store gases. A number of instruments and processes developed after 1650 were subsequently incorporated into gaslight technology. Among the instruments were the retort, Stephen Hales' pneumatic trough, and Lavoisier's gasometer. Among the processes were Joseph Black's description of fixed air's reaction with lime water and Carl Scheele's discovery of sulphuretted hydrogen's absorption by lime water.

The earliest item of chemical apparatus eventually incorporated into the gas plant was the retort. Retorts were closed furnaces used to heat materials while avoiding direct contact with flame (figure 1.2). Retorts in some form, such as closed crucibles, had long been in the alchemists' repertory of tools, but in the seventeenth and especially the eighteenth century, owing to greater interest in the

FIGURE 1.2

A seventeenth-century retort. Johann Glauber, *A description of new philosophical furnaces* (1651), 50. Courtesy of Roy G. Neville Historical Chemical Library, Chemical Heritage Foundation.

ultimate principles of matter, analytic chemistry and hence distillation became
increasingly important, and chemists designed more complex retorts for their analytic experiments.[26] At the same time, retorts were used extensively in pharmaceutical work. In the latter part of the eighteenth century, as chemists began to use
higher temperatures when performing analyses, some began to use iron gun barrels
as retorts, since they resisted higher temperatures better than the retorts that had
been used previously.[27] The gas industry's first retorts were made of iron.

Stephen Hales' pneumatic trough was a truly novel instrument, designed specifically
for experiments in pneumatic chemistry. Its purpose was to facilitate the isolation,
storage, and purification of gases. Before Hales, chemists and alchemists had long used
closed retorts to heat substances and generate vapors, which they collected in attached
receptacles. The vapors were then allowed to condense into fluids. Often the gases
produced would build up pressure in the apparatus and burst the luting, the receptacle,
or even the retort. Often small holes were cut into the apparatus to prevent this.
Many of those who generated hydrogen by dissolving metals managed to contain it
in bottles for a short time. Hales' trough changed that.

Hales, a pastor living near London, became interested in pneumatic chemistry
through his investigations into the mechanisms of plant growth in the 1720s. Having
discovered that plants gave off air, he researched what sorts of air were fixed in solid
bodies of all kinds, without, however, breaking from the tradition of common air as
the only kind of air. He was more interested in the volumes of gases produced. Hales
wanted to wash the air he generated of foreign matter that he thought was mixed in
with it and was somehow reducing the air's volume when it was enclosed in a container. At first, he used a device in which a storage receptacle was in direct communication with the retort in which the gases were generated. Water underneath the
gases served as a volume buffer. The water was forced out as the gas pressure above
it increased. Hales had found, to his dismay, that the volume of air over the water
decreased in the days after the conclusion of his experiments. This was because the
water absorbed some gas, but Hales thought it was because some "acid sulphureous
fumes" produced with the air "did resorb and fix the elastick particles." In order to
avoid this, he devised the pneumatic trough "so that as the air which was raised in
distillation, passed thro' the water up to the top of the receiver . . . a good part of
the acid spirit and sulphureous fumes were by this means intercepted and retained in
the water."[28] (See figure 1.3.)

The pneumatic trough would go on to become ubiquitous in the laboratories of
pneumatic chemists, especially in the latter half the eighteenth century. Some later

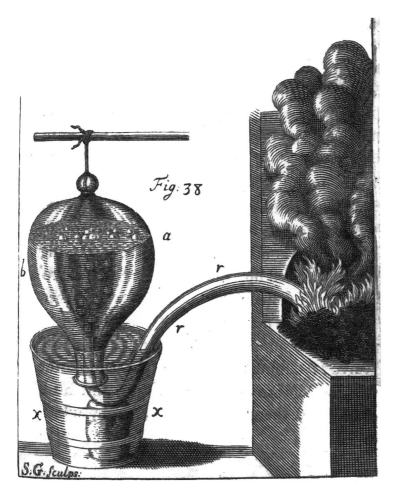

FIGURE 1.3

Stephen Hales' pneumatic trough. Hales, *Vegetable Staticks* (1727), 262. Courtesy of Roy G. Neville Historical Chemical Library, Chemical Heritage Foundation.

incarnations, beginning with those built by Henry Cavendish, used mercury instead of water, since gases are far less soluble in mercury. The trough proved useful for storing gases. Renamed the hydraulic main, it eventually found use in gas lighting apparatus as the first purifying step, removing oils and tars from the gas, and as a way isolating the retorts from the rest of the gas plants. The hydraulic main in a gasworks was used for the very purpose that its ancestor the pneumatic trough had been created to serve: isolating and washing gases.

The work of Joseph Black was crucial for later gas technology. While studying at the University of Glasgow, Black became interested in the medicinal use of alkaline substances such as lime (calcium carbonate) and "magnesia alba" (magnesium bicarbonate or magnesium carbonate) to cure stones and other ailments. In his experiments with these materials, Black produced "fixed air" (carbon dioxide), which, unlike Hales, he saw as a species of air distinct from common air. He observed that this air reacted with lime water by turning it a milky white.[29] This reaction became chemists' standard test for detecting the presence of fixed air in experimentation. The lime water reaction was used in the gaslight industry to purify coal gas of carbonic acid and to increase the luminosity of gas flames.

Lime water was also used to purify coal gas of sulphuretted hydrogen (hydrogen sulphide). Because of its rancid odor and its toxicity, sulphuretted hydrogen was a great obstacle to the use of gas lighting, especially indoors. Most coal contains some amount of sulphur, and sulphuretted hydrogen is produced during distillation in proportion to the sulphur concentration in the coal. Lime water reacts with it, and so the same means used to purify coal gas of carbonic acid served also to remove most of the sulphuretted hydrogen.

Sulphuretted hydrogen was produced as early as 1754 by Hilaire Martin Rouelle (1718–1779).[30] It was, however, Carl Wilhelm Scheele (1742–1786) who, in 1777, first described the properties of what he called "stinking" (stinkende) or sometimes "inflammable" (brennende) sulphureous air (Schwefelluft). Scheele thought it was a combination of phlogiston (the word he used when referring to hydrogen), sulphur, and heat. He produced it, among other ways, by dissolving liver of sulphur (a mixture of potassium-sulphur salts) with spirit of salt (hydrochloric acid). He collected the gas and determined various of its characteristics, including the fact that although it didn't precipitate lime water it was absorbed by it in significant quantities.[31] Scheele's work was fairly well known and was translated into English and French in short order.

The gasometer (figure 1.4) was the final piece of pneumatic chemical apparatus to be incorporated into the gas plant. It was made famous by Lavoisier, who described it in the *Traité élémentaire de chimie* (1789). The third section of that historic work was devoted to chemical laboratory techniques and apparatus, and included a part dedicated to pneumatic chemistry.[32] After describing the pneumatic trough as implemented by Priestley, Lavoisier explained his gasometer in detail. The gasometer, as the name implies, was a complex instrument used to store gas and to measure its volume. The pressure on the enclosed gas was regulated by a counterweight sitting in a pan supported by a balance arm that raised the gasometer's holding tank as it

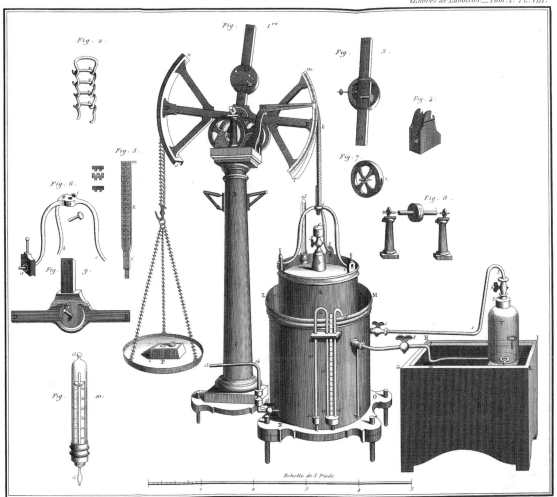

Œuvres de Lavoisier__Tom.I. Pl.VIII.

TRAITÉ ÉLÉMENTAIRE DE CHIMIE

FIGURE 1.4
Lavoisier's gasometer. Lavoisier, *Traité elementaire de chimie* (1789), as reprinted in *Œuvres* (1865), volume 1, plate 8.

filled with gas. The gasometer's origins go back to the years when Lavoisier began investigating gases and their properties. Lavoisier came to regard gases as chemical agents and eventually developed the theory that they represented a third state of matter.[33] At first he used a modified pneumatic trough that had spigots on the gas receptacle to let out gas in controllable quantities. This pneumatic trough proved to be inadequate when Lavoisier began to use gravimetric measurements in his work. He had to find a way to include the mass of gases in his calculations. Weighing a quantity of gas is not easy, but volume, if the density of the gas is known, offers a way to calculate its mass. The gasometer, designed to measure the volume of a gas, was the solution. Lavoisier developed it, over several years, in conjunction with a military engineer named Jean-Baptiste Meusnier.[34] A gasometer built by the instrument maker Pierre Mégnié between 1785 and 1788 was presented to the Académie in March of 1788. A large and expensive instrument, costing 636 livres, it could not have been reproduced by any except the best instrument makers, or purchased by any except the best-funded chemists in Europe.[35]

With the wide diffusion and extensive influence of Lavoisier's work, and with the ongoing interest in pneumatic chemistry, gasometers soon became more common in chemical laboratories. Instrument makers all over Europe began to make them.[36] Mentions of gasometers appear in chemical texts of the period, often referring the reader to the *Traité*.[37] Lavoisier's costly gasometer, however, was not the model that proliferated.[38] Martinus Van Marum, a Dutch chemist who had witnessed Lavoisier's demonstrations on the composition of water in Paris in 1785, wanted to repeat these demonstrations, but owing to the expense associated with the gasometers he was unable to do so until 1791—and then only with gasometers of his own design, in which a number of containers were used to shift water around in order to maintain the gas under constant pressure.[39] Other variants soon appeared, including one—designed by the well-known Parisian instrument makers Louis-Joseph and Pierre-François Dumotiez—that had four separate counterweights supported by pulleys running over vertical supports. The Dumotiez brothers were selling and advertising these by 1795. Their gasometer was the model that became the most popular in chemists' laboratories. A similar one was initially adopted by the gas industry as a result of James Watt's choice of such a design for his own pneumatic apparatus.

By 1790, all the apparatus and all the processes fundamental to the early-nineteenth-century gas plant existed in the laboratories of pneumatic chemists.

1.5   LIGHTING WITH GAS

In view of all the experimentation on inflammable gases, especially in the second half of the eighteenth century, it is not surprising that pneumatic chemists suggested some forms of lighting. In this context three men are particularly noteworthy. The first was Alessandro Volta, who designed a lamp and a lighter. The second was Jan-Pieter Minckelers, one of the canonical inventors of gas lighting, who lit his lecture hall with coal gas. The third was Charles Diller, an instrument maker who made popular demonstrations with his "philosophical fireworks." Volta's and Diller's devices inspired many imitations.

ALESSANDRO VOLTA AND THE INFLAMMABLE AIR LIGHTER

Alessandro Volta, besides being interested in the properties of inflammable airs, was adept at inventing and making scientific instruments that used inflammable airs. The best-known of these was the eudiometer, a device whose purpose was to measure the salubriousness of air. Among the others were an inflammable air pistol and, of the greatest importance in this history, various forms of inflammable air lamps.[40]

Volta had originally mentioned burning inflammable air in a lamp in one of his letters on inflammable air. He developed this idea into two devices: a lamp (lucerna) and a lighter (accendilume).[41] They were similar to one another, but the lamp used marsh gas whereas the lighter used metallic inflammable air.[42] The lamp had a short career and did not inspire direct imitators. Volta abandoned the idea of gas lamps after two or three years because he found that the light was too weak and that too many bottles of gas were needed for only a few hours' light.[43]

The lighter was to enjoy much greater popularity than the lamp, though it was never more than a curiosity. It also inspired a host of imitators. Many of the instruments derived from it were also used as lamps, and were frequently referred to as such.

The lighter (figure 1.5), much like its ancestor the lamp, consisted of two reservoirs stacked vertically, the upper one filled with water and the lower one with metallic inflammable air. A tube connected the upper reservoir to a lower one through a valve. A second valved tube led upward out of the lower reservoir and vented inflammable air to the outside. A sparking mechanism was located immediately adjacent to the mouth of the second tube. To produce a light, the operator let water from the upper reservoir into the lower one, increasing the pressure of the inflammable air. The second valve on a pipe in the lower reservoir was then opened, and the inflammable air now streaming out was ignited with a spark. When this device was used as a lighter,

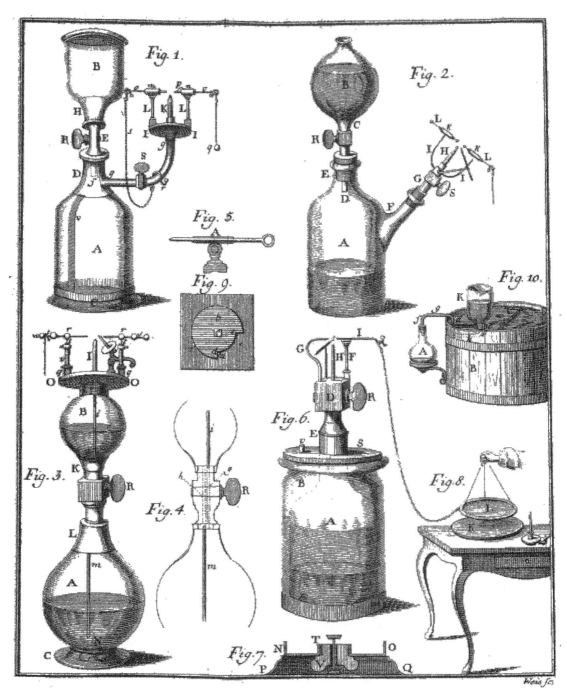

FIGURE I.5

A series of inflammable air lamps. Friedrich Ehrmann, *Déscription et usage de quelques lampes à air inflammable* (1780).

a candle placed close to the exit jet was lit once the flame had been struck. The lower reservoir had to be refilled carefully with metallic inflammable air using a pneumatic trough, a cumbersome and delicate procedure that could be performed only by someone with a chemical laboratory at his disposal and with sufficient skill not to mix any common air with inflammable air.[44] Volta continued working on the lighter at least until 1779.[45]

Thanks mostly to Volta's constant efforts to expand his base of contacts, the lighter was widely diffused in Europe, especially in Germany. Many people either copied it or asked Volta for one, or at least for some basic plans. A number of works soon appeared in print either exclusively dedicated to it or describing it in some detail.[46]

In 1777, as Volta's publications on inflammable gases were spreading through Europe, he also went on a lecturing tour, during which he showed his inflammable air lighter. Among the people who saw it was a certain Fürstenberger in Basel.[47] Fürstenberger was a collector of scientific instruments who eventually made a lighter quite similar to Volta's.[48] Soon lighters were being made by many instrument makers in France and Germany, among them Georg Friedrich Brander of Augsburg and Frédéric-Louis Ehrmann of Strasbourg.[49] One of Volta's friends in Strasbourg, Jean-Jacques Théodore Barbier de Tinan, also made one.[50] He mentioned in his letters to Volta that he used it frequently as a lighter and even as a lamp.[51] Ehrmann stated that all these lighters used inflammable air made from iron and vitriolic acid, with no mention of marsh gas or other sources of heavier inflammable airs.[52] This is corroborated by Jan Ingenhousz, a Dutch natural philosopher who learned of the lighter from Barbier de Tinan during a trip to Strasbourg in 1780.[53] "Inflammable air from standing water cannot be used in this lamp," he later wrote. "The air does not easily take fire from a small electric spark, and even if a flame starts at the tube it is soon extinguished. . . . At least, I have had no success with air from marshy waters."[54] These sorts of problems would occur if the marsh gas wasn't purified of carbon dioxide or nitrogen.

The lighter found its way to Britain via George Nassau Clavering, the third Earl of Cowper, who lived in Florence and was an avid collector of scientific instruments. In July of 1778, having heard about Volta's lamp, lighter, and other inventions, he asked Volta for a collection of them. Though he was thinking of presenting them to the Royal Society,[55] Cowper mentioned in a subsequent letter that he sent them to Edward Nairne, a London instrument maker and fellow of the Royal Society who was interested in electricity.[56] Volta also gave a lighter to Cowper sometime in 1779, and Nairne later presented it to the Royal Society.[57] Nairne was deeply impressed by Volta's work and recommended to the president that Volta be given membership in the Royal Society in 1778, but no new foreign members were being admitted.[58] Soon after Nairne presented the instruments to the Royal Society, Joseph Banks gave a series of public lectures in London partly based on Volta's instruments, especially the pistol.[59]

The lighter was most popular between 1790 and 1810, when many references to Volta's lighter appeared in English, French, and German encyclopedias and texts.[60] The difficulty of making hydrogen and the dangers inherent in storing it while preventing an admixture of atmospheric air, as well as its delicate construction and high cost, meant that the lighter could never achieve adoption outside of a limited group of enthusiasts or men of science.[61] In addition, the limited volume of hydrogen storable in the lighter effectively excluded its use for lighting, as Volta had quickly realized. Although a few people used lighters as night lamps, they could not have been used extensively for that purpose.[62]

JAN-PIETER MINCKELERS

The work of Jan-Pieter Minckelers (1748–1824), a chemist based at the University of Louvain, represents an important attempt to use coal gas for lighting that originated in the context of philosophical pneumatic chemistry. It also demonstrates that by 1783 someone knowledgeable in pneumatic chemistry could implement a version of gas lighting without years of preparation.

Minckelers was born in Maastricht. His father and grandfather had both been apothecaries. After studying in his home town, he went to the University of Louvain in 1764 to study theology. Although ordained a deacon, he decided to teach philosophy at the university. Over time, he become the closest collaborator of François-Jean Thysbaert, the director of the school of arts at the university, where lectures on experimental philosophy were offered. It was under Thysbaert's direction that Minckelers did his experiments on coal gas.[63]

The trigger for Minckelers' efforts was the 1783 ballooning craze, during which inflammable air went from being a fairly mundane and secondary subject of discourse among a select group of chemists to a topic of everyday conversation throughout Europe. In that year, the Montgolfier brothers created a sensation with their inaugural balloon flights. The Montgolfier brothers relied on hot air, although this was not immediately clear to anyone, even the Montgolfiers themselves, since they thought their fires were also producing new gases that contributed to the balloon's buoyancy. Their flights prompted government officials all over Europe to promote research on balloons and inflammable gases, and the leading chemists of France, including Lavoisier and Claude-Louis Berthollet, got into the act. The academic chemists wanted to get involved because the ballooning craze provided an opportunity to secure credit for pneumatic chemistry in the eyes of the public.[64]

Though most of the interest in the ballooning craze was focused on hydrogen, other inflammable gases did not escape attention.[65] Producing hydrogen in large quantities was bound to be expensive and cumbersome. Some balloon enthusiasts "discovered" that inflammable gas could also be produced by distilling coal and other materials, and for a fraction of the cost of hydrogen.[66]

It was at this point that Jan-Pieter Minckelers became involved with ballooning.[67] In 1783, Minckelers, Thysbaert, and Carolus Van Bochaute (another natural philosopher at the University of Louvain) were commissioned by the duc d'Arenberg to investigate ballooning, and specifically how to make inflammable air abundantly and cheaply. That the three of them were well acquainted with pneumatic chemistry is evident from the references in Minckelers' *Mémoire sur l'air inflammable* (1784).[68] Minckelers and company worked intensely on their assigned task, doing many experiments and, after a few months, floating several balloons. In an effort to find the best and cheapest inflammable air, they distilled various substances (including straw, wood, bones, and nuts) and measured the densities of the evolved airs, which they listed in tables included in the *Mémoire*. Because of its narrow focus, the paper included almost no other details about the gases. In the end, Minckelers et al. decided to use coal gas for their ballooning experiments because it was easily produced in large quantities when light coal was subjected to high heat in closed retorts.[69]

Sometime in 1785, Minckelers lit his lecture hall with coal gas; later some students' bedrooms were lit with it too.[70] Few details are known about Minckelers' apparatus or about the success (or lack thereof) of his lighting techniques. Minckelers fled Louvain in 1790 to escape the political turmoil there; presumably he left the gaslight apparatus behind. (It has been said to have survived until the First World War.[71]) His work didn't influence any subsequent development of gas lighting.

Volta is among the possible inspirations for Minckelers' idea of using gas for lighting. It is known that Volta visited Louvain on at least two occasions, and that he gave demonstrations of his apparatus—certainly his eudiometer and perhaps his lighter—to an audience that included Minckelers.[72] Was Volta's lighter the inspiration for Minckelers' gaslight? Lacking contemporary documents, we cannot say for certain, but we do know that Volta had demonstrated his lighter at other times during visits outside Italy, and that it was copied by some who had seen it.

Despite limited implementation and limited subsequent influence, Minckelers' work demonstrates that by 1785 pneumatic chemistry had progressed to a point where someone familiar with the subject had sufficient understanding of the manipulation and properties of gases to be able to construct an apparatus that would produce coal gas in less than a year. Minckelers was no neophyte at chemistry. He had been steeped in the subject, and had been chosen for the ballooning research precisely for that reason. Minckelers and his compatriots knew, for example, that fixed air could be removed from coal gas using lime water.[73] Nor was his work trivial. He had been exposed to all the leading chemical research of the period, and had even been in touch with Volta, from whom he may have heard of lighting by inflammable gas. Minckelers' true achievement was not the invention of lighting by gas, or the discovery of coal gas; it was the first successful application of lighting by coal gas on a scale larger than that of a desktop scientific instrument.

CHARLES DILLER AND THE PHILOSOPHICAL FIREWORKS

In the 1780s, a Dutch instrument maker named Charles Diller invented what he called "the philosophical fireworks." The fireworks, initially conceived as a public display for the purpose of amusement, was also intended to be science on display. Though the expert community lost interest in the fireworks, they persisted as a form of theatrical display and garden entertainment until about 1805. They also inspired a host of related scientific instruments based on lighting with inflammable gases. The fireworks were to play an important rhetorical role in the battle between Boulton & Watt and Frederick Winsor's National Light and Heat Company; they probably also influenced Philippe Lebon.[74]

Little is known about Charles Diller. Based in The Hague, he probably was a student of Jean-Nicolas-Sébastien Allamand, a Leiden professor and publisher.[75] Diller must have been an instrument maker of some note—Stadholder William V patronized him and purchased several of his instruments.[76] Diller also was involved in the ballooning craze—in 1783 he built a balloon 30 feet in diameter.[77]

Diller's philosophical fireworks used bladders of inflammable gas to produce flames of different colors. Moving parts representing stars, dragons, snakes, and other things emitted flame in succession as they moved around. One source reports that Diller's fireworks had 6,000 flames, presumably not displayed all at once.[78] The actual forms such fireworks took on varied, especially as imitators sprang up, and Diller himself used different versions.[79]

Diller's found inspiration from various sources when he designed the fireworks. Among them were the fireworks displays and illuminations in public places common in the second half of the eighteenth century. These displays included fireworks based on gunpowder, but also other sorts of stable illuminants, including lamps of various sorts. The promoters and creators of these displays vied to outdo one another to attract an audience to the pleasure gardens hosting the fireworks, and they increasingly turned to experimental philosophy to inspire the greater variety that drew people to the display.[80] Diller's philosophical fireworks fit into this tradition, and on at least one occasion it was suggested that his fireworks were much better than the "ordinary ones."[81]

But Diller may have taken some inspiration from a number of scientific instruments, including Volta's lighter and lamp. As an instrument maker, Diller would almost certainly have been aware of Volta's family of new instruments based on inflammable air. Indeed, Diller sold an inflammable air pistol and a cannon, two of Volta's creations, to the Stadholder.[82] The lighter was contemporary with the pistol and the cannon, and was receiving attention in the period immediately before and during the fireworks' first appearance.

Whatever the source of Diller's ideas, by 1780 he was making and selling the philosophical fireworks. In 1787 he set off on a grand tour that would take him to Paris and London. In June of 1787, after arriving in Paris from The Hague, he made a number of presentations, including one to the Académie royale des sciences:

Last Monday the 25th, a completely novel spectacle was enjoyed at the Panthéon: a display of fireworks with inflammable air, without smoke, without odour, and which surpasses all previous fireworks. The author, desiring to submit this new kind of experiment to the judgement of the Academy of Sciences, invited this body. Though quite familiar with the phenomena of physics and chemistry, they have singularly approved of the talents of the artist, the elegance of his machines, and the perfection of his means.[83]

Diller was invited to repeat the show for the royal family, and did so a few weeks later; he was awarded a pension by the king for his efforts.[84] The demonstrations created such interest that the Académie commissioned a group, consisting of le Roy, Brisson, Lavoisier, Monge, Berthollet, and Fourcroy, to investigate Diller's "feux d'artifice."[85] Diller did not reveal all the details of the mechanism to the committee, and in particular he did not reveal what the sources of his inflammable gases were, but the committee was certain that it was not inflammable gas taken from metals, as Diller's flames were clearly brighter than what their own experience with hydrogen indicated they would be. Diller even suggested to the committee that this gas could be used for lighting, owing to its brilliance.

The committee prepared and published a description of Diller's apparatus, the most complete available of the philosophical fireworks.[86] It consisted of bladders kept pressurized by screws and set in wooden boxes. Pipes connected the bladders to the flowers and suns serving as burners and visible to the audience. The displayed pieces were moved by the pressure of the gas, or by Diller or an assistant. Finally, the committee was most impressed by the odor-free nature and non-explosivity of the gas, in contrast to wood or coal gas, which always had foul smells.

The Dumotiez brothers made trials of their own to ascertain the nature of Diller's gases. Within a year they reported to the Académie that one of them was probably ether (diethyl ether). Diller had volatilized ether by forcing atmospheric air through sponges soaked in the liquid. The brothers then began to make and sell simple devices based on this design.[87] As late as 1805, an apparatus built in Paris by "M. Dumutier" was used in a London show titled *Pantascopia.*[88] In 1788, another Parisian instrument maker, François Bienvenu, made a "lampe à gaz inflammable" in imitation of Diller's.[89] Bienvenu's inflammable air lamp spread outside of France and references to it appear in texts published in the Netherlands, Germany, and Britain.[90]

Diller's work was reported in Germany and Britain, and Diller himself went to Britain soon after his Parisian sojourn.[91] After private showings to members of the Royal Society, he put on his spectacle in London's Lyceum Theatre; its run began on April 11, 1788 and continued until the middle of July.[92] It generated excitement and interest there as it had in France.[93] The *Critical Review,* reporting the Dumotiez brothers' investigations while Diller's show was playing in England, described them as the "fire-works which have afforded so much entertainment in London." The author of that report was hopeful about the future of Diller's invention: "We have reason to think [the process] will be successful."[94] A spectator at the Lyceum described the display as follows:

We all went to Diller's philosophical fireworks, which is a most beautiful and most ingenious exhibition, as well as a very fashionable one at present. It is an imitation of fireworks, but without any noise or any smoke, and the figures and designs are infinitely more elegant and beautiful, with the additional advantage of a great variety of the richest and finest colours. He represents the growth of plants and flowers, showing first a little stem, which grows gradually, and from which is shot out both leaves and foliage; then comes the flower budding, expanding by degrees, till you have the whole plant in its full growth, with the flowers in full bloom. The proper colours are observed, and the changes of colour take place in their proper order. He represents different insects and animals, and has a most curious chase of a viper after its prey, and of a little flying dragon after a butterfly. I cannot give you any just idea of the beauty and elegance of the show. There is a vast deal of ether employed, and the room smelt so strongly of Hoffman as to add very much to my pleasure and to that of Mrs. Johnston, who has the same affection for Hoffman that I have. Everybody else was loudly complaining of the stench, while we were whiffing it up and agreeing that it was a nosegay, and that it smelt of a *good night.*[95]

Soon thereafter, Diller fell ill and died.[96]

More about the gases Diller used can be found in the writings of Sir George Adams Jr., instrument maker to the king of England in the late eighteenth century. After describing how a bladder can be filled with inflammable gas and used like a torch, he states: "Mr. Dillier [sic] exhibited some very beautiful fireworks of this kind, at London, of different figures and colors. The color varying with the mixture, one third of the air of the lungs mixed with inflammable air of pit-coal, gives a blue colored flame; inflammable mixed with nitrous air, affords a green color; the vapor of ether affords a white flame."[97]

After Diller left the scene, people continued to use his fireworks. James Dinwiddie, a natural philosopher and a collector of scientific instruments, purchased Diller's philosophical fireworks in 1788, although it is not clear if they were a copy or the original that Diller had used on tour.[98] Diller had a partner who perpetuated the performances. While he was still in France, Diller had teamed up with John Cartwright, a musician who was teaching the queen how to play "musical glasses." Diller and Cartwright had gone to London together, and Cartwright provided musical interludes to Diller's displays.[99] After Diller's death, Cartwright used a version of Diller's apparatus and improved it.[100] Over the years, he occasionally put on demonstrations with the fireworks and musical glasses, even as late as 1811.[101]

Diller's public demonstrations and his apparatus reached an eclectic audience. He advertised broadly and attracted many people, including members of the nobility and the gentry (even the king of France)[102] and members of the French Académie and the English Royal Society. Copies of Diller's apparatus were made and used throughout Europe, beginning in Britain and France, but also as far as Italy and Germany, probably through the Dumotiez' incarnation of his instruments, as well as Bienvenu's.[103] Subsequently, even if they did not copy Diller's device, Lebon and Winsor followed in the same tradition of public illuminations in which Diller was situated before them.

By the early 1790s, then, there were many independent versions of gas lighting, and other uses of inflammable gases had been either attempted or suggested. Some of them, like the family of lighters and lamps derived from Volta's, were fairly widespread, at least among instrument makers; others, such as Diller's, were seen by many, copied extensively, and reported on in publications. None of them, however, proved to be the seed of the gaslight technology which eventually became an industry. In view of the number of attempts to make use of inflammable gases, the situation did not seem hopeful to Antoine-François Fourcroy, who wrote in the *Leçons élémentaires* (1782): "Inflammable gas is not of much use. Some have, however, tried to replace other combustible material with it in various aspects of life, such as lighting, fire arms, etc."[104] Fourcroy went on to mention a food warmer built by Neret, Volta's various instruments, and something made by Jacques Bianchi (another Parisian instrument maker) that was similar to Diller's machine.[105] In later years, probably because of the emergence of ballooning in 1783, Fourcroy softened his opinion on the futility of using inflammable gas, merely listing the inventions in subsequent editions without comment.[106] Nevertheless, despite many attempts, neither gas lighting nor any use of inflammable gas on a large scale, other than ballooning, was close to being a reality in 1795.

What the panoply of instruments, public lectures, and demonstrations had achieved, however, was that the idea of lighting with inflammable gases, and various ways of doing it, were spread throughout Europe, both geographically and throughout society. It is no wonder that many inventors and engineers working outside of the bounds of formal natural philosophy simultaneously thought of gas lighting at this time. The idea was circulating widely in the public sphere. Thus, the open science of the Enlightenment, which blurred the distinctions between pure science and the arts as well as those between research, education, and entertainment, contributed to the development of gas lighting. Public science made knowledge of pneumatic chemistry and the idea of lighting with gases available long before Lebon and Murdoch arrived on the scene.

The existence of these forms of gas lighting based in the pneumatic chemistry tradition demonstrates that the invention of gas lighting did not simply stem from the observation that inflammable airs exist, that they can be contained, and that perhaps they can be used for lighting. Volta, Diller, and the many instrument makers who developed their devices were never able to make them more than a scientific curiosity, and never really addressed questions as to how gas lighting could be made practical on a large scale. This in turn helps to clarify precisely what industrial distillation brought to the ultimate emergence of gas lighting. None of the gaslight pioneers mentioned in this chapter had an entrepreneurial mindset, and none sought to commercialize the technology on a large scale.

These versions of gas lighting also illustrate some of the difficulties involved in developing a practical gaslight industry. Specifically, it was one thing to make a scientific instrument or a device for display purposes, but transforming this apparatus into an industrially usable form required a good deal more work. Ensuring a constant and reliable supply of inflammable gas was an important challenge. For this reason, hydrogen could never have been adequate as the inflammable gas of choice for light. As balloon enthusiasts found out, it was quite difficult to produce in industrial quantities.[107] Using large quantities of vitriolic acid and especially iron was never going to be feasible for large-scale production of gases. Iron was too bulky (both as a source material and as waste) and too expensive. And so Volta's lighter was never to be more than a small-scale instrument. These forms of gas lighting also show that using methane—or a mixture of gases with methane predominant—was not an obvious choice either. Volta, Ingenhousz, and others tried to use marsh gas but never succeeded. It was too difficult to light, it went out too easily, and it had to be purified before it could be used. The purification problem was particularly daunting. Although how to separate fixed air from other gases was well known, regularly doing so with lime water was simply not worth the effort. There is no record of anyone doing this, and hydrogen was always the gas of choice. It was only with larger flames that methane-based gas lighting was feasible, but this demanded a much larger supply than could ever be used in a device such as Volta's lighter.

1.6 CONCLUSION

By the end of the eighteenth century, familiarity with inflammable gases produced from coal had developed to the point that facts about their basic properties were easily accessible in the most readily available chemistry texts of the period, including the fact that there were different kinds of inflammable airs.[108] Many techniques for the manipulation of gases had been developed, including production, purification, and storage. The episode of Minckelers' gas lighting demonstrates that someone with good knowledge of pneumatic chemistry could produce and deploy this technology on a larger-than-laboratory scale without years of experimental development, and without having to invent all the apparatus. In the years after Minckelers' work, new chemical texts from Lavoisier, Fourcroy, and others disseminated, expanded, and clarified this knowledge. In addition, the development of pneumatic laboratory apparatus had advanced to the point where all the basic elements of gaslight apparatus could be found in embryonic form in the best-equipped chemical laboratories. The retort had been present for a long time, and the pneumatic trough was more than 60 years old and had undergone many refinements. Laboratory techniques for purifying and separating gases had flourished, including those most important to the gas industry: using lime water to remove carbon dioxide and hydrogen sulphide, and washing and isolating gases with water in the pneumatic trough. The most recent addition to this list was Lavoisier's gasometer, which was further simplified by others after him. Like the readily available knowledge about gases, these laboratory techniques and instruments were all appropriated by the nascent gas industry in the last years of the eighteenth century.

As pneumatic chemists created the knowledge (and the instruments) that eventually made gas lighting a possibility, this knowledge was disseminated widely thanks to the open and public nature of eighteenth-century science. Volta's lighter and related devices embodied the idea of lighting with gases and probably inspired Minckelers and Diller, who put on spectacular and well-attended demonstrations in Holland, France, and Britain. The possibility of gas lighting was widely transmitted to many groups of people—including engineers and inventors working in industrial distillation, who would soon come up with more commercially oriented versions of gas lighting. This connection between pneumatic chemistry and gas lighting will come up repeatedly in the following chapters as the individual gaslight pioneers are examined.

Although pneumatic chemistry was important for the birth of gas lighting, the line between science and technology was not sharp, and there was constant interplay between the two. The retort, for example, was as much a part of practical apothecary culture as of scientific studies. The source materials, reactions, and observations that first led Boyle to study inflammable air came from medical and apothecary practice and from the coal mines of Britain. There was an ongoing back-and-forth between science and the arts, between speculation and practical use.

2.1   INTRODUCTION

The technological tradition that spawned gas lighting by borrowing knowledge and techniques from pneumatic chemistry was industrial distillation, particularly of wood and coal. Distillation had been used for commercial and industrial purposes for centuries, but interest in the process and in ways of improving it and adapting it to different uses was intensifying at the end of the eighteenth century. This was driven by a number of economic and related social factors, some of which also had regional characteristics. In some parts of Europe, deforestation, which had been occurring since the end of the Middle Ages, was reaching crisis proportions by the end of the eighteenth century. Particularly in France and Germany, the constantly rising price of wood was spurring calls for conservation through the enactment of laws, the introduction of new technologies, and the use of wood substitutes, particularly coal and sometimes turf. Distillation was seen as a possible way of achieving either of these ends, and so there were numerous new designs for distillation ovens that made more efficient use of wood or coal, or that coked coal so that it could be substituted for wood as a fuel. For the history of gas lighting, the most important of these new distillation ovens invented on the Continent was Philippe Lebon's thermolamp.

In Britain, somewhat different forces were at work. Coal had already been introduced on a large scale, both domestically (for heating and cooking) and in industry (as the growing substitution of coke for charcoal in iron smelting attested). Rather than a desire to combat deforestation and economize on wood use, what spurred interest in distillation in Britain was a desire to find substitute sources of tar and pitch for maritime purposes. Britain had long been heavily dependent on Scandinavian tar and pitch imports, and commentators had raised the importance of decreasing this dependence on potentially fickle foreign sources. North America had become a new source of tar, but with the coming of the American Revolution that source of tar was threatened. Coal tar made by distillation was a possibility located much closer to home.

In addition to these regional motivations, developments in chemistry were creating a growing awareness for possible new application of distillation. Analytic chemistry was expanding in the eighteenth century, creating a greater knowledge of new materials and their possible uses.[1]

These forces combined to spur interest in distillation in Britain, France, and Germany, leading to the design, construction and use of industrial-scale distillation ovens. Distilling wood and coal also produces gases, of course, so it is not a surprise that many of those looking into distillation thought of using the gases, especially as the awareness of gases and their properties was spreading with the growth of pneumatic chemistry. The acquaintance with pneumatic chemistry of some of the distillation-based gaslight pioneers is quite evident. In the last 15 years of the eighteenth century there were at least six independent instances in which someone interested in industrial distillation suggested using its gaseous byproducts for various purposes, although in none of those cases was the production of gas the original motivation. For William Murdoch, gas eventually became the most important product of distillation, and an entirely different industry, gas lighting, was born. The other people who developed some form of gas lighting after an initial interest in industrial distillation were Philippe Lebon, Jean-Baptiste Lanoix, George Dixon, John Champion, and Archibald Cochrane (ninth Earl of Dundonald). Distillation was the road to gas lighting, but none who started on it had lighting in mind initially.

The early apparatus of the gaslight pioneers was still relatively small, and whether the new technology they were working on was gas lighting or distillation for other products was really a question of emphasis in use, rather than design. Lebon began by looking at all the products of distillation, shifted to gas lighting around 1799 when he thought it had commercial possibilities, and then reverted to distilling for tar after 1802 when he did not find enough investors interested in lighting. The same is true for all the other gaslight pioneers in this chapter: gas was understood as one of the many products of distillation. Only Lebon had the idea of building a new industry solely on lighting, and then only for a few years. In Austria, Zachaeus Winzler, another early experimenter with gas lighting, always gave equal importance to other products in his writings. This was similarly true for Murdoch and Winsor, as outlined in chapters 3 and 4. They did not originally envisage solely a lighting technology, but shifted their emphasis in that direction after working on distillation more generally for some time.[2] The idea of making lighting the primary focus of the technology came only after some work, and had important implications for the design of distillation apparatus.

Over time, however, an important bifurcation in the technological tradition of gas lighting and distillation took place. The British branch increasingly focused on lighting while the Continental one focused on the other products, with the result that they left aside any thought of using the gas for commercial purposes. This culminated after 1818 in Karl Ludwig Reichenbach's design of a distillation oven for the production of various wood derivatives, particularly tar, methanol, acetone, and acetic acid. Though it is true that gas lighting failed to mature on the Continent after Lebon's work, and was later re-imported from Britain, Lebon's invention had in the meantime become the inspiration for a smaller industry. Seen in this light, the Continental failure to exploit Lebon's invention is more correctly understood as a bifurcation in technological traditions than as a failure to develop gaslight technology on the Continent. In effect, the Continental tradition kept more closely to the original distillation model, while the British model became gas lighting.

This bifurcation between the Continental and British traditions had two important causes. The first was the presence in Britain of coal in abundant quantities, so that it was always used for distillation there whereas wood was used on the Continent. The second was the specific skills associated with iron working in Britain, which made the development of machinery designed to handle gases easier. This picture of the development of gas lighting and distillation technology supports the views of the historians E. A. Wrigley and Christine MacLeod, who have argued that the Industrial Revolution was the result of two streams of technological innovation, one of which was common to all of Europe and one of which was confined to Britain and related to its burgeoning coal and iron industries.[3]

A further important theme of this chapter is the commercial orientation or entrepreneurial mindset of all these distillation-based gaslight pioneers. It was of vital importance for the development of gas lighting. All the men discussed in this chapter were trying to commercialize a distillation process, even before they thought of gas lighting, and they continued along the commercial road. Lebon, Winsor, Champion, and Murdoch took out patents. Almost all tried to raise funds to develop their business. This is in contrast to the people discussed in the previous chapter, none of whom took any of these sorts of business-oriented steps. A second consequence of the commercial orientation was their interest in industrial application, and hence larger scales. None of the people discussed in chapter 1, with the possible exception of Minckelers, showed any interest in scale. All of the people discussed in this chapter were interested in working on industrial scales even before they ever got into gas lighting.

This interest in industrial distillation was driven by forces related to politics and economics. In seeking the sources of new inventions, macro-inventions in particular are often seen as exogenous to prevailing economic conditions, almost as random events.[4] At one level, this is certainly true of gas lighting: it was not a search for cheaper sources of light that drove either pneumatic chemistry or industrial distillation. But at another level, prevailing economic conditions did have an important role to play in prodding the invention of gas lighting. The many approximately simultaneous attempts at gas lighting suggest that its invention was not purely a stroke of genius or luck, nor was it something solely produced by science largely independent of economic conditions. Rather, there were existing conditions that made gas lighting's invention more likely, and some of these—the wood and tar crises—were economic in nature. These economic conditions, when combined with the knowledge that pneumatic chemistry had created, resulted in a burst of simultaneous inventions of gas lighting.

Section 2.2 looks at the industrial distillation tradition, and particularly at why there was interest in new forms of distillation in Britain and on the Continent toward the end of the eighteenth century. Section 2.3 considers Lebon's invention of the thermolamp in the context of the distillation tradition. Section 2.4 describes the work of Zachaeus Winzler, who was inspired by Lebon's invention. Winzler's work also demonstrates the dependence of gas lighting on contemporary pneumatic chemistry. Although Winzler had no commercial success, his patron Hugo zu Salm eventually saw success with a new version of the thermolamp designed by Reichenbach. Section 2.5 explores why the Continental gas lighting and distillation tradition eventually shifted back exclusively to distillation, while the British version became gas lighting.

2.2   INDUSTRIAL DISTILLATION

Long before the eighteenth century, distillation of wood had been used to make charcoal for gunpowder and for the smelting of iron. This was done by carefully starving a burning pile of wood of air by covering it with earth, turf, or leaves. When the pitch and tar produced during the process were collected, as was done in Scandinavia, a gutter would be installed running through the base of the pile. This basic process remained largely unchanged until the eighteenth century and beyond in many areas. With the advent of the coke smelting of iron in Britain in the eighteenth century, coal was reduced to coke by a very similar method. Some variations on this basic process were introduced at the end of the eighteenth century, the most important of which was the beehive oven (figure 2.1). It allowed for much smaller pieces of coal to be coked than the hearth process.[5] The beehive oven was not, however, an immediate path to gas lighting. The importance of tar, pitch, charcoal, and coke for metallurgical, maritime, and military uses meant that distillation was very common throughout Europe by the end of the eighteenth century.

DISTILLATION IN BRITAIN

In Britain, it was a shortage of tar for maritime use that stimulated investigation of the new forms for the distillation of coal in the late eighteenth century. The causes of this shortage, however, dated back to the seventeenth century. The Royal Navy had long relied on the domestic production of naval stocks, but in the seventeenth century, because domestic supplies were dwindling, it turned to Sweden as a source

FIGURE 2.1
Cross-section of a beehive oven. Alexander Smith, *Intermediate textbook of chemistry* (1919), p. 439. Courtesy of Roy G. Neville Historical Chemical Library, Chemical Heritage Foundation.

for tar from pine to coat its ships. The price of tar began to rise with the growth of navies. It doubled within ten years after the Swedish state granted the Stockholm Tar Company a sales monopoly in 1689. The war between Russia and Sweden that started in 1699 caused exports to decline, and in 1703 it appeared that Sweden could no longer produce enough tar to meet Britain's demand. In 1705, in an effort to break its dependence on Swedish tar, the British government passed a law encouraging production in the North American colonies, and as a result those colonies (North Carolina in particular) became the dominant supplier of naval stocks to Britain in the eighteenth century.[6] In the mid 1770s, however, unrest in the American colonies endangered Britain's tar supply once more. In 1774 North Carolina threatened to cut off exports. Its total tar shipments dropped from 87,000 barrels in 1774 to 216 in 1777. Britain had to turn once again to Sweden, but the intervening shortages renewed the search for domestic sources.[7]

The production of tar from coal by distillation, which had already been explored in the seventeenth century, became a subject of fresh interest with the tar crisis.[8] In 1779, George Dixon II (1731–1785), an engineer from Durham who owned a colliery, began investigating the process in response to contemporary shortages. After he succeeded in making coal tar, the government paid for a ship to travel to the West Indies to be coated with it as an experiment, which was judged successful.[9] At some point Dixon also thought of using the coal gas he was producing for lighting, but he dropped the idea when he set off an explosion in the tar-making apparatus while trying to extinguish a gas flame. Subsequently he abandoned tar making because of the high cost of transporting tar to nearby ports for sale.[10] At approximately the same time, a group of lamp black manufacturers in Bristol developed and patented a similar distillation process for making tar from coal. The patent was granted to one of their number, John Champion, in 1779.[11] Like Dixon, Champion went on to suggest lighting with coal gas, but it was lighthouses that he had in mind.[12] He moved to London in 1790 to set up a demonstration apparatus in Chelsea, hoping to draw the interest of the Trinity Corporation, which was responsible for the operation of lighthouses. The 86-year-old Champion, feeling that he did not have the resources to make it work alone, wrote to Matthew Boulton in June 1790 suggesting a partnership with profit sharing. Boulton never wrote back despite a further letter from Champion.[13] The project came to nothing after that.[14]

These attempts in the 1770s to make tar from coal failed partly because the cost of making and transporting the tar to ports was too high. Renewed Swedish supplies undercut their prices.[15] Only if the tar works were close to coal mines and to navigable rivers could coal tar be sold profitably.[16] In addition, coal tar proved not to be a direct substitute for pine tar, as it had to be prepared carefully to prevent oils from separating from it during production.[17] The first person to have some commercial success with coal tar was Archibald Cochrane, who had begun experimenting with tar making just before 1780, eventually patenting the process.[18] He found some investors, and by 1785 the British Tar Company had spent £22,400 setting up tar works in various locations.[19] Though the company found maritime customers for coal tar and sold coke to iron foundries, the tar business declined toward the end of the century as a result of the increasing use of copper on the undersides of ships.[20]

Around 1787, Cochrane occasionally gathered inflammable gas from his tar ovens in bladders and used it to amuse guests at his home.[21] As with Dixon, an explosion made Cochrane and his employees far more reluctant to try such things.

DISTILLATION IN FRANCE AND GERMANY

In contrast to Britain, interest in new forms of and uses for distillation in France and Germany was driven by the desire to economize wood use, spurred by progressive deforestation. Although precise figures are difficult to come by, forest cover in France had been declining since the end of the Middle Ages—a trend that would not be definitively reversed until the nineteenth century, when dedicated conservation efforts and new industrial patterns led to significant reforestation in some areas.[22] In the years before the French Revolution, however, there was talk of a wood crisis. Prices were increasing, and erosion of hilly areas was damaging already marginal farmland.[23] The wood crisis seems to have contributed to the discontent of pre-Revolutionary France, especially in the areas most affected.[24] Although public authorities made some attempts to prevent over-cutting and to encourage the substitution of wood with coal, their efforts at enforcement were desultory and largely ineffective.

The situation in Germany was similar. There had been talk of wood impoverishment (Holznot) since the seventeenth century. The sense of crisis had inspired a peculiar genre of pamphlet writing (Holzsparliteratur) dedicated to the art of conserving wood (Holzsparkunst). Most of the pamphlets were very limited editions with only local circulation, but there were many produced over the years. A good number of them provided details of conserving stoves (Sparofen). At the end of the eighteenth century this literature was reaching the crest of its popularity. Many of the pamphlets called for state intervention and for the use of "practical reason" to address the problem. Although the accuracy of this literature's representations of the actual forest cover and the scarcity of wood has been debated, it is clear that there were regional problems. In areas that lacked access to effective and cheap transportation, shortages of wood were becoming acute in the late eighteenth century, particularly for certain industries.[25]

In response to these concerns, some people looked to distillation as a possible means to make more efficient use of wood or as a way to produce coke as a substitute for wood. There was, for example, a large-scale project in Sulzbach, promoted by the prince of Nassau-Saarbruck, in which coal was distilled in clay retorts to produce tar and coke. The prince erected nine furnaces, each capable of holding 2,000 pounds of coal, with a heating cycle lasting three days. The tar and oil were separated and used in miners' lamps and as tar to coat wagon wheels; the coke was used in the smelting of iron.[26] Similar distillation methods of making coke, mostly for iron smelting, were also suggested and tried in the 1770s.[27] It was, however, difficult to use local coal for metallurgical purposes on the Continent, owing to its relatively high sulphur content.[28]

Europe's wood shortage worsened, and in 1782 government authorities in Paris and Lyon took measures to regulate the wood supply. Ways of using coal rather than wood were also sought. To encourage bakers to burn less wood, the regional superintendent of Lyon, a man named de Flesselles, proposed in 1783 to the Société royale d'agriculture de Lyon that a competition be held for the design of a baker's oven that used coal. A winner would not be declared before the oven had baked twelve consecutive batches of good bread. The competition attracted a number of entrants. Jean-Baptiste Lanoix, a pharmacist and "demonstrateur de chymie," won by a unanimous decision of the judges.[29] Lanoix's ovens used coke rather than coal, because sulphur and other impurities in coal would have fouled the bread. Lanoix's oven attracted the notice of the national director of hospitals and barracks, who reported on its advantages to the ministry. Investors formed a company to produce and sell Lanoix's stove, but it didn't attract enough interest from bakers to succeed. In the meantime, Lanoix built a larger version of his coking oven in Lyon, where he continued to distill coal, making use of all the byproducts. Moreover, like his distillation counterparts in Britain, Lanoix began to use the gases generated for illumination—in this case, to light his own house. All these experiments came to an end, however, with the outbreak of the French Revolution.[30]

2.3   PHILIPPE LEBON AND THE THERMOLAMP

The thermolamp was the most important of the forms of gas lighting described in this chapter. Like the work of Dixon and that of Lanoix, it arose from an interest in distillation as an industrial process. The thermolamp, however, had a relatively brief commercial existence as gas lighting, originating with Philippe Lebon's invention and patent (1799) and continuing with its adoption, mostly in Germany and France but also in Russia and the United States, a few years later, by which point it was not much used for lighting. After a short burst of activity, interest in the thermolamp as a form of gas lighting was waning by 1810. Its technological descendants were used for the production of methanol, acetic acid, and other products. Lebon's work was nevertheless important in the history of gas lighting because of the inspiration and competitive impulse he gave to Winsor and Murdoch.

Philippe Lebon (1767–1803) was born in Brachy, Champagne. After his early education in Châlons-sur-Marne, Lebon decided to pursue a career in the French civil engineering corps and enrolled in the corps' École des ponts et chaussées in Paris. He moved to Paris in April of 1787, just before Diller arrived there with his philosophical fireworks.[31] Although Lebon never mentions Diller in his writings, the descriptions he gives of his own lighting follow in the tradition of Diller's philosophical fireworks.[32]

Lebon, an excellent engineering student, graduated at the top of his class in 1792. That same year, he submitted an entry to a national competition, sponsored by the École des ponts et chaussées, for a design of a steam engine.[33] His entry won first place, and he was granted 2,000 francs to continue his research into steam engines.[34] In 1791 Lebon began to investigate distillation for industrial purposes such as the production of pyroligneous acid (a mixture including acetic acid and methanol).[35] The French geologist Barthélemy Faujas-de-Saint-Fond had recently published a treatise, partly inspired by Cochrane's work, on potential uses for the products of the distillation of coal, and that treatise may have stimulated Lebon's interest.[36] In 1793 Lebon was posted to Angoulême in the Charente district of western France as a civil engineer.[37]

Lebon's research resulted in a patent for distillation processes (pour distiller au moyen du vuide [sic] et du froide), granted on September 11, 1796.[38] "Of all chemical operations," he had written in the preface to the patent application, "distillation is without a doubt the most precious and extensively used."[39] Lebon thought he could use distillation as a technique for producing salts, for purifying oils, and for separating the fixed and volatile parts of any substance. He mentioned that the government was seeking ways of improving distillation processes and ovens, similar to what had happened in Lanoix's time.

Lebon followed up his work on distillation by further investigating the possible uses of the products from combustible materials, including inflammable gases. Sometime in the next year or two, he tried using these gases to heat and light. In 1798 he wrote a paper on "New methods of using combustibles more usefully for heat and light, and of collecting byproducts thereby produced," which he presented to the Institut national; he applied for a patent with the same title on September 4, 1798. (See figure 2.2.[40]) Lebon was thinking, as he stated in the 1796 patent, about the efficiency of distillation processes. He thought he had come up with a better distillation process—one that dissolved wood into its constituent principles (one of which was inflammable gas) more efficiently than previous processes. He specifically mentioned in his patent application that inflammable gases produced in this way could be used for balloons as well as for heating and lighting.[41]

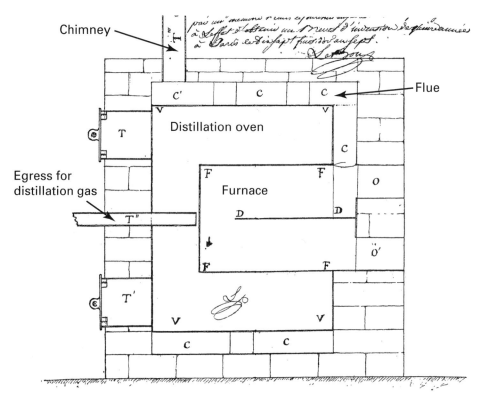

FIGURE 2.2

Lebon's thermolamp. V: distillation chamber containing wood, loaded through doors T and T'. The pipe (T") exhausted gases from the distillation chamber. Fuel was loaded into the furnace (F) through doors O and O'. The hot gases and smoke from the fire were exhausted from the furnace through a flue (C) encircling the distillation chamber before venting through the chimney. Taken from Lebon's 1799 patent application, as reprinted in René Masse, *Le Gaz* (1914), Annexe I, 234–237.

The French government granted Lebon his patent on September 28, 1799, more than a year after he submitted it. Lebon's first public announcements of his process generated some interest, but he decided that more work was needed before he could commercialize it.[42] On August 25, 1801, Lebon submitted an extension to the original patent in which he tried to give more details about the invention proper. (The patent had focused on a process, intentionally leaving vague the details of the apparatus.) His thinking had clearly evolved over the intervening two years. The extension indicated that Lebon was now more interested in heating and lighting: "Since in most cases thermolamps will be used for lighting and heating, I will consider them particularly from this point of view." Reflecting this new emphasis, the term 'thermolampe' had been coined, but not by Lebon: "This is how a number of people are content to call my apparatus."[43]

The patent extension showed that Lebon was aware that the noxious and malodorous products of combustion made it difficult to use wood gas for heating and lighting. He proposed using enclosed burners with chimneys to draw away the smoke and odors produced by combustion from the flame without allowing them to mix with the room's air. He also saw the need to purify the gases by cooling them to remove suspended oils by passing the gas through water.[44]

At this point, Lebon thought he was ready to market his thermolamp and raise funds by subscription. It was not his first attempt to find support. In the time leading up to the 1801 addition to the patent, he had tried to get the French government to adopt his scheme for lighting city streets, but apparently the financial pressures of France's wars made this sort of project a luxury the government couldn't afford.[45] By 1801, his experiments had cost him a good bit of his own funds. With no external source of funding, he was feeling financial strain personally.

Lebon's marketing efforts were successful. He began to garner attention, first in France, when he presented his thermolamp at the Athénée des arts on July 19, 1801 to an audience that included Fourcroy, Guyton de Morveau, and Gay-Lussac.[47] He published a pamphlet, then leased the Hôtel de Seignelay, in an upscale neighborhood of Paris, where he installed two thermolamps, using one to light five rooms and the other to light the gardens and the building's facade.[48] He then advertised public demonstrations there in various newspapers, and many people came, paying three francs for the privilege.[49]

On October 23, 1801, an English traveler visited Lebon and wrote a description of what he saw:

In lieu of fire or candle, on the chimney stood a large crystal globe, in which appeared a bright and clear flame diffusing a very agreeable heat; and on different pieces of furniture were placed candlesticks with metal candles, from the top of each of which issued a steady light, like that of a lamp burning with spirits of wine. These different receptacles were supplied with inflammable gas by means of tubes communicating with an apparatus underneath. By this contrivance, in short, all the apartments were warmed very comfortably, and illuminated in a brilliant manner.[50]

The Treaty of Lunéville had been signed a few months earlier, in February of 1801, ending the war on the Continent, at least temporarily. The peace allowed Lebon's invention and demonstrations to be reported more broadly, and for more foreign visitors to witness them, as revealed by the great interest shown his work in German-language publications.[46]

Despite the attention, Lebon's drive to find investors failed to gain momentum, probably because of the malodorous smell of the combusted gas. As a German newspaper reported, "The greatest inconvenience of this thermolamp, at least up till now, is the unpleasant odor which it produces."[51] The subscription's failure made Lebon's financial situation precarious.[52] He tried to interest some individuals in the thermolamp, and sold at least one, but that strategy too proved futile.[53] He returned to his job as an engineer in the Corps des ponts et chaussées, which assigned him to the Vosges department in May 1802.[54]

Lebon's failure in France has not yet been sufficiently explained. The most commonly cited reason is that the French government was preoccupied with the war, and the defeat of Britain in particular, and was not interested in funding non-military projects such as Lebon's gas lighting. Though there is certainly some truth to this evaluation, there is also more to Lebon's failure than official indifference, as private interest could also have led to larger projects. Perhaps Lebon simply failed to inspire, but contemporary commentators from Germany remarked on the lack of interest in the thermolamp in France relative to Germany. Ludwig Gilbert, professor of physics in Halle and editor of the *Annalen der Physik*,[55] mentions on at least two occasions— once in 1802, during the first excitement, and again in 1806, after the thermolamp had lost its luster—that no French chemical or physics journal carried reports on the thermolamp, and that it was ignored by the Parisian press.[56] Gilbert was certainly overstating the case. The thermolamp was well known, and five years after Lebon's demonstrations a French commentator was able to write: "Everyone remembers the ingenious discovery of the thermolamp, and the experiments done by Mr. Lebon."[57] In addition, at least one report was published in a French academic journal.[58] Finally, and most significantly, there are examples of French chemists or engineers who took inspiration from Lebon. Antoine Thillaye-Platel, who heard of Lebon while working as a pharmacist in Rouen, designed a thermolamp for carbonizing turf.[59] Maxime Ryss-Poncelet tried to adapt the thermolamp to carbonizing coal.[60] But as an editor of a scientific journal that regularly reprinted articles from French and English journals, Gilbert was familiar with the situation in important foreign publications. In contrast to both English journals (which seemed to have passed Lebon by with hardly any comment[61]) and French ones, German-language journals featured many reports on the thermolamp, most of them written by people who took an active interest in the subject. Moreover, by the 1820s French commentators were bemoaning the French failure to commercialize the thermolamp once gas lighting had attained success in Britain.[62]

Gilbert, however, was no great admirer of Lebon or his invention. He dismissed it as "more of a amusing scientific plaything than an invention of great use"—at least in its current form. He also thought little of Lebon's chemical knowledge. Gilbert points out that Lebon did not purify the gas of its carbon dioxide (kohlensaure Gas), as he could have done using lime water, though this is not a simple step for someone who was working with his first prototypes.[63]

After his failure, however, Lebon did not forget his thermolamp. He tried to apply it in some way that might draw the official support he had failed to garner thus far. He wrote to the ministry suggesting that he make tar for naval purposes from wood. He had more success with this tactic, and was given a forest concession in Rouen for this purpose in September of 1803. Lebon's untimely death in Paris in June of the following year, probably of disease, ended this project. Many years later a story that he was murdered began to circulate, but there is no evidence for that.[64] His widow tried to carry on with the thermolamp, but was ultimately unsuccessful.[65]

There are various dates given for Lebon's "discovery" of the inflammability of wood gas, and various stories (in a typical "eureka" style) of how he made the discovery.[66] There seems to be no contemporary evidence that can be used to date the "discovery" or to validate that the "discovery incident" occurred. In fact, Lebon never claimed, either in his patents or in his later publications, to have discovered that inflammable gas could be produced from wood. Even if he had made such a discovery, it would not have been an revelation for the scientific community, which had long been aware of the fact. Had Lebon announced such a discovery in the mid 1790s, or even in 1787 (the year he entered the École des ponts et chaussées), it would have been a laughable statement of his ignorance of the state of contemporary pneumatic chemistry. The story of his discovery of the inflammability of wood gas is certainly a later invention on the part of those trying to strengthen Lebon's credentials as an inventor and in particular as the first inventor of gas lighting, especially in view of the competing nationalistic claims of priority of invention that were made in the nineteenth century. The evolution of Lebon's thought, originating from the study of distillation as an industrial process, also belies this discovery story, which places the emphasis on gas lighting from the very beginning of Lebon's work—something Lebon didn't do in his first patent. This story (like similar ones told of Murdoch) has distorted the understanding of Lebon's inventive process, prejudicing his study of distillation and his understanding of pneumatic chemistry in order to strengthen his claim to be the inventor of gas lighting. As Gilbert wrote in 1806 when commenting on Lebon's invention: "It has long been known that destructive distillation of vegetable matter produces a quantity of gas, and that most of this gas is inflammable. It therefore required no great degree of inventive genius to try burning this gas and using it for illumination."[67]

What, then, is the significance of Lebon's work? His technical work was not tremendously original. The distillation oven he set up was similar in its basic form to earlier devices built by Cochrane, Lanoix, and Dixon. Nor was his technical work substantially different from what Minckelers had done, although Lebon and Minckelers had come to gas lighting in different ways—Minckelers from the point of view of a university lecturer making inflammable gases for balloons, Lebon from work with distillation processes. Lebon, however, was far more entrepreneurial in regard to gas lighting than Minckelers or anyone else before him had been, except perhaps Lanoix. No one had taken out a patent, advertised the invention, or tried to gain public and private support and funds to the degree Lebon had. The greater degree of attention that Lebon attracted shows his success in this regard. This difference proved to be crucial. Although Lebon's projects ended in failure, his publicity influenced Winzler, Winsor, Boulton & Watt, and a host of imitators in Germany and elsewhere, both technologically and commercially. Neither Minckelers, nor Cochrane, nor Lanoix, nor anyone else had such pretensions or such influence. Lebon's successors would, in turn, spur each other on by their competition, successful results ultimately coming in Britain. Thus, Lebon's most important contribution to the development of gas lighting was the injection of an entrepreneurial and eventually competitive element to a field that had already had a few entrants. The importance of advertising and publicity in entrepreneurship in industry in this period will come up in connection with Winzler later in this chapter, in connection with Boulton & Watt in chapter 3, and, most spectacularly, in connection with Winsor in chapter 4.

2.4   ZACHAEUS WINZLER AND THE THERMOLAMP IN GERMANY

Reference has already been made to the decisive impulse that Lebon gave to Frederick Winsor and Boulton & Watt (through Gregory Watt), both of whom were instrumental in bringing about the ultimately successful version of gas lighting that proliferated in the nineteenth century. Boulton & Watt and Winsor were not the only ones, however, for whom news of the thermolamp prompted action. Inventors, instrument makers, men of science, and others—particularly in Germany—were fascinated by the thermolamp. Ludwig Gilbert wrote: "It is known how in Germany novelties receive the greatest eulogies. This was also true with the thermolamp: we had more than one zealot wanting to tear down every oven and turn them all into thermolamps."[68]

The foremost figure in the development of the German thermolamp was Zachaeus Andreas Winzler.[69] Much is known about his work because in 1803 he wrote a lengthy book on the subject (*Die Thermolampe in Deutschland*). Winzler built many large thermolamps for industrial use during the 15 years in which he was interested in the technology. His work and his book are important to the history of gas lighting for two reasons. First, they once again show to what degree the development of gas lighting was based on contemporary chemistry, as Winzler was able to recreate Lebon's work after hearing a description of it based on Winzler's chemical knowledge and apparatus. Second, they show that Winzler followed a development path different from the British one that led to gas lighting. Although both Winsor in Britain and Winzler in Austria were inspired by Lebon, the British path eventually led to the gaslight industry, whereas the German path led to an industry that produced synthetic chemicals from wood derivatives.

Winzler (1750–1830?) was born in Unlingen in Moravia.[70] He studied philosophy in Constanz, and then theology and medicine in Marburg. While at Marburg, he developed an interest in chemistry and metallurgy. He worked in various places, sometimes in mines and sometimes in dye manufactures. He also traveled widely, visiting Britain, Holland, and various places in Germany. He settled in Vienna in 1779 after he was appointed the overseer of saltpeter production for all of Austria. In 1796 he established a saltpeter works in Hungary; after its failure he established another at Znaim (Znojmo), in Moravia.[71]

When Winzler read of Lebon's discovery in newspapers in 1801, he was interested, but not enough to pursue investigating the technology himself. As many reports continued to come in, and after Lebon's small treatise was translated and published in 1802 in Regensburg, Winzler became convinced of the invention's advantages and decided to make a thermolamp for himself, without ever having seen Lebon's lamp or plans of it.[72]

Winzler was, in his own opinion, well prepared for this task because of his lively interest in chemistry: "My grounding in physics and the new chemistry, which has recently been my favorite subject, proved to be very useful in this regard."[73] His book shows that he was well informed about contemporary chemistry—contemporary chemists are cited throughout the text.[74] He was quite up to date on pneumatic chemistry too. His conception of the composition of what he refers to in the book as "inflammable air" (brennbar Luft) was, as was Lebon's, similar to what could be found in major chemical texts of the period. He thought the inflammability of this gas was due primarily to its hydrogen, but he recognized that some elemental carbon (Kohlenstoff)[75] was chemically combined with it, and he thought (correctly) that the carbon increased the lighting potential of the gas. In contrast to what happened when carbon was chemically combined with hydrogen, Winzler also knew that carbonic acid, when mechanically mixed with inflammable gas, lowered its lighting potential. He knew that the carbonic acid could be removed a number of ways, including with water, with dry or wet lime, or with alkali water, but preferred to use pure water, which was simpler to obtain and less messy.[76]

Winzler demonstrated knowledge and ability with chemical apparatus in the process of building his own thermolamp, both in what he chose and in what he rejected. This is of crucial importance because it reveals how much of the knowledge and the apparatus used in constructing a thermolamp—the first commercially oriented gas plant—could be found in the laboratory of an educated and current chemist. When Winzler first decided to try his hand at making a demonstration thermolamp, he turned to the instruments he had available in his laboratory rather than make custom components. Winzler provides clear details as to what pieces of apparatus he used:

I chose an ordinary ypfer retort[77]; filled it with soft cut up wood, only as much as it could hold, and laid it in the free flame of a distillation oven; closed it with its top; stuck a pipe, which was bent downwards, into the mouth of the retort, and luted it there with strong metal; sunk this pipe a few inches into an empty Woulfe flask; connected the flask via a glass pipe to a wooden receiver, which had been previous completely filled pure water; luted everything with fireproof clay; gradually increased the fire, and had the pleasure of seeing my theoretical plan confirmed by a complete success.[78]

The Woulfe apparatus that Winzler mentioned (shown here in figure 2.3) was invented in the 1760s by Peter Woulfe,[79] a London chemist and instrument maker with some

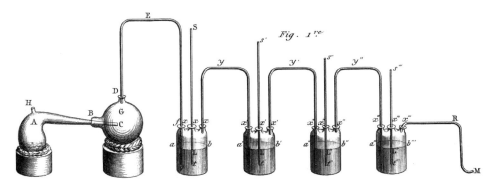

FIGURE 2.3

A Woulfe apparatus. Lavoisier, *Traité elementaire de chimie* (1789), as reprinted in *Œuvres* (1865), vol. 1, plate 4.

very eccentric habits. Its purpose was to facilitate the separation of vapors and gases produced in a retort (A) through distillation by passing the products first through a cooled, empty globe (C), which condensed vapors and collected the newly formed liquids. The gases would continue on through the globe into another cooled bottle (L); there they were bubbled through water, which removed any remaining suspended solids and liquids. The water also acted as a seal to prevent the gases from returning into the retort. Other bottles could be added if further purifying steps were necessary. The Woulfe bottle proved tremendously useful and became an important part of the standard pneumatic apparatus. Such an apparatus was a product of pneumatic chemistry; constructing one required knowledge and understanding of the properties of gases, and knowledge that gases retain their form, in most cases, even after passing through water.

After these first experiments using laboratory apparatus, Winzler was persuaded by friends, who were also to become partners in the resulting venture, to build a larger plant, and to make public demonstrations with a view to commercializing the thermolamp. He finished the larger thermolamp sometime in 1802. Once again, in his book he provided more details about the possible options available in a pneumatic chemistry laboratory. He recommended a retort with two openings to avoid having to re-lute the piping after every charge. He specifically suggested the pear-shaped retort of French origin. A larger Woulfe bottle would do for purification. For an apparatus of this scale, Winzler recommended a gasholder of some sort, which would provide the user with an even stream of gas when needed, rather than the hard-to-control streams generated directly by the retort. Though Winzler knew of the gasometer, he decided to keep his gasholder simple. He suggested using large bags of leather or thick paper soaked in oil, with solid tops and bottoms, to provide pressure. Only if the gasholder was very large did he recommend using exterior guide rails, as found on some gasometers, to provide stability.[80]

Winzler then sought a venue for public demonstrations. Perhaps thinking of Diller-style displays, he decided to eschew public halls and theaters to avoid having to give "a coat of puffery" to his demonstrations. After he was granted permission to use a house belonging to a nobleman, Winzler moved to Vienna from Znaim on April 13, 1802.[81] The noble patron was Archduke Karl, third son of Emperor Leopold II and brother of the reigning Emperor Franz II. The building to be lighted was the k. k. Alsercaserne (imperial and royal barracks).[82]

Having built the apparatus (figure 2.4), the partners decided that the more people saw their experiments, the better; thus, they charged no admission.[83] The first public displays were held in May, running almost daily for about three weeks.[84] Winzler then had to return to his post in Znaim, but his companions continued the demonstration for a few more weeks. In that same year they gave the thermolamp to a Hungarian count, Theodor von Bathyan, who took it down the Danube to Pressburg (Bratislava).[85]

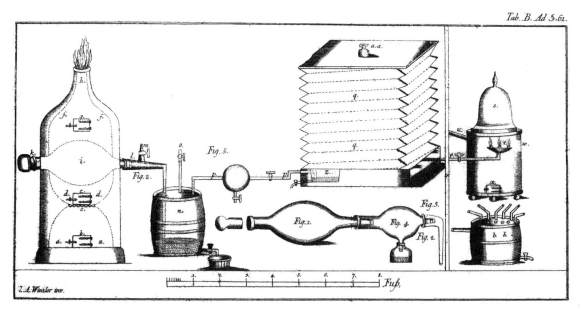

FIGURE 2.4
One of Winzler's thermolamp designs. Winzler, *Die Thermolampe in Deutschland* (1803). Courtesy of Science Museum, London.

Winzler and others[86] reported that, on the whole, the demonstrations were well received. Winzler and his partners thought of asking the Emperor for exclusive rights to manufacture and sell the thermolamp, but they do not seem to have pursued that goal.[87] They had their doubters, however. In 1803 someone wrote a pamphlet attacking the project; Winzler responded with another long-winded book in the same year.[88]

Like Lebon, Winzler did not conceive of the thermolamp simply as gas-making apparatus. His professional background in metallurgy and saltpeter production made him sensitive to the importance of charcoal and acids. As evidenced by his inclusion of coke and distillation in the full title of his book, he thought in much broader terms. He also poured scorn on Giersch, the author of an article on the thermolamp, who had neglected to mention that it could produce coke.[89]

Once back in Moravia, Winzler wrote his book, and proceeded to develop his ideas further. He thought many experiments would be necessary before larger versions of the thermolamp could be built: "The transition to the other extreme can only be made by means of many intermediary steps."[90] Lacking resources, however, he needed to find some form of patronage to fund his efforts. After a failed attempt at lighting the Znaim barracks,[91] he built another fairly small one at his saltpeter works in Znaim. It soon fell victim to the ravages of the Napoleonic wars.[92]

Winzler's search for a patron met with success in 1807, leading ultimately to the largest thermolamp he would build.[93] His new patrons were the zu Salm family, including Count Karl Joseph zu Salm (1750–1811) and his industrially inclined son Hugo Karl zu Salm (1776–1836), who lived at Blansko to the north of Brünn (Brno), about 100 kilometers from Znaim.[94] It was Hugo zu Salm who made the connection with Winzler. Hugo zu Salm had been interested in chemistry from an early age, but after he was captured in Italy by Napoleon's troops and subsequently ransomed he pursued it more closely. He developed a treatment for rabies and investigated the production of indigo dyes, among other things. He also wrote a number of books and articles on technical and scientific subjects, including a paper on distillation processes that was presented to the Berlin Akademie.[95] In 1801 he traveled to Britain to garner further knowledge, particularly about iron founding and other industries. He brought back with him drawings of textile and other machines, eventually having some built for his estates. On his return journey, he passed through Berlin; there Martin Klaproth gave him drawings of Lebon's thermolamp, which piqued his interest in the device. When zu Salm heard of Winzler's work on the thermolamp, he contacted Winzler to explore the possibility of his building a thermolamp on the family's estate.[96]

Winzler's thermolamp for the zu Salms was a large brick oven with a capacity of 80 Viennese Klafter of wood. This is the equivalent of between 180 and 270 cubic meters, depending on the definition of the Klafter used.[97] Construction was completed and the first tests were made around December 1807. As with Cochrane's ovens, a single distillation cycle lasted many days owing to the oven's size. The cooling period was also several days long.[98] Gas lighting, which had been an important part of Winzler's demonstrations in Vienna, was no longer part of the project; although some thought was given to using the gas to decarburize iron, in the end it was simply routed back into the furnace to be burned.[99] During the test, Winzler struggled with the gas pressure in this thermolamp; gas leaked out between the bricks. The leaks also allowed air to enter the oven, causing the wood to burn away. Because of the long cooling time and the air leaks, that oven was abandoned after initial tests.[100] Further work was halted when the thermolamp and parts of the zu Salm properties were damaged in 1809 by the French armies invading Austria during the War of the Fifth Coalition.[101] In July 1808 Winzler's wife died after falling into a vat of chemicals. In 1811, he moved to Vienna, where he then tried to build another thermolamp in which gas was used to fire bricks; that attempt failed because the flames could not sufficiently cure the bricks. Winzler then disappeared completely from the historical record.[102] In 1817, after the war's end, Hugo zu Salm built a smaller thermolamp (figure 2.5); it too was hampered by gas leaks and long cooling times.[103]

Despite their various failures, the Winzler and zu Salm thermolamps inspired a host of imitators and inventors to set up coking and charcoaling ovens in Moravia and Bohemia, making resin and tar for various uses. The most important of these was the chemist Karl Ludwig von Reichenbach (1788–1869). Reichenbach had been a student during the Napoleonic wars, and had been imprisoned for some time for political activities. After his release in 1811, he completed doctoral work on hydrostatic bellows at the University of Tübingen. Marriage brought him significant wealth that enabled him to tour industrial sites in Germany, Austria, and France in the years 1816–1818. During one of these trips, in 1818, he visited the Blansko thermolamps. They inspired him to design and build an oven for wood distillation in Hausach in Baden, but that oven did not achieve any great success. In 1821, he partnered with Hugo zu Salm in building metallurgical plants in Blansko. He also designed and built an oven (figure 2.6) that solved the problems of air leakage that had plagued zu Salm's ovens. It had a double-walled distillation chamber, with the space between the walls filled with sand. The wood inside the oven was heated by iron pipes running through the oven, through which the combustion gases were exhausted. This design succeeded in producing various products from wood, particularly charcoal, tar, methanol, acetone, and

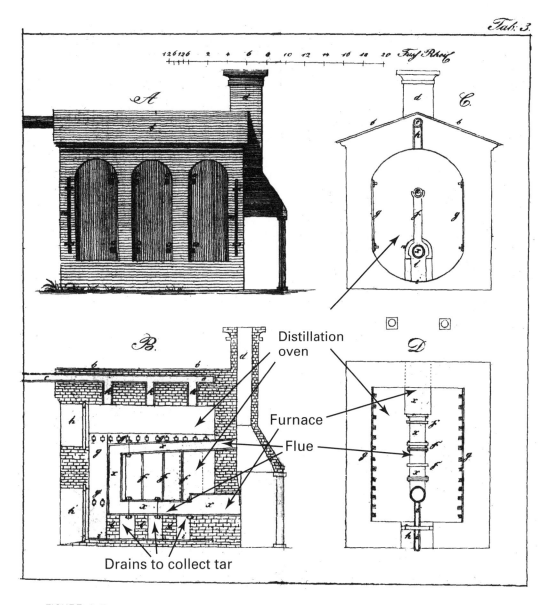

FIGURE 2.5

Exterior side (A), interior side (B), front (C), and top (D) views of zu Salm's 1817 thermol-amp. The furnace heated the wood in the oven primarily by the hot gases passing through the flue. This thermolamp was one-fifth the size of Winzler's earlier model. Christian Hollunder, *Tagebuch einer metallurgisch-technologischen Reise* (1824), plate 3. Courtesy of Bayerische Staatsbibliothek.

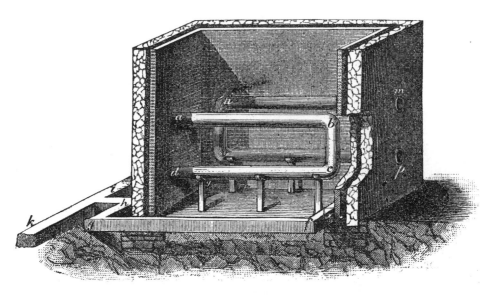

FIGURE 2.6
Reichenbach's distillation oven. The wood inside was heated by the hot gases passing through
the flues (a, b, c, d). Edmund Ronalds, *Chemical Technology* (1856), vol. 1, part 1, p. 344. Courtesy
of Roy G. Neville Historical Chemical Library, Chemical Heritage Foundation.

acetic acid. The wood gas produced during distillation was simply flared. A wood
distillates industry flourished from the 1830s based on this oven, expanding especially
at the end of the nineteenth century as new uses continued to be found for the
products—acetic acid was used in producing stains, and as a solvent for natural dyes;
methanol was used in producing aniline dyes; acetone was used in producing smoke-
less gunpowder, as a solvent for lacquers, and in the production of pesticides.[104] Zu
Salm and Reichenbach prospered, and they established a number of metallurgical and
chemical works in the area, amassing substantial wealth. Reichenbach also pursued
chemical investigations and discovered many compounds in wood distillation extracts,
including creosote, paraffin wax, and pittacal (the first synthetic dye).[105]

Reichenbach's distillation oven represented a complete reversion of the thermolamp tradition back to pure distillation. For Lebon lighting had become the center-
piece of the thermolamp's utility by 1800, but after failing to find investors Lebon
focused on tar production. Likewise, Winzler's early rhetoric and demonstrations
included domestic heating and lighting with gas, but once he left small-scale apparatus
behind his implementations were increasingly associated with other products, including charcoal, tar, and acid.[106] Reichenbach and zu Salm completed the shift away from

lighting, and by the 1840s articles on the thermolamp in technical texts made no mention of gas.[107] A similar shift away from lighting also occurred in France, where, as in Germany, a chemical industry based on the distillation of wood emerged during the nineteenth century.

It was the work of Jean–Baptiste Mollerat that formed the basis for the French wood distillates industry. Mollerat, like many others, had been working on industrial distillation independently of Lebon, and in 1804 he had patented a wood distillation process. Eventually, after 1840, his distillation techniques were deployed throughout France and beyond.[108] As with Reichenbach's oven, the gas Mollerat's apparatus produced was not given any importance, nor was it used in any way.[109]

In the meantime, coal-gas lighting was being reintroduced from Britain back to the Continent. Boulton & Watt's work was first reported on the Continent in journals, especially after 1808, when Murdoch's paper on gas lighting was published by the Royal Society.[110] Experimentation with coal-gas apparatus began in the following decade, with Maxime Ryss-Poncelet's experiments in Liège in 1810. Experimentation with gas lighting increased when the end of the Napoleonic blockade eased the flow of information from the Britain. Frederick Accum, a chemist who worked for the Gas Light and Coke Company for some time before becoming an independent expert, wrote the first sizable book on the technology in 1815.[111] Its publication in French and German in 1816[112] led to some attempts at introducing gas lighting in Paris, Freiburg, and Vienna.[113] It was only after 1820, however, that the industry began to establish itself on the Continent on a scale it had achieved in Britain the decade before, with companies lighting many city streets and buildings.[114]

2.5   TECHNOLOGICAL BIFURCATION

How did this bifurcation in the distillation tradition take place? The technical evolution of the thermolamp on the Continent makes it clear that the interest of those developing it there in the distillation of wood was at least in part a cause of this bifurcation. Specifically, wood gas has much poorer lighting properties than coal gas. Since coal was much less available on the Continent than in Britain, the natural tendency to use wood for distillation there eventually led those working on the thermolamp to lose all interest in lighting and to focus on the other products of distillation. The difference in illuminating qualities had direct effects on the technical design decisions of Continental engineers, who tended to direct their efforts toward the non-gaseous products of distillation. The problems specifically associated with designing practicable gaslight technology, as opposed to more efficient distillation, were never addressed. Most notable among these problems in implementing gas lighting were the impurities in the gas (which caused odors and impinged on flame luminosity) and the transition to a continuous, even flow of gas (required for gas lighting).[115]

Winzler and others knew how to purify wood gas of hydrogen sulphide and carbon dioxide with lime water, the very process that was adopted in Britain.[116] But since using lime water required an entirely new purification apparatus and generated messy spent lime, the thermolamp pioneers consciously decided not to use it. The purity of gas was not an essential desideratum; the other products of distillation were as attractive as the gas was. Over time, they would become more attractive, and the importance of purification faded completely.

With regard to the transition to a continuous process, the gasometer was one of the most important pieces of apparatus adopted by Boulton & Watt in Britain to smooth out gas flow. It allowed gas to be generated in batches, so excess gas could be stored as it was produced, and let out in an even flow to the burners as the gas was needed. Like Boulton & Watt, Winzler and others also knew of the gasometer, but again did not recommend it because it was too complicated, and because it was too expensive if the emphasis was not on using a continuous stream of gas.[117] As with purification, the impetus to use gas was not strong enough. Similarly, the mode of charging and discharging the distillation ovens could be adapted to supplying lamps with gas constantly and evenly. The German thermolamps, like Cochrane's ovens, were all oriented toward distilling large batches of wood or coal in cycles that would last for days and even weeks, with long downtimes for recharging and cleaning the ovens. This was acceptable for producing charcoal, tar, and acid, but not if gas was being generated for lighting. In contrast, Boulton & Watt, who designed their apparatus almost exclusively for gas production, worked on finding ways to reduce the batch times to a few hours, with the recharging of the retorts taking minutes. They did this at first by using cranes and metal baskets to hold a single charge of coal, and later by using horizontal retorts designed to be emptied and reloaded quickly. These rapid-turnaround and continuous-mode solutions were necessary to make gas lighting possible, and for that reason they were only adopted in Britain, where the possibilities for gas lighting were greater because coal gas, with its higher illuminating power, was being used there.

Britain had a relative advantage over the Continent for the development of gas lighting because coal was already being used in large quantities both domestically and in industry. Coal was also available on the Continent; however, owing to coal's relative abundance in Britain and the size of its coal trade, trying coal there for gas lighting was far more likely to occur. In addition, the use of coal for gas lighting was more viable from a commercial point of view. Some contemporary German commentators remarked on how useful the thermolamp could be in Britain in view of the enormous volume of coal present there.[118] A German journal reported that even Lebon thought of going to Britain, because of its coal market and its need for maritime tar.[119] The contrary observation was also made in Britain: carbonizing turf or wood with a thermolamp might be useful on the Continent, but at home, with such an abundance of coal, it was only of value in coal-poor places, such as Ireland.[120] In 1811, Maxime Ryss-Poncelet, who also worked on a coal distillation apparatus in Liège, another coal-producing area, claimed that Lebon would have had more success had he been closer to coal mines, as coal gas had better illuminating properties.[121] And in 1819, Jean-Antoine Chaptal also attributed Britain's lead in gas lighting to its more developed iron and coal industries.[122] As a result of the natural affinity to using coal in Britain, the problem of purification and the switch to a continuous process were solved first in Britain by Boulton & Watt and others. Furthermore, the fact that wood gas was only rarely used for lighting anywhere during the nineteenth century suggests that coal was indeed crucial for the emergence of gas lighting in Britain rather than elsewhere in Europe.[123] Finally, the gas industry remained much larger in Britain as compared to the rest of Europe throughout the nineteenth century because of its large coal industry.[124]

Coal was not, however, the only cause of this technological bifurcation. As will be described in the following chapter, there were other factors, the most important of which was the role of Boulton & Watt, and specifically the skills and experience they had in designing and building large iron machines to produce, manipulate, and store gases. Winzler and zu Salm struggled for years to make large airtight thermolamps, and never really succeeded. It was only Reichenbach's design that proved to be sufficiently sealed for the distillation process to work, and it achieved this largely with brickwork, not with iron as in Britain. Boulton & Watt, in contrast, already had a great deal of experience with making air-tight machines, primarily steam engines. The birth and scaling up of gas lighting at Boulton & Watt then owed a great deal to its pre-existing expertise and abilities with building such pneumatic apparatus on a large scale. Many other gaslight pioneers, including Lebon, Lanoix, and Champion, were unable to make the transition to large apparatus, despite their desire to pursue the possibility.

Political considerations were also important. Specifically, ongoing war influenced the technology's development path on the Continent. Some historians of the Industrial Revolution have pointed out that in this period Britain did not suffer military action on its own territory, whereas much of Europe saw regular outbreaks. Wartime conditions imposed sometimes severe constraints on engineers and inventors on Continent as they tried to develop new technologies.[125] At least two of the Austrian thermolamps were damaged during the wars, and the zu Salm's family's wealth was partially sequestered by Napoleon, hobbling whatever efforts they could direct toward the development of the technology.[126] Lebon's work was also affected by the war: although he failed to get official backing for lighting, he did win it for tar production, a process that had some military application. All these events together suggest that the Revolutionary and Napoleonic Wars had more than a minor effect on the technological development of gas lighting and distillation on the Continent.

2.6   CONCLUSION

The growing interest in industrial distillation at the end of the eighteenth century led to multiple simultaneous inventions of gas lighting, a process catalyzed by the diffusion of knowledge coming from pneumatic chemistry. The interest in distillation processes was being driven by a number of factors, such as the loss of the American colonies as a source of naval tar in Britain and the anxiety over deforestation on the European Continent. Among the inventors of gas lighting, Lebon and Murdoch stand out. Although some of the others, including Lanoix and Dixon, thought of the commercial possibilities of gas lighting, Lebon and Murdoch were each at the head of a lengthy stream of technical development that lasted throughout the nineteenth century. Lebon's technology spurred the growth of gas lighting in Britain through Winsor's borrowing of Lebon's technology and by frightening Murdoch into action, while in Germany and France distillation took a different path into synthetic chemical production. The most important characteristic common to all the people interested in gas lighting and distillation discussed in this chapter, as opposed to those discussed in the first chapter, is their commercial orientation or entrepreneurial mindset. They were all interested in creating a new technology that could be deployed commercially, and they all tried to do so with varying degrees of success. Though the people discussed in the previous chapter may have thought about commercial possibilities, none in fact did anything significant in that direction. This desire to commercialize the technology meant that there was also an interest in scale, which brought with it a host of problems, specifically related to purification and the transition to continuous gas flow, that needed to be surmounted in order to make gas lighting practicable. In France and Germany, these problems were never solved, as gas was judged not to hold enough value to be pursued. In Britain, however, Boulton & Watt started on the road to scaling up gas lighting into a full industry. Finally, the commercial orientation also meant that they had an interest in marketing and publicizing the invention.

This chapter has focused on industrial distillation, but the some of the themes brought up in the first chapter also have relevance here. All the people mentioned in this chapter who worked on gas lighting had some knowledge of pneumatic chemistry. The case of Winzler is particularly revealing in this regard. He was able to recreate Lebon's work from descriptions of it, using his understanding of chemical processes and his laboratory apparatus. The broad range of people who had access to this knowledge is also notable—for example, Lanoix (a French pharmacist), Winzler (a saltpeter manufacturer in Austria), and Dixon (a colliery owner in England). The geographic and professional range is indicative of how relatively easy it was to access and disseminate scientific and technical knowledge, as is the speed with which Lebon's work became known, at least in Germany. In discussing the causes of the Industrial Revolution, historians have sometime referred to institutions, such as encyclopedias, journals, and literary societies, that made scientific and technical knowledge more easily accessible. To use Mokyr's term, the access costs to information were lowered by these various projects conceived in the Enlightenment.[127] Some of the men mentioned in this chapter, including Lanoix, Winzler, and Lebon, clearly had been educated in contemporary chemistry. We know that these three studied at universities or professional schools. Informal training was also important. Cochrane and Dixon were known for their avid interest in chemistry. The knowledge outlined in the first chapter could easily be found in chemistry texts to which these gaslight pioneers had access.

The expansion of industrial distillation in the late eighteenth century and the invention of gas lighting as a derivative technology reveal the wide and deep roots of some technological innovation in the Industrial Revolution. In a paper on innovation in the eighteenth-century textile industry, Patrick O'Brien et al. have argued that historians who want to explain the sources of innovation in the industry should take the long view, looking at patterns of trade, consumption, and culture.[128] This long view is also required for gas lighting. The accumulation of knowledge, instruments, and techniques in pneumatic chemistry that helped transform industrial distillation into gas lighting originated in the seventeenth century, with contributions coming from all over Europe over the course of the eighteenth century. Likewise, interest in developing new forms of distillation was present all over Europe, and depended on economic conditions that had been building for many decades, such as increasing wood prices and the economics and politics of the maritime tar market in Britain. Within this broad background environment, regional and local conditions influenced the particular forms that the new distillation techniques took. In Britain, it was the abundance of coal and the presence of ironworking skills and experience, especially in the area of pneumatics, that helped direct the development of gas lighting technology. Even the relative success of Cochrane's coal tar business depended on locating tar-production sites close to mines and on cheap modes of transportation. In Germany, local interest in wood derivatives meant that inventive energies were directed more to the efficient distillation of wood for other products. By the 1820s, this led to Reichenbach's design of a wood distillation oven, which led to the founding of a small wood products industry. In the meantime, gas lighting had become a flourishing industry in Britain.

II A QUESTION OF SCALE

The creation of gas lighting as an urban network was a two-stage process. The first step was to design a form of the technology usable as stand-alone installations in large buildings; the second was to expand this to the network model. Although the network model dominated the nineteenth-century gas industry, stand-alone plants at large buildings such as factories continued to be used, even well into the second half of the century.[1] Each of the steps in gas lighting's evolution was carried out by different groups, with Boulton & Watt largely responsible for moving it from the workshop to industrial scale (with some refinements added by their former employee Samuel Clegg) and with the Gas Light and Coke Company largely responsible for the second stage. This part of the book is devoted to the first stage in the process.

3 BOULTON & WATT

The first commercialization of gas lighting was no trivial undertaking, as the efforts and failures of the Continental gaslight pioneers attest. It was a long way from having some idea of how the technology could work, and even from building demonstration apparatus, to commercial success—something that that was first achieved in Britain, as with so many other technologies of the Industrial Revolution. Craftsmen, engineers, and entrepreneurs were able to take what in many cases had been invented elsewhere and make the technology practical through a series of refinements, innovations, or micro-inventions, a process that differed from the original invention of the technology.[2] Even in Britain, there were many groups besides Boulton & Watt tinkering with gas lighting up to 1810 in various forms, but in the end only Boulton & Watt (and to a lesser extent Clegg, who drew from their work) had any success. Between 1802 and 1810, Boulton & Watt developed the technology, deploying it in mills in the north of England. In the process, they identified many of the problems associated with scaling gas apparatus up from the first small models, and solved some of them. By the time Boulton & Watt lost interest in the technology, around 1812, they had acquired sufficient momentum that others could capitalize on their work. Although a good deal more work was needed to further transform gas lighting into an urban network industry, the decisive initial momentum came from Boulton & Watt. They had demonstrated that gas lighting was workable on large industrial scales, and it was this work that became a crucial element in the industry's successful birth in Britain.

How did Boulton & Watt establish gas lighting as an industrial technology? Put another way, why did they succeed when so many others tried and failed? First, they were able to draw upon a wealth of technical skills and experience, both within the firm and from suppliers, derived from their background in manufacturing. Second, the firm had access to an extensive pool of resources that enabled it to do many experiments, to promote the technology through an existing network of large industrial customers, and to publicize its utility. It was this combination of technical innovativeness and the careful management of publicity that also marked the firm's work with steam engines, and indeed there are many parallels between the firm's work with gas and steam engines and its work with gas lighting.

In regard to manufacturing skills and experience, Joel Mokyr, John Harris, and many others have argued that British mechanical and artisanal skills in the eighteenth century were important sources of British invention and innovation, particularly in creating usable versions of technologies that may also have been invented or emulated in other countries.[3] Larry Stewart, Margaret Jacob, and others have further argued that there was a greater proximity of the skilled mechanics of Britain to contemporary science that aided them in their development work.[4] These links persisted to the end of the century, when gas lighting developed and when scientific, technological, and industrial development became mutually reinforcing,[5] or, in other words, had reached a point of sustained technological momentum. These mechanical skills had various sources, including the iron founding industry and the scientific instrument trade.[6] Both of these branches were present at Boulton & Watt,[7] and the firm used its expertise in both areas as it worked on gas lighting. In terms of expertise with scientific instruments, the firm had been manufacturing a pneumatic apparatus derived from contemporary scientific instruments for Thomas Beddoes' pneumatic medicine.[8] That pneumatic apparatus provided the basic model of the gas plant. By using its experience with building large iron machinery, Boulton & Watt transformed the pneumatic apparatus into a gas plant usable in industrial mills. In addition, the firm's history with building steam engines meant that it and its suppliers knew how to use iron to build industrial-scale machines to handle gases. Though many of the parts of a large gas plant were different from the parts found in a steam engine, others were not. Boulton & Watt and its suppliers were already manufacturing air-tight pipes, joints, and containers that could be used in gas plants.

The second reason why Boulton & Watt was able to consolidate gas lighting as an industrial technology was that it was able to draw on its extensive resources as a company of some standing. Some historians, including Svante Lindqvist and Donald Cardwell, have identified the period 1790–1825 as a time when technological innovations were increasingly made within institutions, such as companies or military groups. Earlier in the eighteenth century, individuals inventors or engineers (perhaps inspired and fostered by Enlightenment institutions with industrial interests, such as scientific societies) were predominant in technological innovation. But around the turn of the nineteenth century companies and other institutions began to play a more important role, a trend that become more established as the century wore on.[9] Lindqvist suggested that this happened because companies and other institutions were better able than individuals to marshal the greater resources needed to develop and commercialize more complex technologies. It was increasingly beyond the resources of individuals to do the industrial development work needed to deploy new technologies. This pattern is apparent in the case of gas lighting. Its consolidation as an industrial technology by Boulton & Watt depended to a important degree on the substantial resources dedicated to the project by that firm and by some of its steam engine customers, particularly the textile firm Philips & Lee of Salford. Philips & Lee, led by George Augustus Lee, became development partners and took an active part in many design decisions. More important, Lee's willingness to spend enormous sums on a gas plant for his mills provided Boulton & Watt with its first opportunity to install and test a full-scale gas plant.[10]

The resources Boulton & Watt used in developing and commercializing gas lighting were not only financial and material; some were social. In its work on gas lighting, the firm relied on its large network of connections, both among industrial customers and in political and scientific circles.[11] This included its abilities to advertise gas lighting to potential customers and beyond. As Iwan Rhys Morus has shown, display and demonstration were important for inventors and engineers in negotiating the place and acceptance of technical work in this period.[12] Led by James Watt Jr., Boulton & Watt, like Lebon and Winzler, used opportunities for the public display of gas lighting to promote its work among potential customers. Unlike Lebon and Winzler, however, the principals at Boulton & Watt had important connections. Later, when competitive concerns were becoming pressing, they made use of James Watt's contacts in the Royal Society to gain prestige by association with the Royal Society, and also to bring their work to the public. As David Miller and Christine MacLeod have discussed, Boulton & Watt, led by Matthew Boulton and later by James Watt Jr., carefully cultivated the image of James Watt as a natural philosopher, and later as a heroic inventor. It was very important for the firm commercially that Watt's image be established so that prestige could be derived from it.[13] In a similar vein, Boulton & Watt's Royal Society connections (discussed in chapter 4) served as an important means of establishing William Murdoch's status as the "true" inventor of gas lighting in the context of a commercial dispute, as well as the economic viability of the new invention.

Many of the other gaslight pioneers' efforts floundered because they did not have such ready access to resources as Boulton & Watt did. Lebon's efforts ran aground on the first point: he lacked the financial means to go beyond his initial development work and marketing efforts. When Lebon failed to attract outside investment, he was forced to set aside his project and return to the engineering corps, and later to focus on tar (which, unlike gas lighting, was attractive for the military). Likewise, it was not simply by chance that John Champion had written to offer Matthew Boulton a share in his idea of gas lighting if Boulton would help manufacture the apparatus. Champion understood that he lacked the resources to take the idea of gas lighting from the simple apparatus he had constructed to a new scale, and thought that Boulton & Watt's financial and manufacturing resources could provide an important impetus for his technology. Boulton declined to act, and Champion's idea died.

Historians of gas lighting have debated Boulton & Watt's place in the industry's early years. While most histories celebrated the firm's work,[14] Malcolm Falkus argued in one of the few critical papers on the early history of gas lighting that its role had been exaggerated because the dominant form of gas lighting in the nineteenth century was as an urban network, not as stand-alone units.[15] Falkus argued that Frederick Winsor and the London-based Gas Light and Coke Company were the true originators of this model, not Boulton & Watt, and that they developed their version of gas lighting independently of Boulton & Watt.

Though the Gas Light and Coke Company's creation of the network model was significant and difficult, the firm of Boulton & Watt, particularly under the direction of James Watt Jr., made three crucial contributions to this new industry despite its relatively short-lived involvement in gas lighting.

First, it took an idea (lighting with inflammable gases) that Murdoch had conceived in a form similar to many others of the period and transformed it into a functioning technology deployed in many large mills—something no other gaslight pioneer even approached. This entailed recognizing and in numerous cases solving non-trivial problems associated with scaling up gaslight apparatus, including how best to charge the retorts (horizontal settings); extending apparatus life span and robustness (keeping retorts constantly hot); dealing with gas purification and supply (using water to purify and gasometers for storage); flame efficiency; and assessment of economics. Boulton & Watt did not solve all the problems, and Watt Jr., who had been the main driving force behind the project, lost interest before the technology was sufficiently mature. It did, however, address a good number of them.

The second achievement was a social one: the firm demonstrated the viability of the technology to the public. A major element of this was the medal James Watt Jr. contrived to have the Royal Society give William Murdoch in 1808 to strike a blow against the Winsorites in Boulton & Watt's battles with them. The paper and all the publicity surrounding the battle linked the prestige of the Royal Society and that of Boulton & Watt to gas lighting, effectively sanctioning it as a valuable technology in the public consciousness. Even if Boulton & Watt exited the field, it left the technology firmly rooted in public awareness. The impression the firm created included a positive assessment of the technology's economic viability, one that was not firmly based on actual results to date.

Boulton & Watt's third achievement was to train and give experience to many craftsmen and engineers in the area of gaslight technology. Many of its employees and its contractors, as well as the employees of its customers such as Philips & Lee, gained knowledge of how to operate and build gaslight apparatus. The range of experience of their customers, especially Philips & Lee, was extensive because they were not merely passive purchasers. They were actively involved in design decisions. Many of the people who gained experience through Boulton & Watt's work went on to participate in the further development of the industry. The most important of these was Clegg, but there were also unnamed craftsmen who were hired from Manchester by the gaslight industry.

Section 3.2 tells the story from Murdoch's original invention of gas lighting to when Boulton & Watt began to install its first large gas plant at Philips & Lee in Salford in 1805–06. Section 3.3 looks at how this installation generated interest and orders for gas plants among the owners of many industrial mills in the north of England in 1806, and how Boulton & Watt responded. Section 3.4 examines Boulton & Watt's design work from 1805 to 1809, looking particularly at issues related to the large size of their gas plants relative to the workshop results of other gaslight pioneers, and how they tried to determine the economic efficiency of gas lighting. Section 3.5 recounts Boulton & Watt's last major sales of gaslight apparatus, in 1808–1810.

3.2   FROM INVENTION TO THE FIRST PILOT PLANT

WILLIAM MURDOCH AND DISTILLATION PROCESSES

William Murdoch (1754–1839; the name was later changed to Murdock) was born in Old Cumnock, Ayrshire, Scotland, the third of seven children. His father had been a gunner with the Royal Artillery, and was employed by the local Lairds, the Boswells, as a miller and millwright. He also did various engineering jobs, such as installing and repairing pumps at local coal mines and repairing bridges. In addition to whatever he learned at his father's side, William received an education in the Scottish school system of the eighteenth century, which was usually better than the English system. Through family connections, he was hired by Boulton & Watt in August of 1777 at the age of 23. He soon became Boulton's chief pattern maker, and began working as an engine erector for the firm. He was posted to Cornwall in the fall of 1779 to work as an engine erector for the firm's many mining customers. Cornwall was to be his home base until 1798, when he moved to Birmingham to work at Boulton & Watt's Soho Foundry and Manufactory.[16]

FIGURE 3.1
William Murdoch, c. 1800. Courtesy of Science Museum, London.

The firm of Boulton & Watt was a propitious environment for this talented engineer to develop new ideas. The firm had been founded in 1775 as a partnership between Matthew Boulton, an adept manufacturer in the Birmingham metal trade with a flair for marketing and business, and James Watt, the brilliant engineer whose work improving the steam engine provided the foundation of their collaboration. The firm had flourished over the years by successfully marketing engines, mostly to mines for pumping water, but it was also expanding by manufacturing stationary steam engines as power sources for industrial mills. Boulton and Watt were quite ruthless in defending Watt's steam engine patent, suing people who did not pay them license fees and retarding the development work of others on the steam engine with their pugnacity. Their business skills extended beyond marketing, design, and competition, and their firm was more sophisticated than many others in its management and accounting practices.[17] Boulton and Watt were also very well connected with some of the local scientific and industrial elites, most importantly through their membership in the Lunar Society, an informal group of luminaries, mostly from the Birmingham area, who corresponded and gathered regularly to discuss scientific and technical matters, beginning in the 1760s.[18] Joseph Priestley, Erasmus Darwin, James Keir, and Josiah Wedgwood were members. In addition to these local connections, Boulton and Watt had many international contacts.[19] By joining their firm, Murdoch had come close to an important nexus of natural philosophy, the arts, and industry where new and important scientific and technical information circulated easily, and in some cases was generated locally. Examples of this range from Watt's discovery of the composition of water to Priestley's extensive work on the chemistry of airs. Watt's friendship with Thomas Beddoes, who had an extensive library of chemical and medical texts from all over Europe, was to be very important for determining what form gas apparatus would take.[20] In addition, the firm and the people around it were capable of important technical and industrial achievements, including Watt's letter copying machine, James Keir's soap manufacturing plant, and Wedgwood's pottery works.[21]

Murdoch was never a member of the Lunar Society, but his talents as an engineer were soon recognized by his employers. In 1778 Matthew Boulton wrote to James Watt that Murdoch "is now a good engineer," and in 1784 Boulton characterized Murdoch as the "best engine erector I ever saw."[22] Murdoch had an active interest in invention, sometimes tinkering with and improving the steam engines he was responsible for erecting (much to Watt's annoyance). He was responsible for a series of important inventions. In 1781 he designed the sun-and-planet gearing mechanism that was eventually used to transform the reciprocal motion of a steam engine's piston into the rotary motion that was necessary to drive most factory machines. He also invented the D-slide valve, for which he received a patent in 1799. This valve was the most frequently used method to regulate steam flows in locomotive engines in the nineteenth century. Between 1783 and 1785 he experimented with and even built a number of steam carriages. In addition, he was a competent inventor of machine tools, as he showed once he had moved to Birmingham to set up the works at the Soho Foundry.[23]

Murdoch, like many of his contemporaries who worked on gas lighting, came to the invention through the investigation of distillation processes. Though his primary interests lay in mechanical engineering, Murdoch also maintained an interest in chemistry that led him to produce two minor inventions: an iron cement—a mixture of iron and sal ammoniac (mostly ammonium chloride)—that was used to fuse steam engine components, and a new clarifying agent for beers based on codfish skins.[24]

In 1790 Murdoch became interested in the chemical processes by which various chemicals agents, dyes, paints, and other products were made from materials containing sulphur—typically pyrites from the Cornwall mines—by distillation processes similar to coking processes. Like the other British inventors of gas lighting, he was at least in part seeking a replacement for maritime tar, as his patent of 1791 indicated: "a method of making . . . copperas,[25] vitriol, and different sorts of dye or dying stuff, paints, and colors; and also a composition for preserving the bottoms of all kinds of vessels, and all wood required to be immersed in water."[26] In his very brief patent, Murdoch gives a cursory description of the process, which was to "put the [pyrites] into a kiln, house, oven, cone, or heap, covered nearly close, and then set the same on fire, admitting no more air than is sufficient to cause the said pyrites, mundick,[27] or other minerals or ores, to burn." The smoke produced was collected in a receiver and condensed. The patent did not cover any specific apparatus, being about the process of distilling pyrite generally.

There is no evidence that the aforementioned patent led to any commercial activity, but it is likely that Murdoch at some point chose to distil coal or wood in the stoves he was using, like so many others at this time, and that he thought about how the inflammable gases produced in the process might be used.[28] He performed several basic experiments with gas lighting in the period 1794–1798, mostly while he was living in Cornwall.[29] The apparatus he used, described in his 1808 paper, was simple: "My apparatus consisted of an iron retort, with tinned copper and iron tubes through which the gas was conducted to a considerable distance; and there, as well as at intermediate points, was burned through apertures or varied forms and dimensions."[30]

James Watt Jr. and others saw the experiments and apparatus, but they generated no particular excitement at the time. Indeed, there are hardly any references to Murdoch's gas lighting in any archives before 1801. In the many surviving letters and journals, some directly concerning Murdoch in Cornwall, references to gas lighting are fleeting. James Watt Jr. later claimed that he discouraged Murdoch from getting a patent at the time because of the difficulties the firm was having over steam engine patents. Watt's awareness of previous uses of inflammable gases, such as Archibald Cochrane's, made him think a patent would not be defensible.[31]

After Murdoch moved to Birmingham in 1798 to help with the Soho manufactory and the newly opened foundry, the firm of Boulton & Watt began some experimental work, casting a retort for the purpose of testing various coals.[32] James Watt Jr. stated that the apparatus "was applied during many successive nights, to the lighting of the [Foundry] . . . experiments on different apertures were repeated and extended upon a large scale. Various methods were also practised of washing and purifying the air, to get rid of the smoke and smell."[33] However, they decided not to pursue further work with gas lighting for the moment because of ongoing steam engine work and litigation.[34] James Watt Jr. specifically discourage Murdoch from working on the invention, as he wrote to Robinson Boulton in 1799: "With respect to Murdoch's lamp, I am very doubtful whether it can be made to pay & have dissuaded him from thinking about it at present. My father is to have the apparatus over to Heathfield & try some experiments with it."[35] Murdoch then turned to designing steam engines.

DEVELOPMENT AT BOULTON & WATT, 1802–1805

The firm of Boulton & Watt did its first serious work on gas lighting in 1801. By then, the firm had changed significantly. The two founding partners, Matthew Boulton and James Watt, were less active in the management of the firm, and by 1800 Watt had effectively withdrawn from day-to-day affairs.[36] Their sons James Watt Jr. (1769–1848) and Matthew Robinson Boulton (c.1772–1842) were now in charge.[37] James Watt Jr. had received a classical education at a local school, then had begun a practical education at John Wilkinson's iron foundry at the age of 15, studying carpentry, drafting, and accounting. He was then sent to Switzerland and Germany for further education in French and German and in matters related to trade and metallurgy. Returning to England in 1788, he joined a fustian manufacturing firm in Manchester, where he helped promote and sell his father's rotary steam engines to the local textile firms. He became part of the local intellectual community as the co-secretary of the Manchester Literary and Philosophical Society. He also developed an affinity for radical politics, and in 1792 he visited revolutionary France. Returning somewhat less radical in 1794, he turned his attentions to the family business. He largely took over direction of the steam engine works, became active in prosecuting steam engine pirates for patent violations, and helped established the new Soho Foundry.

The nature and the activities of the firm of Boulton & Watt changed over the years. Originally, the firm did not manufacture many components for engines, relying instead on suppliers for most parts and producing only some precision parts in house. Its focus was on the engineering work involved in erecting the engines. During the 1790s, the firm began manufacturing more steam engine parts at the Soho Manufactory, and by 1793 it was making half of the components of its engines. Beginning in 1796, when a new partnership called Boulton, Watt & Co. was formed to run the recently completed Soho Foundry, nearly all manufacturing was done in house.[38] The partners in the new firm were James Watt Jr., Matthew Robinson Boulton, the elder Boulton, and the junior Watt's half-brother, Gregory. The elder Watt was not a partner, but did supply capital to the firm. Although Boulton, Watt & Co. manufactured most components of the gas plants Boulton & Watt sold, it was still the original partnership of Boulton & Watt that sold gas plants to customers. It sent orders to manufacture most parts on to the Soho Foundry.[39] "Boulton & Watt" is used here to refer to both companies, as in practice they were closely related and the distinction between them in the gas lightning context is evident only in accounting matters. The senior Boulton died in 1809 having played no part in the gas business. Gregory Watt died in 1804 and had played no role in running the firm before that. Thus, the management of Boulton & Watt in the context of gas lighting lay entirely with James Watt Jr. and Robinson Boulton, with the former clearly taking the lead and the latter largely uninterested in the gas business.[40]

As described above, the sparks that set off the serious work on gas lighting at Boulton & Watt, with a clear purpose of commercializing it, were Gregory Watt's visit to the Continent in 1801 and his letter to James Watt Jr. mentioning Lebon's ambitions.[41] The idea that they could be beaten to gas lighting by this Frenchman who was actively promoting his scheme was sufficient motivation for them to begin work in earnest. Lebon had, in effect, through the threat of competition, injected entrepreneurial energy into the gaslight project at Boulton & Watt. James Watt Jr. responded to Gregory in December 1801: "Murdoch is going to Cornwall upon his own affairs . . . & upon his return here some decisive experiments are to be made which will determine whether we shall proceed upon his plan or not."[42]

An opportunity to demonstrate the new form of lighting publicly came when Napoleonic France and Britain signed a cease-fire agreement and a preliminary peace treaty in London in October 1801, with a view to finalizing terms in upcoming negotiations. This marked the first time since 1793 that the two countries were at peace, and the event was celebrated with fireworks and "illuminations." The final treaty was to be settled in Amiens, north of Paris. Boulton & Watt decided in November 1801 to have illuminations of their own, and ordered "84 dozen illuminating lamps" of various colors, six dozen candles, and 14 gallons of oil.[43] Gaslights would make a small but nice addition.

Murdoch designed a new retort to be made of cast iron (figure 3.2). A vertical pot that could hold about 15 pounds of coal, it was approximately one foot deep and six inches across.[44] As a preliminary test, the foundry was lit on February 20, 1802 with the retort placed in a stove.[45] When the Peace of Amiens was signed on March 25, 1802, Boulton & Watt set the date of the celebratory illuminations for March 31. Clegg later described these gaslights as two copper vases, one at either end of the manufactory. These "Bengal lights" were fed from retorts placed in fireplaces.[46] Clegg's presence at the firm at this period is crucial because he witnessed important development work which he would then use after he went off on his own beginning in 1805, and later when working for the Gas Light and Coke Company beginning in 1813.

After this first small demonstration, the firm tackled the problem of building larger plants, a fitful process of design and experimentation lasting several years. Indeed, this work was actually a continuation of experiments going on at Boulton & Watt for some time in the context of the chemistry of gases and the design of pneumatic

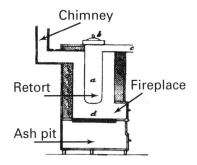

FIGURE 3.2
One of Murdoch's first retorts. Creighton, "Gas-Lights," *Encyclopaedia Britannica* (1824).

apparatus. The work had been begun by James Watt; it went back at least to his study on the nature of water. More directly, however, Watt had become very interested in the medicinal possibilities of gases after the untimely death of his daughter Jessie in 1794 from pulmonary tuberculosis. He was in contact with Thomas Beddoes, a medical doctor who had been thinking about the medicinal uses of gases when, in collaboration with Watt, he began to practice pneumatic medicine in earnest. Watt designed and built a pneumatic apparatus whose purpose was to produce and deliver doses of gas to patients suffering various ailments. The gas, typically hydrocarbonate (a mixture of carbon monoxide and hydrogen), was produced in a distilling apparatus called an alembic.[47] In the case of hydrocarbonate, this was done by dropping water onto heated charcoal. The gas was then washed and cooled with water in a refrigeratory and stored in a gasholder or gasometer until the patient was ready to inhale a dose, diluted with atmospheric air in a ratio of 20 to 1.[48] The subsequent transformation of the apparatus to industrial gas plants was then the continuation of tradition of experimentation in chemistry that spanned theoretical concerns, such as the nature of airs and steams in Watt's water experiments, to applications, such as pneumatic medicine, and now finally to industrial machinery.[49] In commenting on Watt's experiments, Larry Stewart has argued that they were representative of the continuities between experimentation in natural philosophy and in industrial application that were often present in the late eighteenth century in the same physical location. For Stewart, Watt's experiment were "partly exploratory, partly an attempt to assess theoretical explanations for the hotly contested phlogiston, and sometimes an effort to determine whether any practical benefits might be achieved, as in the assessment of the quality of airs for human life."[50] These experiments now culminated in the creation of a new gaslight industry, although the experimental work was now taken up by Watt's heirs.

Stewart's claims about the wide range of experimental work done in specific sites is part of a broader argument about the "spread of experimental enthusiasm" in this period that ultimately contributed to the technological innovativeness found in the Industrial Revolution. Many historians have argued that the experimental techniques learned from natural and experimental philosophy proved important for technological innovation in the late eighteenth century as a culture of experimentation became entrenched.[51] Indeed, Watt himself acknowledged that he learned from Joseph Black "correct modes of reasoning and of making experiments, of which he set me the example, certainly conduced very much to facilitate the progress of my inventions."[52] The style of experimentation Watt brought to his business enterprises persisted after he gave up an active role, to the extent that there was continuity in the context of pneumatic chemistry. Watt designed the original pneumatic apparatus, and others stepped in after 1801 to create industrially useful versions of the gas plant. Watt's experimental heirs in this matter were the Boulton & Watt engineers Henry Creighton, John Southern, and William Murdoch, as well as James Watt Jr., together with some of their customers, and especially George Augustus Lee.

The experimental work was primarily aimed at increasing the scale of the plant. A retort the size of the Amiens one could keep a single lamp burning for four hours or so,[53] and sizable buildings such as the Soho Foundry or Manufactory could easily have hundreds of candles and lamps lit on any given night. The first big challenge lay in finding a way to recharge the retorts quickly. It would take them some years to find an adequate solution, and the first plant at Philips & Lee's mill in Salford was problematic. Specific issues included the expansion of the coal as it was distilled into coke. This created a heavy encrustation on the walls of the retort, which had to be broken up and scraped out before the next load could be introduced. A second issue was speed. For workshop use or for producing gas in small batches such as the Amiens illuminations, recharging the retort quickly was not very important. It was simply not necessary to produce gas for more than a few hours in a row, and one load sufficed. For industrial gas lighting, however, downtime for recharging had to be minimized, and the question of how to recharge the retort became important. Winzler, with his thermolamps, and Cochrane, with his tar ovens, didn't have this problem, because their installations weren't expected to produce gas every day for hours. Boulton & Watt needed to supply gas constantly for hours at a time.

The first attempted solutions involved retort designs that simplified recharging by having two openings, one at the top and a second toward the bottom. These models were quickly abandoned as the increased complexity and associated manufacturing costs proved to be too much of a barrier. Retorts had a short service life, typically a few months, burning out fairly quickly with constant heating cycles. They had to be simple to keep replacement costs and times down, and complicated retort designs would make the cost prohibitive. The experience Boulton & Watt gained in 1802–1805 meant that by the time the first plant was installed at Philips & Lee's the retort design had reverted to a vertical pot much like the Amiens retort, although much larger. (See figure 3.6 below.) These retorts were recharged with cranes that lifted wire frame baskets of coal. Boulton & Watt never tried complicated retorts again.

Another important problem in creating a form of the apparatus for more general use was ensuring the availability of a smooth flow of gas. As was noted in chapter 2, gas was generated in a batch process, meaning that it was produced unevenly over the course of a few hours as the coal was heated; production then ceased once the coal had been exhausted and was no longer capable of emitting gases. The workers

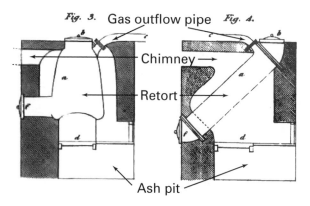

FIGURE 3.3
Early retorts, designed to simplify recharging. Creighton, "Gas-Lights," *Encyclopaedia Britannica* (1824).

(called stokers) then had to empty the retorts and reload them with fresh coal. The gas, however, was consumed evenly from early evening onwards, and there was then a mismatch between supply and demand schedules. In the case of gas, Boulton & Watt used the gasometer as a buffer to store production to achieve the shift from the unevenness of gas generation to the smooth continuous flow required for lighting. As Winzler and the German gaslight pioneers realized, the gasometer would be an adequate means of guaranteeing an even supply of gas. They had eschewed the solution because of the instrument's expense, occasionally using large canvas bags instead.

Gasometers had been introduced at Boulton & Watt when Watt read Lavoisier's *Traité*, and Beddoes later sent him a sketch of one in 1791.[54] Watt then built one with a single counterweight to balance the weight of the gasometer's movable upper section. This gasometer's design was less about measuring the volume of the air than about storing it for later use, but the name stuck. After 1794 Boulton & Watt got into the business of manufacturing pneumatic apparatus of this kind for sale. The volume of the gasometer was quite small: half a cubic foot for the personal model and one cubic foot for the hospital edition (figure 3.4). Sales of the pneumatic apparatus were never vibrant, but Boulton & Watt had enough success with it over the years 1795–1803 that they sometimes made a profit.[55]

Pneumatic medicine ultimately proved to be a failure, as Beddoes despairingly admitted in one of his last letters,[56] but the legacy of pneumatic medicine was not lost. Its apparatus was inherited by the fledgling gaslight technology within the firm.

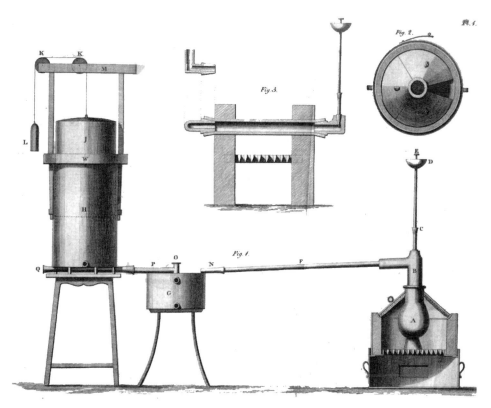

FIGURE 3.4
Thomas Beddoes' and James Watt's pneumatic apparatus. Beddoes, *Considerations on the Medicinal Use, and on the Production of Factitious Airs* (1795), part 2, plate 1. Courtesy of Roy G. Neville Historical Chemical Library, Chemical Heritage Foundation.

Experiments on gas lighting began at just the time when pneumatic medicine began losing its currency. When research was conducted into how to scale up the gaslight apparatus, attention was naturally given to the gasometers that Boulton & Watt had been manufacturing for almost a decade. The gasometers were far too small, of course, as a cubic foot of gas was sufficient for only one burner for one hour's light. A larger model, 8 feet deep with a diameter of 6 feet and a capacity of 300 cubic feet, was built and in use by March of 1803.[57] No sketch or design drawing of this gasometer survives, but it was probably similar to the earliest ones installed at first purchasers of the gaslight apparatus (figure 3.5). These had a design almost identical to the one used in Watt's pneumatic apparatus: the single counterweight was attached to the center of the movable section, with the chain passing over two wheels to the attached

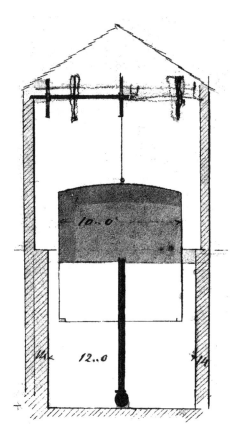

FIGURE 3.5
Sketches of gasometers prepared for Philips & Lee in 1805 or 1806. The basic design, including the counterweight mechanism and the location of the pipes, follows the pneumatic apparatus closely. BWA-MS-3147/5/804. Copyright Birmingham Archives and Heritage.

counterweights. The feed pipes entered from the bottom. With the exception of the feed pipes, this design was the prototype for nearly all of the gasometers Boulton & Watt manufactured from 1806 to 1815. As the pioneer manufacturer of gaslight apparatus, Boulton & Watt also served as an inspiration for many early gasometer designs. In fact, the Royal Society's report on the safety of gasworks in 1814 described the gasometers as being "in effect the same thing as the receiver of the pneumatic apparatus."[58]

Watt's pneumatic apparatus provided the bridge between the technology as it was implemented at Boulton & Watt, and then in the rest of Britain, and the scientific instruments derived from Lavoisier's gasometer. It is rarely possible in this period to find such a clear connection between an industrial technology and a scientific instrument. It is, however, likely that even if Boulton & Watt had not been manufacturing apparatus for pneumatic medicine, people at the firm would still have been aware of Lavoisier's gasometer, as the German chemists had been. It was hardly a secret after all, being described in the *Traité élémentaire* and other works. But the connection is nevertheless of some importance, since Boulton & Watt's experience with manufacturing and using the gasometer certainly made it easier for them to absorb it into gas lighting, a step other gaslight pioneers were more hesitant to take, and points to the importance of Murdoch's invention of gas lighting in a context of strong scientific and technical networks.

PHILIPS & LEE AND THE FIRST WAVE OF COMMERCIAL INTEREST

The period from mid 1805 to late 1806 marked the first time Boulton & Watt turned outward and installed a large gas plant at a customer's mill. In doing this, they began to draw on their wide base of industrial customers for support for their new project. Up until then, the development work was mostly at Soho, but in 1805, spurred on by the emergence of possible competitors closer to home, they solicited orders, and built and installed their first apparatus at Philips & Lee's textile mill in Salford near Manchester. This first installation, and Lee's enthusiasm for it, produced interest and a series of orders from other industrial mill owners. Boulton & Watt did not, however, feel that the technology was mature enough to build more quite yet. Despite the many orders, no further plants were built while analysis and experimental work on Philips & Lee's installation continued during 1806. The firm originally was intent on filling new orders, and its engineers even surveyed many mills, but by the end of 1806 the firm was focusing on other matters, mostly marine steam engines. Work on gas lighting stopped until late 1807. This section tells the story of this first wave of sales and development work from mid 1805 to late 1806.

George August Lee (1761–1826) was the son of John Lee (a theater manager and actor) and his actress wife. He was well educated, and became particularly interested in commerce and the textile trade, working as a clerk and later as a manager at a mill. He joined the Salford Engine Twist Company from 1792 as a managing partner. In 1807 the firm was renamed Philips & Lee when other partners left. The Salford Mill was one of the largest in the Manchester area and was described in Murdoch's paper as possibly the largest in the country. It was equally a travel destination for industrialists seeking inspiration and the curious looking for a landmark. Lee was always interested in improving the mill, and was a technophile who rarely passed on something new. He introduced fire-resistant cast iron frames for his machines, as well as steam power and heating at the mill. Under his direction, the firm grew and prospered.[59]

Lee knew James Watt the elder and James Watt junior and had been a customer of Boulton & Watt since the late 1790s. He had visited Soho with some frequency and would have seen the gaslight experiments going on there. His natural affinity for improvement and novelty undoubtedly made him think about using gas lighting at the Salford Mill. By 1803 he was already considering installing a gas plant, and he sent a letter to Boulton & Watt urging action.[60] Though his enthusiasm could not be accommodated that year, a small experimental apparatus was built at his house in 1804.[61] In early 1805 he was again thinking of gas lighting and became convinced of its viability. After a visit on the part of Murdoch to Manchester, where they discussed the subject, Lee was resolved. On March 27, 1805, he wrote to encourage James Watt Jr. to develop it into a salable product and effectively ordered one for his mill:

I have intended every day since W. Murdochs departure to write you upon the subject of the new mode of lighting by inflammable gas. Is it not an object of attention for you to undertake to prepare the retorts, air-holders pipes & other apparatus with directions & drawings for erecting them, which could afford you a profit & him a recompense for the invention? I am convinced it [will] be as generally introduced as your engines here & that you will have the same pre-eminence & preference. In case you think it eligible to undertake it you will please to prepare the requisite apparatus for our mills as early as possible.[62]

This letter set in the motion the design work and finally the manufacturing of a gas plant for Philips & Lee in late 1805.[63] It was not only Lee's interest that was moving Boulton & Watt, however. More and more engineers and inventors were also trying out the new mode of lighting and making it better known. Hints that others were working on gas lighting had reached the firm by the end of 1804, but in 1805 it became a constant stream that created real pressure to commercialize it quickly or potentially risk losing any head start they might have. Winsor had by then been on a very aggressive campaign for almost a year, and although he was based in London, he sent flyers all over the country—even to Boulton & Watt, being in ignorance of that firm's plans.[64] In another note about gas lighting, one that appeared in the *Monthly Magazine*, John Northern of Leeds described his own small apparatus: "I have the great hope that some active mechanic or chemist will, in the end, hit on a plan to produce light for large factories."[65] Other examples of gas lighting cropped up from still closer quarters. Peter Ewart, another Manchester cotton spinner and a former Boulton & Watt engineer who still acted informally as its agent, wrote to James Watt Jr. in June 1805 about a local shop that had begun to use gas lighting.[66]

Gas lighting was also being used in Scotland. On a trip there, James Watt Jr. wrote in a letter from Glasgow that Murdoch was becoming quite worried "from having learned that the new light is here a subject of general enquiry & admiration, but that a competitor has started up to supply the public demand."[67] The excitement and interest surrounding gas lighting was such that in late 1805, when James Watt Jr. visited Glasgow again, he wrote back to Boulton:

The new lights are much in vogue here; many have attempted them, and some have succeeded tolerably in lighting their shops with them. I also hear that a cotton-mill in this neighbourhood is lighted up with gas. A long account of the new lights was published in the newspapers some time ago, in which they had the candour to ascribe the invention to Mr. Murdoch. From what I have heard respecting these attempts, I think there is full room for the Soho improvements, though, when once they see one properly executed, it will have numerous imitations.[68]

One of these Glasgow gas pioneers soon came to ruin when his store burned down.[69] The other shops using gas in Glasgow and in Edinburgh persisted only for a year or two before the heat, the smell, and the constant maintenance forced most of them to abandon gas lighting, demonstrating that the problems associated with practicability were not easily solved.[70] Finally, there were also a number of reports of shops and buildings in London being lighted with gas in 1805.[71]

Anxious about possible competition, Boulton & Watt were actively soliciting orders from among their steam engine customers. Among mill owners, excitement was building. In August 1805, a Boulton & Watt engineer, Eidingtoun Smeaton Hutton, wrote from Manchester that either James Watt Jr. or Murdoch was needed to deal with the various mill owners who wanted gas lighting.[72] James Watt Jr., meanwhile, was in Scotland spreading the word; he received expressions of interest.[73]

While new orders were being solicited and collected, the preparations for installing a gas plant at Lee's mill were proceeding apace. Hutton, still working in the Manchester area, was asked in September 1805 to survey the six-storey Old Mill and to lay out plans for apparatus with a capacity to replace 3,000 candles.[74] The entire apparatus was to be nestled in among a number of building within Lee's yard, with the retorts set against some old sheds and the gasometers next to the wall of the Old Mill, close to an interior staircase, about 100 feet from the retorts. The pipes were to go from the gasometers into the stairwell and up to the mill's many floors. A further complication was that gasometers and retorts had to be located at a level below the lowest lamps. The primary constituents of coal gas—methane and hydrogen—are lighter than air and could not be pumped downhill except under pressure, something Boulton & Watt was not prepared to do. The only solution was to locate the apparatus below the burners and allow the gas to rise. All of Boulton & Watt's installations required gasometer pits to be dug out, which sometimes created water problems as most mills were located next to rivers. It also made accessing the lower part of the gasometer challenging, a serious problem if the pipes leading gas into or out of the gasometer became clogged, as they frequently did because of the many impurities in the gas. Boulton &Watt eventually gave up the having the gas pipes enter the gasometer from underneath for this reason. The solution the firm came up with was a complicated mechanism involving movable pipe joints (figure 3.10).

The initial plans called for four gasometers sitting in pits of approximately 8.75 feet in outer diameter, the gasometer itself having an inner diameter of 7 feet and a depth of 4½–5 feet for a volume of 170–200 cubic feet.[75] There were to be three retorts 3.6 feet in outer diameter and about 5 feet high.[76] The retorts were to have a capacity of about 25 cubic feet, and were to hold 16 hundredweight (cwt) of coal (figure 3.6).[77] In addition to the retorts, Boulton & Watt designed and built cranes and grapnels to lower wire frame baskets into the retorts. This solution proved cumbersome in the long run because the extra expense associated with the equipment. The drawings showed no purification device of any sort.

The manufacturing of the Philips & Lee apparatus began in September 1805 and proceeded slowly.[78] A retort and a gasometer were ready in early December, and Murdoch went to Manchester to oversee the assembly of the apparatus. When the rest of the apparatus did not arrive promptly, he and Lee grew impatient and began pushing to have it shipped. They hoped to have it on site in order to start lighting by Christmastime.[79]

Murdoch's anxiety was increased by his discovery that Clegg, his former co-worker, was well advanced in his own operations at another mill. He wrote plaintively back to headquarters that "none of Mr Lees lighting apparatus is yet arrived. . . . It is of no use to thinking of taking orders here for your old servant Clegg is manufacturing them in a more speedy manner than it appears can be done at Soho."[80] Clegg bested his former employers by a couple of weeks, at least according to his son, when he managed to start the lighting of Henry Lodge's mill in Halifax, about 50 kilometers northeast of Manchester.[81]

The components that had arrived at Philips & Lee were installed. Despite Murdoch's initial resolution not to do so, he began operations with only one gasometer

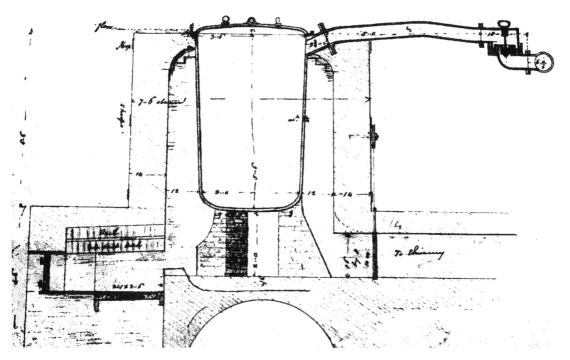

FIGURE 3.6
Vertical retort used by Philips & Lee. BWA-MS-3147/5/804. Copyright Birmingham Archives and Heritage.

and one retort. On New Year's Day 1806 he was able to write the following to Boulton & Watt:

> We were obliged to set to work with the light without the second receiver as Mr Lee was very anxious to determine which of the burners should become general an order for them and many other material will be forwarded by next post. . . . We have lighted 50 lamps of the different kinds this night which have given the greatest satisfaction to Mr Lee & the spinners. There is no Soho stink has yet offended them. Mrs and Misses Lees have visited it this night & their delicate noses have not been offended.[82]

Philips & Lee was the only firm to receive anything in 1805–06.[83]

3.3   FINDING A NEW MARKET IN 1806

The transformation of the pneumatic apparatus into the industrial gas plant involved more than experimental work associated with a different scale and a new purpose. It also included changing the audience for the technology. The pneumatic apparatus was marketed primarily by means of its association with Thomas Beddoes' pneumatic medicine and later his Pneumatic Institution in Bristol, becoming entangled with Beddoes' radical politics to its commercial disadvantage.[84] In its new form, the technology was being offered to a completely different group: the industrial mill owners who were Boulton & Watt's customers for their rotative steam engines. Boulton & Watt's marketing and publicity activities had been an important part of the firm's work for many years. Even before Watt had teamed up with Matthew Boulton, Boulton had established himself as a consummate businessman with a keen sense of finding buyers for his wares.[85] He famously saw a world market for Watt's steam engines and drove the business end of the partnership. In 1784, to publicize Watt's newly developed rotative steam engines, the firm built a showpiece factory, the Albion Flour Mill in London, to put the new engines on display.[86] They had settled on the horsepower as a quantitative standard for the measurement of power. This helped them to convince potential steam engine customers of the value of their products through comparison with other power sources.[87] Their introduction of pressure gauges and later the indicator diagram for steam engines provided a means of measuring the pressure and the power developed by the engine as it cycled. This allowed them to ensure that the engine was closely matched to its load, and thence to give assurances to their customers that they were neither paying for an over-capacity engine nor at risk of needing to add more power capacity soon. As with the unit of horsepower, the gauge and diagram could be used to argue that the engine represented an important savings on other sources of power, such as horses.[88] A similar pattern of reaching potential customers and of quantitatively demonstrating the advantages of gas lighting over other forms of lighting was evident in this new case as well.

Marketing began in 1805, when Boulton & Watt began to work through their network of customers for possible sales of apparatus. They chose to supply the first plant to Philips & Lee, in large part because of Lee's enthusiasm. Boulton & Watt then used the new plant to reach more potential customers, beginning in early 1806. At that time, Robinson Boulton traveled to Manchester to survey the situation at Lee's mill. Lee was ebullient and wanted to show off his new technological wonder. With the plant up and running, he installed some of the first lamps in his own house, and organized a series of showings for the elite of Manchester and his industrialist friends, which Boulton described to James Watt Jr.: "Lee & Co are entertaining all the cognoscenti of Manchester with the wonders of the new lights which is displayed in the mansion as well as the factory."[89] Lee had had pipes laid under the street to his house, which was 300 feet from the factory.

Work continued to bring the installation up to the planned-for 3,000 or so candle equivalents with the arrival of a second gasometer and retort.[90] The lights were being installed in two rooms in the Old Mill (probably the first two floors). These were almost complete when, in early February 1806 Murdoch thought that Lee's effusive talk and demonstrations were creating a real opportunity, and that one of the partners was needed on site to capitalize on it: "It appears a great deal may be done in the lighting way here which will require the assistance of one of you."[91] James Watt Jr., who was more of the promoter of the gaslight project than Boulton, was soon on his way to Manchester.

Once on site, Lee and James Watt Jr. arranged for a grand meeting of potential customers to take place on the night of Tuesday, February 18, 1806. James Watt Jr. described the preparations to Boulton:

There is to be a grand meeting of the Illuminandi at Lee's tonight, to see the wonders in his house and to have the general distribution of the apparatus explained to them, but whether the whole, or any of them are to be admitted into the mill, seems not yet determined. I had a glimpse of it last night; one room with the cock[spur] burners all lighted was nearly free from smell; the other room, in which part of the people had left works and perhaps had not turned their cocks very accurately was somewhat unsavoury. The parties have however seem to make nothing of it and I suppose there will be several orders.[92]

The big night proved successful, with two major orders placed on the spot. Most of the people present at the event became Boulton & Watt's customers for gaslight apparatus, and they represented many of the major industrial mills in the area:

The exhibition upon Tuesday night was completely successfful. Messr McConnel & Kennedy, Mr Jas. Kennedy, Mr George Murray, & Mr Wilkinson of the Chorlton Twist Co. were abundantly gratified. The former has given a positive order for a photogenous apparatus, and the others seem all determined to have it. Peter Marsland has also given an order for the whole of his mills which will take some miles of small pipe and some half dozen retorts & gazometers. Radcliffe & Ross seem also strongly inclined, as indeed seems to be the case with numbers of others. H. Creighton will have full employment for sometime in preparing plans of the mills to show the situation of the retorts & the disposition of the pipes.[93]

Many expressions of interest became confirmed orders within days, and James Watt Jr. wrote to Boulton a few days later with five more firm orders.[94] With this volume coming at a single stroke, Watt Jr. surmised the Soho works would have to be rethought to handle the expansion: "As soon as I can get a little leisure I shall write what occurs respecting the arrangements which it may be necessary to adopt at the foundry & Soho for meeting these orders."[95] Murdoch was sent back to Soho to begin making preparations to fill the flood of orders, but Watt Jr. wanted him back with him as soon as possible to consolidate further possibilities in Leeds and Derby. The hectic pace and the sustained salesmanship were not to Watt Jr.'s liking, but he thought he had to seize the opportunity before it slipped away: "It is very evident that whatever is to be done, must be done now, as the whole tribe of engineers, founders, braziers & c will become our competitors, as soon as they know how."[96]

In a letter to his father written the next day he observed that "the Photogenous business . . . goes on very prosperously most of the large mills having given their orders, to which however for the most part, I have been obliged to eat my way; a mode of solicitation which is not very congenial either to my habits or obligations."[97] Watt senior responded briefly in encouraging tones, but he was not much enthused about the lighting project: "I am glad to hear of your success in the new light way. I doubt your circuit will [?] and wish you soon home however business must be looked after."[98]

Murdoch arrived at Soho on March 1, 1806 and worked with Southern to make preparations for the new orders. In the meantime, James Watt Jr. received confirmation that Clegg had indeed set himself up as a competitor and had lit Lodge's mill. Clegg's father had been a Boulton & Watt customer for many years, but had run into some financial difficulties recently with slowing trade, with the result that he had not been able to make timely payments on his accounts. Watt Jr. reacted to the gaslight news by shutting down the senior Clegg's credit with the firm: "As I find the report is true that Lodge of Halifax has had a photogenous apps. erected by Clegg, I think it would be but reasonable that he should no longer be a trespasser upon our liberality, and therefore propose that his acct. should be sent in immediately with a letter from Soho urging payment."[99] It was not long before the older Clegg was having serious financial problems.[100]

In the first two weeks of March 1806, Watt Jr. continued his sales tour, confirming more orders as he traveled around the mills in the vicinity of Manchester at a frenetic pace.[101] With the large number of orders and consequent preparations to fill them going on at Soho, led by Southern and Murdoch, Watt Jr. was enthusiastic, but Boulton, emotionally and physically more distant from Watt's marketing campaign, wrote back to him with a more sober assessment of their prospects: "My expectations of profits from the photogenous orders are very moderate compared with your prospective statement; so much so that if the results should not fall short of one half of the anticipated profits they will not be disappointed."[102] He also feared the effect the large volume of orders would have on their business. The steam engine line was their profit-maker of long standing, and he did not want to divert many resources away from it as this new business was threatening to do, and indeed had done in the last few weeks. Watt Jr. and Boulton had discussed how to handle the manufacturing of all the parts, agreeing that they should be farmed out to other manufacturing firms to the extent possible. In this letter, Boulton recognized that they would effectively transfer knowledge and skills to these other firms, which would be their competitors after two years or so, but he thought this price was worth paying to protect their existing steam engine business and would not hurt them in the short term. Boulton & Watt was generating expertise in gas lighting, and it was trickling down the supply chain and to the firm's employees. In effect, Boulton & Watt was training many of the first generation of gas engineers and workers.

On March 14, 1806, James Watt Jr. set out with Lee and Murdoch, who had come up from Soho five days before, to visit Leeds and Preston and collect what orders they could among mill owners there, including Benjamin Gott, another major mill owner.[103] They returned to Manchester four days later, and Watt Jr. was finally satisfied. He had secured three more orders; thus, he had commitments from almost all the large mills worth having as customers.[104] He wrote to his father the day after his return, satisfied with the fruits of his efforts of the last three weeks: "This concludes my labors in this vocation for the present, as the orders for all the large mills, which are thought tolerably safe, are now secured. Many of the smaller ones will necessarily drop in the course of another year & the remainder will fall to the lot of our competitors, unregretted by us."[105] Watt Jr. turned his attention to other matters, leaving Manchester the next day and the work of gas lighting to Murdoch, Southern, Creighton, and Hutton.[106] He did not mention the business much in any surviving letters from this point until 1807, when Winsor and the National Light and Heat Company burst on the scene threatening to upset the status quo and forced Watt Jr. to pay a great deal of attention to gas lighting once more.

Preparations began to survey the mills that had placed orders after Lee's demonstrations even as Watt Jr. was on his sales tour. Creighton arrived in Manchester at the end of February 1806, and began making drawings. These drawings usually consisted of only ground plans with some basic estimation of requirements, but even this work was so voluminous that Creighton was overwhelmed. Watt Jr. had to call in Hutton from Scotland to help, as well as to learn more about gas lighting so as to be able to do engineering work himself.

Creighton and Hutton prepared preliminary sketches of the mills, including McConnel & Kennedy's cotton mill at Manchester and Benjamin Gott & Co. at Leeds.[107] They worked until June, surveying at least sixteen mills.[108] With this flood of drawings coming in, Watt Jr. instructed Creighton and Hutton at the beginning of July 1806 to send no more preparatory surveys for gaslight apparatus, although a few more trickled in.[109] Despite all this, very little was done with the surveys.[110] Boulton & Watt did not even manage to get estimates out to the mill owners that had placed orders. Customers, however, were anxious to move ahead despite the delays in receiving quotes. In June 1806, Creighton wrote to the head office communicating the desire of some of these to begin work at least on the gasometer pits while the summer days were still long, but nothing was forthcoming.[111]

This mid-1806 batch of drawings represents the densest cluster of order taking by Boulton & Watt in the firm's history as a gaslight manufacturer. At no other time were preparations made for such a number of installations. Despite this, letters from Creighton and Hutton from May and June of 1806 indicate that they were more taken up with steam engine work than with gas lighting. Boulton's desire to maintain the firm's focus on steam engines was effective even in this period.

A long pause in collecting new orders now set in.[112] Some customers expressed their unease at the delays. Benjamin Gott wrote in July 1807 to chide Watt Jr. a little over the lack of activity in matters gaslight: "Perhaps Mr. Winsor has eclipsed your new lights—or perhaps you do not wish to give him any insight into the way it by at present proceeding with your new works. We shall wait patiently until you find it convenient."[113] Hutton wrote in October 1807 to say that another anxious customer was "much disappointed in not having the . . . appt this season." Hutton added that he was unwilling to take any more details about mills until he had received any definitive news about timing.[114] In the end, the next survey for a major order was for made for James Kennedy in May of 1808.[115]

3.4 DESIGN WORK

Boulton & Watt's design work for gas lighting involved solving a host of technical questions about the structure of the apparatus, as well as identifying and refining many parameters related to the operation of the gas plant, such as what sorts of coals produced the most gas of the highest illuminating quality, and what was the ideal retort temperature. Apparatus structure and mode of operation also influenced one another, and so Boulton & Watt refined each in a series of design iterations and sets of experiments between 1805 and 1808.

This design work was divided into three phases. The first phase, described above, concluded with the first basic design including gasometers and vertical retorts. The second phase began in 1806, after the Philips & Lee apparatus was operating, and centered on the relative sizing of the retorts and gasometers. The final phase, a prelude to further sales in 1808, led to a move to horizontal retorts, and to changes in gasometer piping to avoid having pipes deep in the gasometer pit.

The refinements in the second and particularly the third phases came from learning by using, reflecting how Boulton & Watt understood and responded to Philips & Lee's experience in using gas lighting. Boulton & Watt was clearly desirous of providing the cheapest possible solution, as well one that was easy to maintain. It was less concerned with other points in the design, particularly purification. The purification of gas the firm provided was basic, little more than passing the gas through water. Although Boulton & Watt could have found solutions suggested by others at this time to purify gas, such as lime, it never made much effort in this regard. The economic question, rather than what immediate users of gas lighting experienced, was the essential design parameter for Boulton & Watt. This dynamic of designing for the one who paid the capital costs of plant was a feature of installations at industrial mills, where the mill owner was fundamentally interested in the economics of gas lighting over candles. With the advent of gas as an urban network, however, the dynamic changed completely, as the final user had no direct engagement with the capital costs or even the operation of the plant. The design of gaslight technology was profoundly transformed by this change in users, as will be described in chapter 5. The Gas Light and Coke Company, in addition to controlling the habits of its users, had to pay far more attention to purification.

ASSESSMENT OF PHILIPS & LEE'S PLANT, 1806

Simultaneous with the marketing and sales campaign of February to June 1806, Boulton & Watt engaged in research and development work at Philips & Lee and at Soho to refine the design of their gas plants. The first part of the Philips & Lee apparatus, that serving the first two floors, was completed in February 1806.[116] James Watt Jr. was feeling confident about the project, and he described how gas lighting fared at the mill to his father in optimistic tones:

Mr. Lee's answers extremely well in the mills, when the upper part of the windows are generally a little open, which gives it a complete ventilation. Indeed the pipes seem so tight, and
the people manage the lighting & putting out so well in general, that this precaution scarcely
appears necessary, and in fact had always been practised when candles were used. The Argand
lamps are all discarded & cockspur burners only are in use. . . . This will be universally adopted
in the mills, which have hitherto been all underlighted. . . . The air which first comes over
has a bad smell in burning, but this does not last long, and that which [subsequently?] succeeds
is the most brilliant. Towards the end of the operation they are obliged to open the cocks
much wider to obtain the same portion of light, which occasions the variation I have mentioned in the consumption of gas.

Chimnies are put over the burners in Mr Lee's house to carry off the consumed air & effluvia,
which now answers very well, but have not been brought to bear without some alterations
and occasional offense to the olfactory organs of the Gentleman & Lady & of their
visitors.[117]

Despite this optimism, the Philips & Lee apparatus was still very much a work in
progress. As orders began coming in 1806, Boulton & Watt didn't yet have an idea
what size a gas plant should be to feed a given number of lights.[118] Experiments were
done in February 1806 at the Philips & Lee mill to try to determine this. Watt Jr.
wrote a long letter to Southern back at Soho summarizing the results and giving his
opinion as to what they meant the sizing of the retorts and gasometers. He acknowledged that the experiments were limited in scope. The gasometers were of insufficient
capacity to hold the entire daily production of the two retorts, and hence complete
production figures could not be obtained directly. He thought, however, that although
no general rules for gasometer and retort size would be adopted, it was likely that
gasometers would be designed to hold three hours of gas production from the retorts.
He then observed that "it is a combined question of first cost; conveniency of room
& comparative wear & tear."[119]

Southern responded to James Watt Jr. in two letters.[120] In one dated March 9, 1806,
Southern identified what was to become the determining issue for the relative size
of the retorts and gasometers: the added cost of every heating and cooling cycle.
Southern argued that it was best to have the retorts always hot so as to avoid wasting
coal in reheating the ovens.[121] His argument was based on the economy of coal alone.
But as it turned out over the long run, there was even more cost associated with each
heating and cooling cycle: retorts lasted longer if kept under constant heat and not
allowed to cool. The more frequently a retort went through a heating and cooling
cycle, the more quickly it burned out and the more susceptible it was to cracking. As
the gaslight industry matured, retorts, which at first were replaced on the average

every nine months in the early years, had extended life spans of twelve or fifteen months by the middle of the century. It was soon standard practice to keep them under constant heat. If a retort had to be taken out of operation, it was allowed to cool slowly over the course of a week.[122] Boulton & Watt did not think in these terms in 1806, but by the end of 1807 the retorts were always under heat, except on Sundays when the mills shut down.[123] Gasometers, on the other hand, were not subject to any such operational strains, and therefore they would be less liable to future replacement costs. In general, the larger the gasometer was, the smaller the retorts could be, with the limit set by the condition of retorts producing gas 24 hours per day.

The drawings Southern prepared for the final configuration of the Philips & Lee plant (figure 3.7) show that he used a mode of calculation similar to what he had suggested to Watt Jr. in his letter.[124] The drawings propose eight gasometers with approximately 8,000 cubic feet capacity in total, supplied by six retorts, approximately twice that required if they were run 24 hours a day.[125] In fact, Southern's idea of running them constantly had not yet been adopted at Lee's, and the plan was to run them 12 hours a day.[126]

NEW DESIGN WORK, 1807–08

The Philips & Lee installation was largely finished in 1806. Boulton & Watt did nothing further with gas lighting until late 1807.[127] In December 1807, a second review

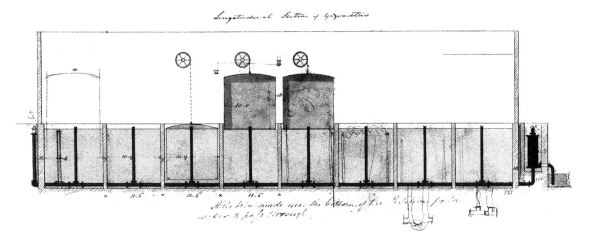

FIGURE 3.7
Side view of the gasometers at Philips & Lee. BWA-MS-3147/5/804. Copyright Birmingham Archives and Heritage.

of the Philips & Lee plant was done by Creighton, working with Lee in Manchester. Watt Jr. and Murdoch remained at Soho, and a series of letters, some quite lengthy, were exchanged during December 1807 and January 1808. With two years' experience using the technology, Lee and Creighton were able to give more details about how gas lighting was functioning, what was working, and what needed to be changed. Although there was much discussed during the review, little was finally changed except that the retort layout was changed to a horizontal setting, and that the gasometer piping was moved from beneath the gasometer to the top. The economics and efficiency questions raised at this time are explored in the next section.

The first letters from Manchester discussed a number of problems, the most important of which was the foul odor of the gas. The rancid smell had been the most evident objection to gas lighting from the outset, and this problem had plagued Lebon. Watt Jr. and others had been aware of the smells gas lighting could produce from their own experiments in Birmingham, referring to the "Soho stink."[128] They had been hopeful that odors would not be a problem with expanded apparatus, and in their first comments from Philips & Lee they dismissed the odors as not being strong enough to be of much concern. Now, however, Creighton addressed the matter again. He mentioned that cannel coal was less malodorous than other types. Lee was also trying to purify the gas by washing it with running water. This worked to some extent, but since the water had to be pumped into the mill from the river, it could only be done when the steam engines were running.[129] In Lee's opinion, the purification problem was the most pressing one, and would be for some time. At the end of 1808 he wrote to Watt Jr. to ask "have you consulted Murdoch about forcing the gas thro' a head of water or other means of purifying it? It is the great desideratum remaining."[130]

Another factor affecting the odors was the completeness of combustion in the lamps. Uncombusted gases smelled worse than those completely consumed in the flame, and so it was paramount to ensure that the burners' apertures were set correctly. Unfortunately, this solution depended on constant gas pressure, which was in turn a function of the gasometers' predictability, but, as gas engineers found with time, the desired constancy was hard to reach. The counterweight system of pressure regulation was eventually abandoned in favor of valve regulators. This was not yet the case at Philips & Lee, however. Creighton explained that "the friction of the gasometer pulleys, shafts & c occasion a considerable difference at times but the gasometer being loaded when the lights are running is a principal cause of this variation."[131] In fact, the counterbalance weights were being shifted every day. When the gasometers were being emptied, an extra 70 pounds was lowered onto the gasometer to drive out the gas, a procedure that completely negated the purpose of the counterweights.[132]

The initial reports Creighton sent to Soho prompted some more serious reflection on the part of Murdoch and Watt Jr. They responded with an extensive list of 27 questions for Creighton and Lee, which they forwarded on January 19, 1808.[133] The questions showed that Murdoch and Watt Jr. were not intimately familiar with how the gaslight apparatus was functioning at Philips & Lee or even with the details of its physical arrangement, as the list included questions such as these: "How do the grapplers last?" "What number of men are employ'd in attending to the retorts & c?" "Is there any use made of the tar?" "Are there any of the retorts worked with separate chimnies?" The nature of these questions indicates that Philips & Lee were doing considerable design and maintenance work, probably the great majority, in keeping gas lighting going at their mill. They were not mere users whose employees were only involved as far as day-to-day operation of the apparatus were concerned. Rather, they maintained and repaired the apparatus, and made decisions about matters of design and implemented them. The question about the chimneys in particular demonstrates that Murdoch and Watt Jr. had not had access to schematics of the final configuration of the apparatus. Murdoch may not have been at Philips & Lee since the spring of 1806.[134]

Creighton's answers to Murdoch's queries, which were mostly about how the apparatus was operated, indicate that many changes had been made since the beginning of 1806, mostly in order to minimize the effects of the wear and tear on the apparatus and the intense heat. The retorts were now always under heat except on Sundays, when the mills shut down. Whereas in the past the retorts were allowed to cool at the end of each cycle, they were now charged under heat.[135] Two men ran the apparatus, each working a 14-hour shift. They recharged the retorts in the two hours they were at work together. For this they received £3 a week.[136]

Gas leaks were also a problem, and would be for the gas industry throughout the nineteenth century. The pipes immediately connected to the retorts were luted with lead to prevent the gas from leaking, but it melted easily and the joints needed to be resoldered every second charge, no easy matter considering the heat. Philips & Lee were going through two hundredweight of lead per month.[137] The heat was still affecting the grapnels, and they continued to bend or break frequently. The cranes were likewise showing signs of heat strain.[138] These problems with the grapnels and cranes, combined with the lower efficiency of vertical retorts as compared to horizontal ones (according to Lee), prompted Lee to recommend that Boulton & Watt redesign the retort.[139] In short order the vertical retorts would be abandoned in favor of the horizontal version. The new retort layout was the most important outcome of this design review.

The distillation byproducts, especially the tar, were becoming a serious issue. The original plan for the plant had included tar traps to separate the tar from the gas. These were large-diameter pipes filled with water through which the gas was bubbled. The tar and spoiled water were removed and the water refreshed via valves, but in some cases these traps were too close to the furnace flues, causing the water to evaporate rapidly. The tar then became thick and consequently the valves moved with difficulty. The only solution was to replace the water every half hour.[140] The tar collected from the traps was allowed to pool in an excavated cistern, with the idea of finding a buyer for it at some point.[141] The tar was obviously leaching into the soil, and Creighton observed that ground water was mixing with it.

Despite the traps, tar was getting into the pipes. Creighton had mentioned earlier in 1808 that the burners were becoming encrusted with tar,[142] and he now explained that the pipes closest to the retort mouth needed to be cleaned out regularly, which entailed similar challenges from the intense heat as the reluting. The pipes leading to the burners and the main pipe from the gasometers were almost choked up with "a matter resembling hardened coak mixed with tar."[143] If the tar or other liquids accumulated in the pipes closer to the burners, the flames were liable to be extinguished or to "vibrate greatly," and the burners were "apt to get partially or wholly stopt with a hard substance which is probably occasioned by the iron's oxidating in some degree & the tar there growing thick & mixing with it."[144] This could only be removed by "hammering or other considerable force which may endanger the soldering," and Creighton recommended redesigning the burners to make them easier to disassemble.[145] As difficult as these problems were for Boulton & Watt, they would take on an even greater degree of complexity for the Gas Light and Coke Company, whose network encompassed a city.

With the recent design work completed, Lee was of the opinion that the gaslight business was ready to grow rapidly: "Depen'd the demand for the apparatus will one day be as great as sudden, & you have now an oppt. of more leisure to determine all difficult or dubious points," of which the most pressing was purification.[146] Boulton & Watt never solved that problem.

EFFICIENCY

The question of the economic efficiency of gas lighting was paramount for Boulton & Watt because Watt Jr. in particular believed it would be financial reasons that would drive mill owners to use gas lighting. To make a financial case for its steam engines, the firm had used the horsepower as a unit of power, and later had used pressure gauges and the indicator diagram. These allowed the firm to compare its engines with other power sources, and to demonstrate quantitatively to its customers that they were saving money. Similarly, to make a financial case for the adoption of gas lighting, it wanted to compare it to other forms of illumination, primarily tallow candles. When the firm began to build its first full-size apparatus for a mill in 1805, the intention was that a gaslight installation would provide as much light as the candles it would replace in a given mill. As the Philips & Lee plant reached completion in early 1807, Boulton & Watt began to do the numbers to see how much money gas lighting was saving the firm.

Boulton & Watt's focus on the economics of gas lighting reflected the industrial context in which the firm was working. Watt Jr. believed that mill owners would only adopt the technology if it would save them money, much as was the case with mines and steam engines. In this, however, he was mistaken. In the first years of the gas industry, what drove the adoption of gas lighting on a large scale was its brightness, safety, and ease of use relative to candles, advantages which came out most clearly when gas lighting was deployed as a network and the users did not have to concern themselves with running gasworks. It was in the urban context that this network form of the technology thrived, and urban users of gas lighting were in fact willing to pay more for gas lighting than for other forms of illumination. Boulton & Watt's mill owner customers had to run their own gasworks, however, and gas lighting was not necessarily any simpler and could in fact be a great deal more complicated. For them, the economic case had to be compelling. After Boulton & Watt gave up on gas lighting, mill owners would adopt the technology to some extent, but they would represent only a small fraction of the industry, with large urban companies dominating.

In regard to economics, the lengthening of the workday has often been suggested as a reason why mill owners adopted gas lighting: they could make more money by getting more use from their buildings and machines.[147] This was not, however, taken into consideration by Boulton & Watt, for a number of reasons. One was that the option of long work hours was already open to the mills with other forms of illumination. In fact, none of Boulton & Watt's customers appeared to have changed their working hours with the adoption of gas lighting, and some were already working 24 hours. When calculating the financial benefits of gas lighting, Boulton & Watt never included these sorts of considerations, and always assumed that a mill's hours of operation would remain the same. Finally, many mills did not use lighting at night for the half of the year with more hours of daylight.

Why did the early adopters among the mill owners choose to implement gas lighting? They clearly had hopes that they could save money, and the greater brilliance of the lights certainly drew them. Candles were messy because they melted and ran, and had to be trimmed regularly lest they create fire hazards. Gas lighting offered an opportunity to mitigate these problems. Part of what motivated the early adopters was technological enthusiasm, something most evident in the case of George Augustus Lee, the central figure among Boulton & Watt's customers. His affinity for new technologies has already been mentioned. Willing to spend enormous sums on the project, he maintained his zeal, almost without regard to how much it was costing him. Other mill owners also showed the same enthusiasm in the first two or three years of Boulton & Watt's work.

THE FIRST TESTS, 1805–06

In making economic comparisons between gas lighting and candles, Boulton & Watt used three experimentally calculated parameters. The first was the ratio of the volume of coal gas required to provide light equal to one pound of tallow candle. The second parameter was the volume of gas needed to provide one candle's luminosity for one hour. The third parameter was the gas yield from a given quantity of coal, which, when combined with the first parameter, gave a sense of how much coal was needed to replace candles on an ongoing basis.

The first such efficiency experiments were done in 1805 to prepare for the Philips & Lee installation, which was meant to provide light equivalent to the 3,000 candles Lee used. Boulton & Watt adopted Count Rumford's method to compare the luminosity of different light sources. It consisted of moving two light sources so that the shadows they cast on a screen were of equal intensity. Their relative distances from the screen corresponded to their relative luminosities. This was to be the standard test for the first few years of the gas industry, but misgivings were expressed about its validity, and it was eventually abandoned.[148]

The 1805 tests (table 3.1) were first done with a small retort (15 pounds capacity) and then a larger one (8 hundredweight) using different types of coal, including cannel coal on the recommendation of William Henry, a local chemist who had researched and published on gases for some years. In 1804 he had formulated what later came to be known as Henry's law of partial pressures. The 1805 tests showed that gas from cannel coal burned brightly and left relatively little ash.[149] It also contained little sulphur. Cannel coal earned its name because in Lancashire, where it was abundant,

TABLE 3.1

Results of luminosity and efficiency tests performed by Boulton & Watt in Birmingham in 1805.

Date	Flame luminosity (candle)	Coal	Tallow equivalent (ft³/lb)	Coal yield (ft³/cwt)	Flame efficiency (ft³/candle)
2 Mar 05	5.5	Wednesbury	17	411	0.56*
2 Mar 05	8.5	Wednesbury	17	411	0.36*
13 Jul 05		Cannel	17	420	0.35*
23 Sep 05	4	Wednesbury	16	370	0.36*

*These numbers are calculated from Boulton & Watt's experimental results. The rest are directly from the archives. For sources, see note 151.

the local accent made 'candle' sound like 'cannel', and it would become one of the favorite coals of the gas industry. Henry had become familiar with the coal based on his own 1804–05 experiments in the nature of coal gas,[150] and this constitutes another example of how the nexus between natural philosophy and the arts was effective in work on gas lighting.[151] Watt Jr. decided from these tests that they would need 750 burners for Philips & Lee if the 3,000-candle estimate was used.

These experiments were expanded (table 3.2) once the first part of the Philips & Lee apparatus was operating in early 1806. Murdoch and Creighton ran the experiments, and Lee also added some figures of his own from their first month of operation.[152] In these tests, the gasometers were allowed to fill to close to capacity, and were then isolated from the retorts. The mill's 203 burners were all set to the same aperture, and they concluded that the efficiency figures were in the range given by the 1805 Soho experiments.[153]

NEW EXPERIMENTS, 1807–08

Experiments restarted in early 1807 (table 3.3) when James Watt Jr. become aware that Winsor and company in London had gaslight plans of their own. Watt Jr. decided to write a paper on gas lighting which would include some firm numbers about the economics.[154] In the preparation, a new set of efficiency experiments were run at Soho, giving results broadly in line with previous ones. Watt Jr. calculated the cost of the completed Philips & Lee apparatus, which was by now running with eight gasometers and six retorts, at £5,000, including all pipes, burners, excavation work, and new buildings. This was £625 per year, assuming a depreciation of 12.5 percent.

TABLE 3.2

Results of luminosity and efficiency tests performed by Henry Creighton, William Murdoch, and George August Lee in Salford in 1806.

Date	Flame luminosity (candle)	Coal	Tallow equivalence (ft³/lb)	Coal yield (ft³/cwt)	Flame efficiency (ft³/candle)
1 Feb 06	1			415	
1 Mar 06	3.5	Clifton	16		0.36*
1 Mar 06		Clifton	19		0.43*

*These numbers are calculated from Boulton & Watt's experimental results. The rest of the figures are taken directly from the archives. For sources, see note 153.

TABLE 3.3

Results of luminosity and efficiency tests performed by Boulton & Watt in Birmingham in 1807.

Date	Flame luminosity (candle)	Coal	Tallow equivalence (ft³/lb)	Coal yield (ft³/cwt)	Flame efficiency (ft³/candle)
26 Feb 07	8.9	Wednesbury	16.25	217	0.76*
26 Feb 07	9.3	Wednesbury	16.25	250	0.83*
6 Mar 07	7.7	Wednesbury	29.25	286*	1.16*
7 Mar 07	9.75	Wednesbury	25.5	328	1.05*
9 Mar 07	12.47	Wigan cannel	14.5	331	0.74*
10 Mar 07	13.1	Wigan cannel	13.2	375	0.77*
11 Mar 07	12	Wigan cannel	14	226	0.24*

*These numbers are calculated from Boulton & Watt's experimental results. The rest of the figures are taken directly from the archives. For sources, see note 157.

Philips & Lee were burning about £104 worth of coal a year, meaning a net yearly expense of £729. The cost of labor was not included in the calculation because it was "certain that the cost of attendance upon the candles would be more than upon the apps." Watt Jr. reckoned that Lee would have had to spend £2,600 to light his mills with tallow to get the equivalent illumination, using the recently calculated 20 cubic feet per pound for the equivalence. This meant that Philips & Lee were saving £1,871 per year with gas lighting. In fact, these calculations proved to be optimistic, not least because the retorts were assumed to last for approximately five years, and no allowance was made for the cost of capital.[155] At this point, however, Watt Jr. once again dropped the entire affair because he found out from his father, who was in London at the time, that it would be too late to have anything published by the Royal Society at that point.[156, 157]

The idea of the paper was picked up again in December 1807, when the firm did the most wide-ranging experiments on efficiency yet. They began with a series at Soho using a small apparatus (table 3.4).[158]

Creighton, who ran the Soho tests, then went on to Philips & Lee to run experiments on the full-size plant there (table 3.5). After three weeks he sent a preliminary three-page report to Boulton & Watt, to which Lee added a page of his own comments.[159] They had tried various types of coal in Lee's main apparatus, measuring how much was produced and how much was needed to feed burners of various luminosities. The results were that the figure hitherto used for the gas yield of coal (360 cubic feet per hundredweight) was approximately correct, and perhaps even a little

TABLE 3.4

Results of luminosity and efficiency tests performed by Henry Creighton in Birmingham in 1807. For sources, see note 158.

Date	Flame luminosity (candle)	Coal	Tallow equivalence (ft^3/lb)	Coal yield (ft^3/cwt)	Flame efficiency (ft^3/candle)
1 Dec 07		Wednesbury		340	0.36
1 Dec 07		Wednesbury		367	0.46
1 Dec 07		Wednesbury		305	0.46
1 Dec 07		Wednesbury		360	0.62
1 Dec 07		Merthyr		448	0.71
1 Dec 07		Merthyr		400	0.89
1 Dec 07		Wedgwood		475	0.36
1 Dec 07		Wedgwood		475	0.38
1 Dec 07		Wedgwood		438	0.43
1 Dec 07		Middleton Rotheswell		396	0.38
1 Dec 07		Middleton Rothwell		440	0.33
1 Dec 07		Wednesbury		530	0.32
1 Dec 07		Wigan cannel		500	0.25
1 Dec 07		Wigan cannel		567	0.25

TABLE 3.5

Results of luminosity and efficiency tests performed by Henry Creighton in Salford in 1807.
For sources, see note 160.

Date	Flame luminosity (candle)	Coal	Tallow equivalence (ft^3/lb)	Coal yield (ft^3/cwt)	Flame efficiency (ft^3/candle)
28 Dec 07	4	Clifton		390	0.7
28 Dec 07	4	Blackrod cannel		337	0.52
28 Dec 07	2	Clifton		390	0.9
28 Dec 07	2	Blackrod cannel		337	0.5

optimistic. The various sorts of cannel coals were judged roughly equivalent, but Creighton thought their value had been overestimated in the past.[160]

The flame efficiency figures were, however, lower than figures from 1805 to 1807, even the most recent ones run at Soho on December 1, 1807. Whereas in the past most tests, though not all, had given flame efficiency figures around 0.3–0.5 cubic feet per candle, these latest figures from Lee's full apparatus placed the range between 0.5 and 1 cubic foot per candle. Lee's judgment on these results was they "afford no flattering prospect of extending the oeconomical principle beyond our former practice & expectations."[161] He thought part of the cause of the lower efficiency value was the poor gas quality that came from over-distilling the coal, as the gas given off at the end of the cycle was of lower quality.

When James Watt Jr. and Murdoch received Creighton's report, they were disappointed by the results and by Lee's opinion of the financial and business prospects of gas lighting. How flame size affected the illuminating efficiency of gas lighting became a central question in these discussions, and Creighton and Lee resumed their experiments in January of 1808 (table 3.6).[162] This time, they looked at tallow equivalence and flame efficiency, and found that the equivalence values of the small flames used at the mill were much higher than with large flames (i.e., the new values showed more gas was needed to replace a single candle). This meant that the previous figures for the tallow equivalence of gas lighting of 14–20 cubic feet per pound were only correct for the large Argand burners. There had been a tendency to favor the cockspur burners (figure 3.8) over the Argands because they were simpler and cheaper to manufacture and much easier to keep clean. The Argands had cylindrical glass chimneys and corresponding mounts, whereas the cockspurs were effectively tubes.[163]

TABLE 3.6

Results of luminosity and efficiency tests performed by Henry Creighton in Salford in 1808.

Date	Flame luminosity (candle)	Coal (lamp)	Tallow equivalence (ft^3/lb)	Coal yield (ft^3/cwt)	Flame efficiency (ft^3/candle)
25 Jan 08	1	Clifton (argand)	24.4*		0.61
25 Jan 08	1	Cannel (argand)	20.8*		0.52
25 Jan 08	1	Clifton (cockspur)	36*		0.9
25 Jan 08	1	Cannel (cockspur)	20*		0.5
25 Jan 08	1	Cannel	57		1.43
25 Jan 08	4	Cannel	20		0.5
25 Jan 08	4	Clifton	28		0.7
25 Jan 08	6	Cannel	16		0.4
25 Jan 08	6	Clifton	24		0.6

*These numbers are calculated from Boulton & Watt's experimental results. The rest of the figures are taken directly from the archives. For sources, see note 162.

These new reports Creighton sent to Soho prompted more serious reflection on the part of Watt Jr. and Murdoch. Watt Jr. responded with questions trying to reconcile the better luminosity and efficiency figures from previous years with Creighton's new figures.[164] In his response, Creighton attributed most of the difference to flame size.[165] With regard to the 1805 Soho experiments in particular, the "great differences" with which Watt Jr. found "extraordinary," both in terms of quality (illuminating power) and quantity produced,[166] Creighton pointed to a number of other differences that could explain the puzzle. At Soho, they had used even larger flames than the four candle Argands, typically 10, 12, or 14 candles, and extrapolating his results to flames of that size accounted for most of the difference in gas quality. In terms of quantity of gas generated, Creighton thought the scale of the experiments accounted for the divergence. At Philips & Lee they never got more than 350 cubic feet per hundred-weight, versus the 560 recorded at Soho, but whereas at Soho they had used a small retort containing 10 pounds of coal, heated very rapidly, at Philips & Lee they heated 1,700 pounds of coal for hours, to the point of exhaustion. The circumstances were simply too different to achieve identical results, and the Soho results were based on invalid extrapolations.[167] The testing and design of gaslight apparatus then depended to an important degree on experiments done with a full production-scale plant. The small experimental machinery used by all gaslight pioneers gave results that differed

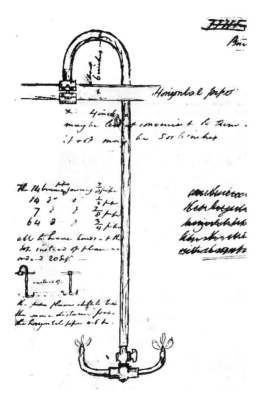

FIGURE 3.8

A cockspur burner designed by Boulton & Watt. BWA-MS-3147/4/115, p. 286. Copyright Birmingham Archives and Heritage.

in important ways from what Boulton & Watt's large plants were now producing. A further change in scale would come once the Gas Light and Coke Company started operations, and there the company was able to do even more extensive experiments. Learning by using was essential in the design of better gas apparatus, and hence sizable institutions which could support experimentation with production apparatus, such as Boulton & Watt and Philips & Lee, were essential to the design process.

With regard to the economics, Creighton went over Watt's analyses and figures of June 1807, which had determined that gas lighting saved Philips & Lee £1,871 per year.[168] He sent in a new report that formed the basis of the Murdoch's 1808 Royal Society paper, with almost all the figures in it identical to those found in this letter. He reported that at this point there were 633 cockspur and 271 Argand burners in the Old and the New Mill, the counting houses, and Lee's home. This was equivalent to a total of 2,500 candles.[169] These consumed 2,500 cubic feet of gas per day, requiring 7 hundredweight of cannel coal. The yearly cost of distilled coal was £125, plus £20 for fuel, less £93 for coke sold, for a net coal expense of £52. The depreciation on £5,000 for the apparatus at 12.5 percent was £625 (the same figures as Watt Jr.'s in 1807), for a total yearly cost for gas lighting of £677. This replaced the equivalent of £2,000 of tallow per year (versus the £2,600 Watt Jr. had estimated), for a net saving of £1,323, or 30 percent less than Watt Jr.'s 1807 number of £1,871. When the Royal Society paper was published on February 25, 1808, the depreciation expense was further reduced to £550. The justification given in the paper was that Lee's apparatus was oversized, which was true.[170] This revision brought the net advantage of gas lighting back to Watt Jr.'s £1,800. At no point in his calculations did Watt Jr. make an allowance for the cost of capital, although when the paper was published it claimed that the cost of capital was included in the £550. Likewise, the 12.5 percent depreciation figures did not include a realistic estimate of retort replacement frequency. In making this change, however, Watt Jr. had some justification in contemporary accounting practice. Boulton & Watt, like some other large firms, depreciated steam engines at 8 percent and buildings at 5 percent per year, whereas after 1800 textile mills often used 5 percent for both.[171] And so, although the reduced depreciation expense did not reflect the reality of retort replacement frequency, the rate Watt Jr. used was similar to what the firm employed when calculating depreciations in its other lines of business.

Gas lighting was clearly not easily moved from the workshop or even small industrial scale to a production industrial context because the industrial apparatus had sufficient different characteristics that design by simple extrapolation did not work. Lee wrote to Watt Jr. about these results: "Your experiments warranted more favorable conclusions because small retorts appear to disengage more gas, and large Argand burners more light."[172]

James Watt Jr. received Creighton's and Lee's letters and wrote back on January 23, 1808, saying "I confess myself somewhat disappointed by the oeconomical statement, for although I did conceive we had rather overdone it before, I did not expect so great a difference as appears, and I think the causes are not yet fully explained."[173] In his response, Creighton and Lee appended another economic analysis, this time assuming gas lighting was used for three hours a day rather than the two actually used at Philips & Lee.[174] Watt Jr. had requested this because it would strengthen the economic case for gas lighting, although it did not correspond to the actual situation.[175] Despite Watt Jr.'s doubts, Lee was unwavering in his opinion: "I feel fully confirmed in the correctness of our oeconomical statement, and as it is founded upon extensive and correct experiments."[176]

James Watt Jr. was still not satisfied, and hoped to find ways to improve the financial analysis. He considered that "the difference between the present & former expts is not sufficiently accounted for." The only possibility he could see was that somehow the cockspurs at Philips & Lee were dimmer now, being around 2.5 candles, than they had been in March 1806, when Murdoch had estimated 3.5 candles.[177] To this, Lee responded that he did think the lights were dimmer now than in the past, but in that case the assumption about how many tallow candles they were replacing would also need to be modified, and so the economics would not be much improved.[178] He later added: "Our photogenous apparatus produces nearly uniform results & confirms most completely our statement."[179] This missive put an end to the discussion and Murdoch and Watt Jr. went with Creighton's figures in the Royal Society paper; but, as was mentioned earlier, they found new economies by arbitrarily changing the depreciation expense.[180]

3.5   THE END OF GAS LIGHTING AT BOULTON & WATT

NEW ORDERS AND DELIVERIES, 1808

Boulton & Watt's renewed interest in gas in 1807–08 was accompanied by another sales drive.[181] As they had done nothing about any of their 1806 orders, some of their customers had lost interest. Peter Marsland was interested enough in 1806 to have drawings prepared, and asked for another estimate in 1809, but never ordered.[182] Others, such as Horrocks & Co. in Stockport, went bankrupt in the meantime.[183] In the quiet year of 1807, some other mills expressed interest but were met with silence or vague assurances. One mill owner asked: "Might I without being deemed too troublesome again ask what would be the expence of the apparatus?"[184] Boulton & Watt even had inquiries about home gaslight apparatus. In 1807, Josiah Wedgwood wrote to Watt Jr. asking to get a quote for his house, which he was about to renovate.[185]

The hesitations and delays stopped in 1808. When Creighton was at work at Philips & Lee, he met with McConnel & Kennedy and James Kennedy, both of whom had begun making preparations for their installations. He wrote: "MK is resolved to proceed with [the lighting apparatus] immediately. . . . I will endeavor to get the whole finally settled as well these as at Mr Jas Kennedys who is now digging the pits for his gasom. as he means to have part of the apparatus ready for next winter." One challenge at James Kennedy's was that the ground was quite damp and so the gasometer pits could not be sunk very deeply, with the result that Creighton worried about whether "the lower rooms of the mill would be deprived of much of their light."[186] This problem of lower area receiving less gas was fairly easily dealt with by Boulton & Watt; but once the Gas Light and Coke Company began operations, balancing large areas with different elevations became much more difficult. Creighton also noted that James Kennedy hoped to purify his gas much the same way as Philips & Lee. Kennedy wanted to draw water from the neighboring canal continuously and pass it through a tar trap so the gas could be washed before it reached the gasometer.[187] The tar and waste water would be drawn off and washed down a drain back into the canal.[188]

Creighton also mentioned that Birley & Co. of Oxford Street in Manchester were also desirous of finalizing their purchase.[189] Unfortunately for Birley, he received no faster service than Boulton & Watt's earlier customers and Creighton wrote four months later that "Mr Birley has made repeated enquiries respecting his estimate for lighting apps. & seems very anxious about it—you probably could furnish him with it soon."[190] Nor was James Kennedy getting his apparatus more expeditiously. It was only in September and October 1808 that the drawings were finished for Kennedy and Birley, as well as additional ones for McConnel & Kennedy.[191] The components were then manufactured quite quickly.

There were some design changes. The drawings indicate that Boulton & Watt had heeded Lee's advice about rethinking the retorts, and they were now all horizontal (figures 3.9). The retort's cross-section resembled the D-form that became quite common in later history of the gaslight industry. Boulton & Watt were using a complicated set of pipes to get the gas into and out of the gasometers. Previously, the pipes had entered from the bottom, but this made them difficult to access in case of blockage, a not infrequent event. Their new design called for pipes that entered the gasometers at the top, and that had movable joints immersed in water to make them airtight. The assembly consisted of two pipes and three movable water joints to allow for the gasometer's rise and fall during operation. This design was ingenious but proved impractical. It was little used outside of Boulton & Watt.

The retorts were shipped off to the three mills before the end of the year, and payments are recorded from Birley (£779) and James Kennedy (£537) before the year was done.[192] Gillespie & Co. in Glasgow were also interested. Wormald, Gott & Wormald of Leeds, who had been among the 1806 orders, restarted their order, and fresh drawings were prepared before the end of 1808.[193] Birley & Co.'s apparatus was running during the winter of 1808–09, although they had some initial problems (such as a poorly seated retort which subsequently cracked, lead accumulating in pipes, and

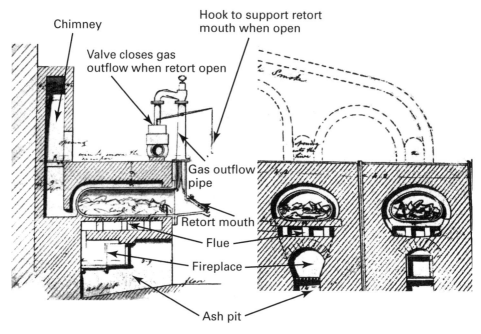

Chimney

Hook to support retort
mouth when open

Valve closes gas
outflow when retort open

Gas outflow
pipe

Retort mouth

Flue

Fireplace

Ash pit

FIGURE 3.9

Retorts used by James Kennedy, side view (left) and front view (right). Note the use of fire brick to protect the retorts. BWA-MS-3147/5/817/8. Copyright Birmingham Archives and Heritage

burners made of thin material).[194] James Kennedy's excavations were completed in February 1809, and he had his apparatus running that year.[195] He also made additions in 1809 and in 1810.[196]

Philips & Lee's mills continued to expand their gas lighting over these years, and became well known for the sheer scales of their works. In 1815, two Austrian Archdukes who were touring England stopped at Philips & Lee when they were in Manchester. The Archdukes kept a journal and described what they saw as Lee ("you would take him for a Swiss of the first distinction, if his way of thinking did not shew the British merchant") showed them around:

We were invited to visit the manufactory of Mr. Lee, one of the greatest in Manchester: and the evening was chosen for the purpose, that we might at the same time see the building lighted with gas. On entering the court-yard, we saw the first gas lamp which thoroughly lights it. The buildings make a very handsome appearance: one of them is seven stories high, and has forty-six windows in a row ; an adjoining building is a story lower, but of the same length. This brilliant illumination, and the noise of the machines, which resembles that of a

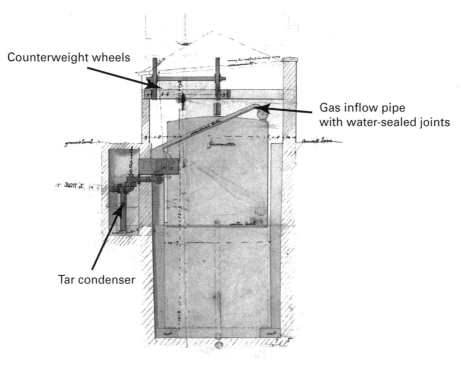

Counterweight wheels

Gas inflow pipe
with water-sealed joints

Tar condenser

FIGURE 3.10
Side of gasometer used by James Kennedy. BWA-MS-3147/5/817/10. Copyright Birmingham
Archives and Heritage.

considerable waterfall, all together makes, as you enter the court-yard, a new and extraordinary
impression. . . . The apartment in which we were received, was lighted by a chandelier, in
which gas burnt; it is conveyed through several pipes, from the ends of which it issues, generally by three openings; the flame from the middle one burns perpendicularly, and from the
two lateral ones in an oblique direction. These lights, which do not at all offend the eye by
their brilliancy, seemed to us rather unsteady. Mr. Lee then shewed us a plan of the building
which contains the great spinning manufactory. . . . The counting houses are also lighted with
gas; and, as we wished to see how it was prepared, we were conducted into the laboratory:
ten stoves stand in it, in a semicircle; each of them contains a retort of thick iron, in the form
of a chest, about five feet long, a foot and a half broad, and a foot high; from these proceed
pipes which all unite in one large common pipe, through which the gas is conveyed into the
receivers, of which there are ten or twelve. No coal is used for the preparation of gas, except
Cannel coal, from the mines of Wigan. The receivers are large gasometers, the wooden
balls of which are mostly borne by counter-weights, which produce a pressure that may be
changed at pleasure. The establishment was one hundred tons [*sic*], or 2000 cwt. of coal per
week.[197]

FIGURE 3.11
A cotton mill lit by gaslights. Andrew Ure, *The Philosophy of Manufactures*, (1835), p. 308. Courtesy of Roy G. Neville Historical Chemical Library, Chemical Heritage Foundation.

OTHER MILLS AND DENOUEMENT, 1809–1812

After Boulton & Watt took on new customers at the end of 1808, they picked up a few new ones, but the business was peaking quickly: in 1809 there were four new customers, in 1810 another four, and only two in 1811. Their income peak lagged their customers by a year as equipment was manufactured and delivered and existing apparatus was expanded and repaired. In 1812 they got no new customers and had total sales of £250, a pittance in comparison to the six previous years. The last of the 1812 sales is recorded in February of that year. Boulton & Watt had clearly lost interest in the manufactured gas business.

The largest among the new post-1808 customers of gas lighting were Wormald, Gott & Wormald and the associated firm of Benjamin Gott & Co. Wormald, Gott & Wormald's gaslight apparatus for the Burley Mill was first prepared at the end of 1808 for £650.[198] Benjamin Gott & Co.'s Armley Mill followed that summer, with an apparatus that cost £496. Gott was very happy with his new lights, writing that they were "excellent" in a letter to Watt Jr.[199] The Park Mill, the third of his big three mills, came next in 1810, for £490.[200] This was expanded in 1811 for £224; but other than a few replacement components, no more major work was done for any of Gott's mills.[201]

The other 1809 customers were Nielson & Co. of Kirland in Scotland (an order for £503, later expanded by £205[202]), Marshall Hives & Co. of Leeds (£890[203]), and Birley & Hornby of Chorlton (£1006[204]). The 1810 customers were Lister Ellis & Co. near Otley (£610[205]), Thomas Coupland & Sons of Leeds (£500[206]), and J. Thomas & E. Lewis of Manchester (£620[207]).

The 1811 customers were Benyon, Benyon & Bage near Shrewsbury (£970[208]), Marshall Hutton & Co. of Shrewsbury (£1150[209]), and Huddart & Co. of Limehouse, London (£487[210]).

After 1812 there are a few minor signs that the gaslight business was still barely alive, such as an apparatus sold in 1814 for £219, probably to William Jones,[211] and a survey prepared by Creighton for John Maberly & Co. in Aberdeen in late 1814 and early 1815,[212] but it was clearly moribund. Gaslight was hardly mentioned in any letter sent within the firm of Boulton & Watt after 1809. There is no mention of any article related to gas lighting in the accounts of the Soho Foundry for 1816–20.[213]

The drawings and plans of the period 1810–12 show that development work on gas lighting at Boulton & Watt had effectively ground to a halt. The retorts did not change in design, and drawings prepared in 1812 are almost identical to those of 1808. The gasometers from the later period were somewhat larger than the earlier ones, reaching 5,000 cubic feet, but this was not a giant step.[214] It was not that gas lighting was judged sufficiently mature. Rather, Boulton & Watt simply weren't developing this line of their business. Watt Jr. had even written in April 1809 that he felt unacquainted with it.[215] Lee's last remaining desideratum, the more effective purification of the gas, had been solved by Clegg (among others[216]) during 1805–08 with the use of lime to remove the hydrogen sulphide. Lime was, however, never used successfully by Boulton & Watt.[217] Perhaps they were unaware of this solution in 1808, but they could not have remained ignorant over the next seven years. They were not sufficiently interested to put much effort into it.

George Augustus Lee had always been a major driving force behind Boulton & Watt's development of gas lighting. He had certainly not been the only one: both Murdoch and Watt Jr. had been very much taken with the project for some time. But they lost interest over time. Robinson Boulton had never been enthusiastic about the technology, and his coolness affected the resources the firm was willing to dedicate to gas lighting. It probably rubbed off on Watt Jr. over time. Watt Jr. certainly developed a case of "gaslight fatigue" after the battles of 1807–09 with the Winsorites.

Even if he never consciously gave up the business, he conceded that he could not keep out serious competitors.

Lee, however, remained as enthusiastic as at the outset. In 1813, after the Gas Light and Coke Company had been established, Lee received a visitor from London. He was the architect for the new General Post Office, and he was investigating using gas lighting. Having heard of the famous gaslights at Philips & Lee, he made a trip to Manchester to see the lights in action. After the visit, Lee wrote to Watt Jr.:

It appears to me that with a little exertion there is a very extensive field opening for the extension of the gas lights and that after expending so much money & brains you will let the quacks in & out of the metropolis derive all the advantage of the dyscovery and our experience.

Being satisfied that coal gas can be effectually purified by agitation in the cream of lime, the time is near at hand for its more general introduction. A gentleman from the general post office in London who is their architect came expressly for the purpose & by their directions to inquire about it and I am since informed means actually to introduce it there. He says the present expense of oil & candles is no less than £3000 pannum. I have endeavoured to guard him against the quacks in London. This & more general but important considerations induce me to press the subject upon your early attention, & if you can give it the due exertions I shall either hear or see you about it.[218]

But James Watt Jr. did let the "quacks" derive all the advantage of the discovery. His mind was now elsewhere: he had been devoting his energies to developing a line of marine steam engines. In 1817, he even purchased a steamship, which he refitted with two Boulton & Watt engines. He crossed the English Channel with it, perhaps the first such crossing for a steamship. The firm's marine engine business continued to increase. Watt Jr. prospered, and died unmarried in 1848. For his part, Murdoch was well paid by the firm for his work over the years, with an annual income of over £1,000. He built himself a gentleman's home near Birmingham in 1817, and was forced to retired by Watt Jr. and Robinson Boulton in 1830, dying a wealthy man in 1839.

3.6   CONCLUSION

Between 1801 and 1810, Boulton & Watt consolidated an invention that had been around in various forms since the late 1770s. Although many people and groups had tried to develop a commercial implementation of gas lighting, none had succeeded.

Of the many groups scattered around the Continent and in Britain, only Boulton & Watt built a version of gas lighting that could be used on an industrial scale. When Boulton & Watt lost interest in the technology, sometime in 1810, the technology had been implemented in many of the largest mills in Britain—in the case of Philips & Lee, a few years earlier. The possibility of using gas lighting was clear. Many scaling issues had been solved by Boulton & Watt, whose proposed solutions endured through most of the history of the gas industry. These included using gasometers for even flow, maximizing gasometer volume to save on retort capacity, running retorts as nearly constantly as possible, horizontal retort configurations with D-shaped profiles, and condensing and washing gases with water. Other challenges, such as pressure regulation and purification, had been identified but would be solved by others.

This story has a number of implications for the historiography of the Industrial Revolution relative to why Britain created industries while other regions failed to do so, even though the basic technology was present everywhere. Boulton & Watt succeeded largely because of their and their suppliers' skills with ironworking. Specifically, they had experience in building large airtight machines. Furthermore, they had a background in making scientific instruments, particularly Beddoes' pneumatic apparatus. This experience with ironworking and pneumatic chemistry meant that when Murdoch thought of lighting with gases in the 1790s he was with a firm that had at its disposal many of the resources that were needed to transform it into a viable industrial technology.

The continuities between the pneumatic apparatus and the gas plant at Boulton & Watt and the engineering experimentation done by the firm to create an industrial form of gas lighting are both representative of a culture of experimentation within the firm. This culture had at least some roots in James Watt's work on the steam engine. It has been suggested that the experimental culture of the late eighteenth century helped bring about the technical innovativeness of the Industrial Revolution, and that this culture was learned in part from natural and experimental philosophy. In the case of gas lighting at least, the culture of experimentation had roots in James Watt's work in pneumatic chemistry and his design of the pneumatic apparatus, and also in the methods and attitudes he had infused in the firm more generally.

The scaling up of apparatus to industrial sizes was a procedure that came to be repeated more frequently as the nineteenth century progressed, whether with chemical processes or physical models for experimentation. The rise of chemical engineering education in the late nineteenth century represented a formalization and systematization of the process of scaling up chemical reactions produced in the laboratory for industrial applications.[219] Likewise, the widespread acceptance of scale models in engineering experimentation later in the nineteenth century required the use of engineering theory, as opposed to the rules of thumb that predated it.[220] In both these cases, the use of theory was important for the scaling process. Boulton & Watt's work with scaling a chemical technology did not, however, rely on any theory of similitude. For them, scaling was an empirical process.

Establishing gas lighting also meant finding a new base of customers, and this involved marketing. As with the steam engine, Boulton & Watt engaged in a number of publicity-seeking activities to create a place for gas lighting. These included public displays at Soho in 1802 and at Philips & Lee in 1806. As will be described in the next chapter, they used the Royal Society (through the paper in the *Philosophical Transactions* and the resulting medal) to attempt to crush their competitors in the Winsor camp, all with Joseph Banks' full complicity. By 1809, gas lighting, William Murdoch, and Boulton & Watt had become a well-known grouping. Many publications had reprinted Murdoch's paper. In a further parallel with steam engines, and indicative of their proclivity to privileging quantification, Boulton & Watt gave great importance to demonstrating that gas lighting was an economically attractive form of lighting.

In London one finds in many parts of this gigantic city large plants where gas is produced and from where it is sent into streets and houses. There is no factory, no meeting place, no theatre, and in many streets no private dwelling where gas is not brought in via pipes. And no one can go along the main streets without being aware that there is in the ground beneath a labyrinth of iron pipes which conduct inflammable air even to the highest rooms. The pipes terminate in valve openings with stylish peaks, and in an instant, darkness is dissipated by the brightest light. Whoever has seen the smallest thermolamp will easily understand how this works.[1]

—August Niemeyer (1819)

If Boulton & Watt thought of gas lighting only in terms of stand-alone gas plants for mills, how and why did gas technology assume the network form that came to define the industry in the nineteenth century? The roots of the network form date back once again to Lebon's Parisian demonstration of his thermolamp. Besides Gregory Watt, Lebon received another visitor during his marketing days in Paris in whom his thermolamp struck a chord. This was a German merchant named Friedrich Winzer (changed to Frederick Winsor during his early residency in Britain) who had traveled from Frankfurt to Paris with the sole purpose of seeing Lebon's invention, which had been reported on in German journals. That Winsor made such a trip is reflective of his somewhat eccentric and dogged nature, especially in regard to technology. Although he did not—and indeed, lacking the education and a technical mind, could not—understand the principles behind the thermolamp, he tried to purchase one from Lebon through an intermediary after the end of the marketing campaign, but without success. This failure did not deter Winsor, and from this point forward until his death in 1830 Winsor devoted all his professional and personal energies to creating a new industry based on the thermolamp, at one point styling himself the "second inventor and improver" of the technology. In all that time, Winsor contributed nothing of value to the design of gaslight technology in any of its details, despite spending many years and much money in experimentation. His many pamphlets and notices related to the technology were full of wild exaggerations that betray an ignorance and misunderstanding of technical details, as well as occasionally a personal invective that bordered on the paranoid. This colorful character, so different from engineers and inventors like Lebon, Murdoch, and Winzler, nevertheless was a central figure in the initiation and unfolding of a new strategy for the gas industry: as a network urban infrastructure, with a large company generating gas in a few large gasworks and selling it to consumers via a network of mains.

The network strategy was very different from Boulton & Watt's vision for gas lighting, according to which each large building would have its own gas plant, but even Winsor did not conceive of this strategy in a single step. The full vision unfolded slowly, some elements of it coming more by force of circumstance than through foresight or planning on the part of Winsor or his growing number of backers. The seed of the vision, however, came from Winsor. Sometime in 1806, it occurred to him gas could be distributed through a network of pipes, much as water was. But this vision did not originally include large gasworks, relying instead on many gas plants similar in size to Boulton & Watt's, distributed throughout the city, each feeding 100 or so lamps. Nor was Winsor's company originally meant to concentrate on producing gas and other products of distillation, as he thought it would also manufacture apparatus for sale, much as Boulton & Watt were doing. A first refinement to the network strategy came with a political battle with Boulton & Watt over the incorporation of the National Light and Heat Company, renamed the Gas Light and Coke Company once incorporated. The compromise reached by the warring parties saw the Gas Light and Coke Company abandon all plans to sell machinery and focus exclusively on selling the products of distillation, particularly gas. Further strategic refinements came once the GLCC began operations in 1812. The company's managers switched to having a few much larger gasworks supplying the whole city once it became clear after a year or so of operation that acquiring land for many small gas plants wasn't going to feasible.

The shift to the network strategy for the technology involved much more than specifically technological problems, although there was no shortage of these. Boulton & Watt's small-plant model required little novelty over what was required to deploy their other technologies, such as steam engines, in terms of finance, business organization, or legal structures. Boulton & Watt did not, for example, have to go to extraordinary lengths to find large quantities of capital to develop the technology. While they spent lots of money, they and their customers had sufficient resources to make the transition from workshop experiments to mill-sized plants without financial strain ever appearing in the years they worked on it. Neither did they or their customers need to change the way they administered their business because of the technology, and the nature of the technology did not create any legal issues for them. The network model, however, went to the limits of the standard for the time in many ways, and the transformation of gas from mill-sized apparatus to a city-encompassing technology was entangled with many business, legal, and political issues. The somewhat haphazard adoption of the network strategy involved a simultaneous deployment of a more structured approach to business management and the use of a legal form (made possible by an act of Parliament) that allowed the company to incorporate as a limited-liability entity with the power to dig up city streets. Gaslight technology was also shaped by the how the company interacted with its customers. The mill owners who purchased Boulton & Watt's gas plants were in control of the technology, and there is little evidence that their employees had much impact on how the gas lighting was deployed. In contrast, the Gas Light and Coke Company's customers had a decisive influence on the technology, and the company had to design ways of controlling them in order to ensure the network's stability.

The development of gas technology into large network infrastructure was novel in some of its technical aspects, but it found some precedents in other infrastructure developments in the Industrial Revolution. From the middle of the eighteenth century through the nineteenth century, Britain and to varying degrees other European countries witnessed the expansion of transport, communication, water, and sewage infrastructure that played an important role in organizing a modern economy and in creating a new urban environment. In Britain, the growth of the road network through turnpike trusts started earlier in the eighteenth century, and was accompanied by the expansion of coastal shipping and the improvement of navigable waterways.[2] The canal booms of the 1760s and the 1790s added a new network of artificial water routes that significantly decreased transportation costs and increased regional economic specialization.[3] Railways and urban infrastructure growth largely lay in the future, but cities were growing rapidly, and their existing resources for water, sewage, and transportation were overwhelmed, creating insalubrious living conditions, especially for the poorer inhabitants.

Gas emerged as a new infrastructure before the railways and most new urban development. Only the water companies preceded it in the city, and for that reason water, and to a lesser extent the canals, provided something of a legal, business, and financial model that the builders of the gas networks could follow. This included the creation of a joint-stock company via an act of Parliament that also gave the company the right to bury its pipes under streets with the permission of the local authorities. As was the case with other large infrastructure projects, internal financing was not an option for building a gas network. It was sufficient for the Boulton & Watt model, but not for the network model. This transformed the nature of the project completely.

This final part of the book tells the story of how the network strategy was adopted and how it was made to work. Chapter 4 focuses on Winsor's work in founding the Gas Light and Coke Company. Chapter 5 describes that company's history after it began operations.

4   FREDERICK WINSOR AND THE NATIONAL LIGHT AND HEAT COMPANY

4.1   INTRODUCTION

The dominant form that gaslight technology took on in the nineteenth century was the network model that originated with Winsor. Despite his pretensions, Winsor was devoid of almost any useful technical knowledge or talent, but he possessed an intensity of entrepreneurial energy and vision that none of the other gaslight pioneers had. The contrast with Boulton & Watt is important: although Boulton & Watt had tremendous resources at their disposal in form of business experience, money, skills, prestige, and a network of industrial customers, Watt Jr. and others at the firm lacked the almost prophetic commitment that marked Winsor's approach to the technology, and they finally lost all interest in gas lighting. Winsor's entrepreneurial dynamism went far beyond the bounds of the rational, a characteristic he shared with many other entrepreneurs in history.[4] It is not clear whether he was self-delusional, willfully deceptive, or hopelessly unrealistic, but during his drive to create the National Light and Heat Company he made a series of wild prognostications about the profitability and benefits of the new technology, such as that expected yearly profits were on the order of £100 million and that the thermolamp was "the most prolific source for the wealth of nations, that ever was recorded in the history of the world."[5] His commitment to the possibilities of the technology was even self-destructive, as he ran up debts and neglected basic accounting in promotional efforts. He had to flee Britain to escape his creditors just as his vision was being realized.

Winsor was one of the most successful entrepreneurs of the early nineteenth century from the point of view of creating social, business, and political momentum for gas lighting through publicity. When he arrived in Britain in 1802, gas lighting was a novel invention. Some people had heard of Lebon's and Murdoch's work, but it was still mostly an idea. Winsor soon started a campaign of public demonstrations and advertising which created a frenzy of interest in gas lighting. By 1807, Winsor had formed a sizable group of investors, among them members of the nobility and politicians, who were willing to support him with hundreds of thousands of pounds of capital investment and to fund a drive to win an act of Parliament creating a limited-liability joint-stock company. So powerful was his vision that, although many of his backers never completely trusted him or his wild rhetoric, they came to believe that it was possible to set up an enormous company based on gas lighting technology. Winsor's grand ambitions collided with Boulton & Watt's plans, and a battle in Parliament ensued over the act of incorporation, leading to the bill's defeat in 1809. This battle, however, was also important in establishing gas lighting because of the publicity it generated, not least through Murdoch's Royal Society paper and medal. A compromise between the warring parties was reached, and the act was passed easily in 1810.

Winsor's first presentations of the new technology were decidedly in the Lebon distillation/gaslight tradition. In 1804, he marketed a patented stove that could make all sorts of products, the most important of which was coke. He hoped to incorporate a company that would sell his stoves and other products. Sometime in 1806, however, he took inspiration from the water companies for a distribution model of generating gas in plants and selling it to consumer through pipes running under the streets. It was just at that time that a number of new water companies, such as the West Middlesex Waterworks Company, were founded, and existing ones, such as the New River Company, were expanding.[6] The water companies, however, provided more than just a distribution model for the Gas Light and Coke Company. They also had a legal and financial form that the GLCC could follow, although no form specific to an infrastructure model yet existed. In this regard, water supply and subsequently gas provision were treated by Parliament and public opinion in this period as ordinary business ventures, to be placed in private hands, with the assumption that competition would protect the public interest.[7] The exception to this attitude was that Parliament was willing to pass acts of incorporation for large companies granting limited liability for shareholders to encourage the pooling of capital, as well as creating special rights to take up streets in order to lay pipes with the authorization of the local paving authorities, provided that the paving was repaired.[8] In exchange for granting this special right, however, Parliament demanded some concessions. The GLCC was mandated to provide street lighting for interested local authorities at lower rates than what oil lamps would cost.[9]

The use of joint-stock finance by the GLCC marks a departure from the financing trends for new technologies in Britain, another indication that gas lighting was at the beginning of the second wave of new technologies. Very few if any of the new technologies of the Industrial Revolution up to this point were financed through stock issuance by incorporated companies, although there were instances of unincorporated companies. Textile machines, coke smelting, iron puddling, and steam engines required relatively little capital, and their deployment could be supported by capital generated within the firms that developed them or by local business networks.[10] The securities markets had, as yet, little direct effect on the industries based on new technologies.[11] For this reason, institutional and cultural innovations have often been presented as more important than capital accumulation for new technologies in the Industrial Revolution.[12] This would change dramatically with the rise of the railways beginning in the 1820s, but gas lighting led the way. The industries in which stock-market financing was important in the late eighteenth century and the early nineteenth century were insurance and large infrastructure projects that exceeded the funding abilities of individuals or small partnerships. Many commercial docks, water companies, river navigation corporations, and canals raised funds through primary stock issuance following parliamentary acts of incorporation, the stockholders typically being local and regional players more or less directly benefiting from the new canal, although central markets in London increasingly participated.[13] The financing of large infrastructure projects through the creation of a joint-stock company via parliamentary act was becoming increasingly standardized and commonplace during canal booms in the late 1760s and again in the early 1790s, with the result that the business community was accustomed to it by 1800.[14] The incorporated joint-stock route was the natural one for Winsor to follow once he had conceived of the large network vision for gas, and his project would join another joint-stock boom that raged from 1805 to 1811.[15]

It is possible to document the process of the GLCC's incorporation drive, including the initial stock offering, in great detail, because, although no records survive from the company's archives before its actual incorporation in 1812, Winsor published many pamphlets and notices in papers, making it possible to track his progress on an almost weekly basis. This provides a valuable overview of the initiation of a large joint-stock company in the period, of which few examples exist at this level of detail.[16]

Another important theme of this chapter is the cultural value of science and technology in the context of creating new companies and industries in the early nineteenth century, especially when intense rivalries were present, such as that between Boulton & Watt and the Winsor group. Both sides were willing to use science and technology to make arguments about the validity of their own business interests. Patents are an example of this. Christine MacLeod has described how patents were taken out during the Industrial Revolution for a variety of motivations, including for the sake of marketing to consumers, attracting potential investors, and gaining prestige.[17] Always the publicist, Winsor took out a number of patents for these reasons, and used them extensively in his public statements, frequently using the moniker "the patentee" to sign his pamphlets and notices. He almost always referred to his devices as "the patent stove" which produced "patent coke"; he at first wanted to found the "New Imperial Patent Company"; he threatened competitors with infringement lawsuits; he tried to use patent rights to keep himself at the center of the gaslight project. While Winsor probably thought that his various patents were also ways to protect his technology, he never in fact tried to use his patent rights in any legal process, nor did they have any effect on the subsequent development of gas lighting. By the time the Gas Light and Coke Company received its charter, the question of patents had disappeared.

Winsor was not, however, the only one to marshal the cultural value of science and technology for business purposes. Boulton & Watt, through close connections with Sir Joseph Banks, the president of the Royal Society, used the society's prestige to wipe away Winsor's claims to priority based on his patents. As was discussed in chapter 3, Boulton & Watt portrayed James Watt as a natural philosopher, and later as a heroic inventor, because of the commercial advantages that could accrue from such a reputation.[18] In a similar vein, the Royal Society connections served as an important means of establishing William Murdoch's status as the "true" inventor of gas lighting. By having a paper putatively written by Murdoch read to the Royal Society and printed in the *Philosophical Transactions*, and then by having the Royal Society give Murdoch a medal for his invention, Winsor was painted as a fraud and impostor seeking to steal credit from the true inventor. The battle then moved to Parliament, and there too men of science were called in by both sides to defend rival claims on very narrow bases. Humphry Davy was even courted by both sides before finally choosing to defend Boulton & Watt. Despite all the technical arguments, the battle was in the end settled in a political way: through a negotiated compromise. Once both sides came to agree on a demarcation separating their two businesses, all the scientific arguments and priority claims disappeared.

Section 4.2 describes Winsor's initial exposure to the thermolamp and his first efforts in London, which eventually failed. Section 4.3 describes his second, much larger effort to found a company, which began in 1806 after Winsor adopted an early form of the network strategy. Section 4.4 details Boulton & Watt's initial response to Winsor's efforts, particularly Murdoch's Royal Society paper and medal. Section 4.5 explores the 1809 battle in Parliament over the bill to grant a charter. Section 4.6

FIGURE 4.1
An 1830 portrait of Frederick Winsor by John Alfred Vinter. Courtesy of Science Museum, London.

describes how the Winsor group and Boulton & Watt reached a compromise that limited the GLCC's activity to selling products of distillation.

4.2   WINSOR AND THE THERMOLAMP, 1802–1804

Friedrich Albrecht Winzer (1763–1830) was born in Braunschweig and worked as a merchant. Little is known about his early life, but he had a degree of interest in natural philosophy, which may explain his presence in 1784 in Paris to witness some of the first balloon flights there, and again in London in 1785. He thereafter made small balloons which he floated using inflammable air.[19] He took up residence in London, worked for the merchant firm of Weinholdt & Co., and was granted English citizenship by an act of Parliament in 1795.[20] With the outbreak of the French wars, he left England for the Continent when his possessions there were threatened, return-ing in 1799 a widower and somewhat poorer.[21] The year 1802 found him in Frankfurt, where he read of Lebon's thermolamp in journals. He traveled to Paris to see Lebon, and visited his demonstrations on a number of occasions, but Lebon would not show him (or anyone else) the details of his thermolamp.[22] Winsor, however, was nothing if not determined. He tried repeatedly to learn more about the thermolamp from Lebon, as well as to purchase one by exchange of letters.[23] He was turned down, but his enthusiasm was not dimmed, nor would it be for the next 30 years. He left Paris in April of 1802. By September he was back in Braunschweig, making many trials with a thermolamp of his own construction.[24] He also published a short memoir containing a trilingual edition of Lebon's 1801 thermolamp pamphlet and an account of his visit to Paris.[25] He then went to Bremen and Hamburg, where he again dis-played his apparatus to scholars, who could see, in Winsor's words, no great difference between his and Lebon's thermolamp, including the offensive smell.[26]

Winsor finally decided that he would have better chances with the thermolamp in England. He moved there, probably in 1803, because he thought an easy market existed for coal tar for maritime use, and because of his facility with the language and his connections there.[27] His activities at this point are obscure, but it is known that he resided in Shrewsbury House in Dover, Kent.[28] Whatever else he may have been doing, Winsor spent a good deal of time and money developing his version of the thermolamp. In late 1803 or early 1804 he moved to Cheapside in London, where he convinced a wealthy coach manufacturer named Kenzie to fund further experi-ments.[29] Winsor then filed for a patent on his new stove, which was granted on May 18, 1804.[30] The patent and Winsor's first advertising make it clear that gas lighting was not his only priority at this point. The patent included inflammable gases among a long list of other useful products, including wood acid, ammonia, and coke, all of which the "improved Oven, Stove, or Apparatus" was able to produce from either coal or wood. Winsor was clearly trying to make the patent as broad as possible in terms of uses, but it did overlap with Archibald Cochrane's earlier patent.[31] Although there was no sense of a network model, the patent included the notion of conducting gas "through tubes of silk, paper, earth, wood, or metal to any distance in houses, rooms, gardens, places, parks and streets to produce light and heat."

At about this time, Winsor printed the first of what turned out to be quite a lengthy series of verbose pamphlets. One of these was titled *Account of the most ingenious and important National Discovery for some Ages. British Imperial Patent Light Ovens and Stoves, respectfully dedicated to both Houses of Parliament, and all Patriotic Societies; and recommended to all the learned in Physics and Chemistry.* The pamphlet provided a brief history of Winsor's dealings with Lebon, and then went on to describe in prolix terms the theory and uses of his patent stove in his characteristically bizarre manner of thinking. The stove, in addition to bringing "great œconomy, utility, convenience, cleanliness, and delight, in human life," would also provide great security and lower fire insurance premiums because no sparks would be emitted. It would improve health by doing away with "the great blackening soot and dust continually inhaled by our lungs." Winsor claimed that inflammable air from his stove could supersede the "dangerous and expensive steam-engines" because "the *azot* [by which Winsor meant inflammable air, not nitrogen] it produces by a mixture with the atmospheric air, is capable of tenfold powers, if employed in the freezing point only" [*sic*]. The list of useful products from the stove included tar, in which he hoped to outdo Cochrane. Other products included "pyrolignous and coal acidities," which could be used to tan skins, to cure meat, to produce whitelead, verdigrease [*sic*], copperas, and alum, to resolve metals, and to fix dyes so that they "would never wash out again." He marshaled the work of famous chemists—Lavoisier, Count Rumford, Fourcroy, Neuman—to demonstrate how the stove was based on solid principles, all the while presenting a quasi-Aristotelian diagram of the four elements. Winsor finished with some of the fantastic and wildly exaggerated financial statements that would earn him much ridicule in the years to come. He suggested that, according to his estimates, the stove represented a 1,000 percent saving over conventional stoves, and that even with this figure it "appears still undervalued." His arguments in favor of adopting the stove also included a humanitarian claim that it would improve the lot of "the most miserable and unhealthiest profession among the human race, that of the poor chimney sweepers" because the diminution in the quantity of soot produced would obviate the need for this trade.[32]

With his pamphlet ready, Winsor began making arrangements for public lectures and further development work. He hired some assistants, among them Edward Heard, who had advertised for a place as a chemist's assistant, having formerly worked at Lardner & Co. of Piccadilly, a London merchant and chemist.[33] Lardner & Co. would in the following year start using some sort of gas lighting at their building, and Winsor would later accuse them of industrial espionage.[34] In regard to Heard, Winsor presented his hiring as a great act of charity on his part, almost as if he were taking in a vagrant. Heard and Winsor got to work in early 1804 by buying supplies and within six weeks had a falling out which turned acrimonious, something that seemed to happen to Winsor with some frequency. Winsor later claimed that Heard, whom he relied on to read his lectures while he demonstrated his apparatus, one day did not show up for his duties, apparently claiming to have mislaid the lecture notes.[35] Winsor summarily dismissed Heard, who went off on his own trying to develop gaslight technology, later embarking on a promotional lecture tour outside of London, and finally settling in Bristol, where he tried to incorporate a gaslight company.[36] In 1806, Heard was the first to take out a patent for lime purification.[37] Winsor rushed out and got a patent that included purification a month after Heard's patent was published in *The Repertory of Arts*, and later claimed Heard had stolen this idea from his use of lime "many years before."[38]

Winsor's promotional campaign now began in earnest (figure 4.2). In May of 1804, he began advertising his Lyceum lectures in newspapers, promising to light a local garden come summer. The lectures at the Lyceum included "a chandelier in the form of a long, flexible tube suspended from the ceiling, communicating at the end with a burner, designed with much taste, being a cupid grasping a torch with one hand and holding the tube with the other."[39] He established an office at the City Coffee House in Cheapside,[40] and his *Account* pamphlet was now on sale.

Over the course of the summer, Winsor's ambitions grew. He hoped to establish an "New Imperial Patent Company" by finding investors to sign up for 100 shares, with no price specified.[41] Winsor claimed that 100 was the legal limit for the number of shares without an act of Parliament.[42] He retained the joint-stock bank of Matthew Bloxam & Co. to run the share subscription, and Bloxam himself eventually became an investor and a leading backer of Winsor's efforts. Winsor printed posters and notices which he mailed out in large numbers to all the prominent people he could. He must have had substantial financial backing for the expenditures he had incurred so far, although it is not clear who his backers were. By September, Winsor was claiming on his posters that he had had expenses amounting to £2,000 in the last three years—

Profit! Information! Amufement!

NOW EXHIBITING,

Moſt Brilliant Illuminations

FROM

SMOKE.

A NEW PATENT DISCOVERY,

AT THE

LARGE THEATRE in the LYCEUM.

EXAMINED AND SANCTIONED BY

The DUKE of RICHMOND, Sir JOSEPH BANKS,

And a numerous Party of their noble and ſcientific Friends.

SMOKE, which evaporates in Millions of Value, is condenſated ſo as to yield the whole Weight and Subſtance of Fuel in Coke, Tar, Oil, Acid, and pure Gas; to produce Heat and Light at any Diſtance from the Stove; whereby unprecedented Advantages are gained, becauſe this Invention is applicable to every Conſumption of Fuel in all Buſineſs, from a Kitchen-Grate to a Steam-Engine.

Books deſcriptive of this Diſcovery, dedicated to both Houſes of Parliament, and all the learned Societies, are ſold at Meſſrs. RICHARDSONS', Royal Exchange; and at the LYCEUM.

The LARGE THEATRE is open every Day from ELEVEN till FOUR o'Clock.

Admittance Two Shillings.

LECTURES, illuſtrated by Experiments, are given every Evening from EIGHT till TEN o'Clock.

Boxes 4s. Pit 3s. Gallery 2s.

*** *Society and Family Tickets of Admittance, to be had at lower Prices, in proportion to the Number of Perſons.*

☞ SUBSCRIPTION BOOKS are opened at the LYCEUM for PATENT STOVES and APPARATUS.

FIGURE 4.2

A bill advertising Winsor's demonstrations. BWA-MS-3147/3/539. Copyright Birmingham Archives and Heritage.

not an unbelievable sum, in view of the scale of the promotional campaign and his experiments.[43] On the advice of the patent office, Winsor generally didn't publish the details of his stoves or show them off in public.[44]

Winsor was always sensitive to establishing the prestige of his projects, and his posters claimed that Sir Joseph Banks of the Royal Society as well as the Duke of Richmond had been to see his demonstrations and had been much impressed.[45] Winsor had not taken sufficient care (if he took any care at all) to consult with people he named in his notices, and some of them would take exception to his public use of their names.[46] In September, Winsor tried to interest Trinity House, the trust responsible for the lighthouses along the English coast, in using his stoves for light. The Deputy Master and some other "brethren" went to visit Winsor at the Lyceum, and were apparently impressed with the technology. Nothing came of it in the short run, but the visit was reported in the local newspapers.[47]

The last four months of 1804 saw Winsor push his campaign to a new level. He wrote and had printed a second pamphlet, *The Superiority of the New Patent Coke over the Use of Coals*, in which he argued that the coke produced in his stoves was far superior to ordinary coal. He listed many of the same arguments he had used in his earlier pamphlet, such as the humanitarian treatment of chimney sweeps and the greater efficiency of his flames, but made no mention of gas lighting. This was not, however, because he had given up on the possibility, but because he thought he could attract more investors by emphasizing the coke aspect of his work rather than the unproven lighting side of it. He continued to speak about the lighting uses of his stove in his lectures and newspaper advertisements. He also lit the front of the Lyceum with gaslights.[48]

Winsor used public lectures to build awareness of his project and to bolster his scientific and commercial credibility. Scientific and technological public lecturing in the early nineteenth stood at a crossroads between the open scientific ideals of the eighteenth-century Enlightenment and the professionalized science of the later nineteenth century.[49] While the most elite members of the scientific community, such as Humphry Davy, were able to build positions of expertise through their public lectures, less elite lecturing was commonplace and had more uncertain effects. Many public lecturers were trying to increase their stature in the scientific community and in society more broadly. Winsor's lectures were aimed at establishing him as an inventor with commercial respectability. An account of the impression made by Winsor's demonstrations was printed in *Kirby's Wonderful and Eccentric Museum*:

The numerous discoveries resulting from the spirit of philosophic research, so generally diffused within these few years, throughout the most civilized nations of Europe, have undeniably contributed to promote in a high degree, the comfort and conveniencies of society. None however promises to be more beneficial, or of more general utility, than a discovery first exhibited at Paris, in 1802, and lately introduced into this country by an ingenious artist who obtained a knowledge of the secret, and who has for several months exhibited it to the curiosity of the public at the Lyceum in the Strand. The object of this discovery, which will doubtless form an important epoch in the annals of domestic œconomy, is to produce light without the aid of wax, oil, tallow, or any combustible now employed for that purpose. The expence of illumination both to the community in general, and to individuals in particular, is most sensibly felt at the present moment, when the materials employed for that purpose have attained to an unprecedented price. The public must therefore feel more deeply interested in a discovery which tends to reduce that expence to a mere trifle, and to supply them with a light infinitely superior to that to which they have hitherto been accustomed. . . . The extraordinary advantages of this method of producing light must be obvious to the most superficial observer.[50]

Winsor was beginning to wrestle with the intricacies of forming a company under English law. The Bubble Act of 1720, passed as the South Sea Company was being formed, prevented the creation of new joint-stock companies with more than five partners except by act of Parliament or royal charter.[51] While it remained possible to create non-incorporated joint-stock companies, their stockholders did not enjoy the privilege of limited liability, and were at least in theory liable for the debts of the company whose shares they held. The interpretation of both the limited liability aspect of incorporated companies and the formation of non-incorporated companies varied over the eighteenth century, but by the beginning of the nineteenth century the Bubble Act was not acting as a brake on the formation of non-incorporated companies. Despite this, the desirability of limited liability was increasing for stockholders, and investors in many prospective companies sought acts of incorporation from Parliament for precisely this reason. Although some succeeded, the most formidable difficulties and largest costs came not from the legal process of getting the act, but from the opposition that interested parties could mount in Parliament.[52] The Bank of England, for example, effectively blocked the incorporation of new London-based banks, and landowners would frequently resist the incorporation of canals with rights of eminent domain that threatened their holdings. These patterns held true for Winsor as well: his investors were interested in an act of incorporation, and the most vociferous opposition came from vested interests—in this case, Boulton & Watt.[53]

As was noted above, Winsor originally planned in 1804 to sell 100 shares. By late September, he was claiming that only a few of the 100 shares originally available were left, and by the end of the year all 100 shares had been sold. A committee of five shareholders was established to validate Winsor's claims.[54] They met in November 1804, and, at least according to Winsor, were satisfied that Winsor's stoves were sound and his claims valid.[55] The subscribers then voted to apply for a royal charter via an act of Parliament, and for this purpose to build a large demonstration furnace that could carbonize a chaldron (32 bushels) of coal. They also voted to give Winsor five guineas each for his expenses to date.[56] Winsor wound down his lectures and demonstrations at the Lyceum, closing the show off toward the end of December 1804.[57]

The subscribers constituted another committee of five who would act as trustees. Winsor legally entered into a partnership with them and licensed his patent in order to proceed with the construction of the large furnace. Among the five was a Doctor Clarke who, as Winsor put it few years later, "seemed to be a Newton in theory." Clarke persuaded the group to allow him to build the large coking stove, and Winsor later claimed that he spent much of the partnership's capital without producing a working model. Winsor variously put the figure between £300 and £1,000. Winsor and Clarke were soon in a dispute over the direction of the Coke Company, with Winsor resisting Clarke's plan to become a co-patentee with Winsor. They reached an impasse and Winsor felt compelled to dissolve the partnership some time in 1805. The entire venture disintegrated.[58]

Winsor returned to his developmental work, focusing this time on gas lighting, and spent the greater part of the next two years on it. Although the committee had dispersed, Winsor seemed to have retained the backing of some important shareholders who were willing to fund further work. He needed to supplement this, however, and began running up personal debts, owing £4,000 by the beginning of 1806.[59] Winsor had completely halted his public activity in 1804. He placed no advertisements and wrote no pamphlets from 1805 until mid 1806.

4.3 THE CAMPAIGN OF 1806–07 AND THE FIRST GENERAL MEETING OF 1807

The period from mid 1806 to mid 1807 marked a new chapter in Winsor's drive toward establishing a company. In 1805 and 1806, he and his backers devised a business plan which they brought to the public toward the end of summer 1806, when they launched a campaign that had two distinct phases. First, they engaged in a wide-ranging canvassing to raise more funds for the National Light and Heat Company, as they now called the venture. The lobbying of Parliament to grant a charter for the new company was the second phase of the campaign, and it proved to be far more difficult. The two-phase campaign was remarkable in its scale and for its success.

Advertisements for the company were placed almost daily in many London newspapers and journals, and Winsor printed bills which he cast as widely as his advertisements. He advertised outside London, with country bankers collecting deposits for shares.[60] From the investment perspective, in the space of a few months in 1806, the company managed to find buyers for 5,000 shares at £50 each, with £5 payable immediately. The shares sold reached 10,000 in early 1807, and 20,000 by May.

The ability of Winsor and company to attract so much investment in such a short time is a remarkable testament to how much money was available in London for a new venture, and how willing people were to place their money with a company based on a new technology, and finally how effective Winsor was at generating interest and keeping the venture in the public eye. There were broader trends that made this moment particularly propitious for Winsor's drive to incorporate a large joint-stock company. The amount of savings was growing in the years around the turn of century, and the stock market was functioning as more efficient means of channeling capital to new companies.[61] In addition, the promoters of the National Light and Heat Company were able to ride a boom in joint-stock companies that was sweeping Britain in the early years of the nineteenth century. The Canal Mania of 1791–1794 had subsided, but a new boom that ran from 1805 to 1811 had started, with tens of companies being promoted. In January of 1808 alone, there were 42 new joint-stock companies formed that had been promoted over the previous years.[62] Unlike the canal boom, the current one featured a broader range of companies, including water companies, and did not come to a crashing halt. It also featured much larger swathes of the population buying shares than had previously been the case.[63]

Early in 1806, Winsor leased a house at 98 Pall Mall and moved his residence and offices there. The choice of Pall Mall indicates how much Winsor's ambitions had grown even since 1804. The location placed him close to exclusive gentlemen's clubs, to the seat of the government at Whitehall and Parliament, and to a number of houses and palaces of nobility and royalty, such as Marlborough House, Carlton House, and the official residence of the king at that date: St. James' Palace. He resumed lecturing, and with the backing of investors he continued to experiment.[64] Wanting to make his demonstrations even more extensive, he placed an order for 5,600 feet of earthen tubes during the summer of 1806, with the intention of laying them under a number of streets the following year, perhaps intending this to be a decisive prelude to the permanent operations of the new company.[65] As the year progressed, Winsor contracted Frank Hardie to take over giving his lectures at Pall Mall in order to free himself for his other activities.[66] Around this time, he leased the neighboring building at 97 Pall Mall for more space, making the new building the official address of the company while keeping his personal address at 98.[67] Hardie was a lecturer in natural philosophy, specializing in electrical phenomena. He had been giving lectures in his own quarters in 1804 when the demand became such that he moved them to another hall of the Lyceum theater at the same time Winsor was there. Hardie's lectures in 1804 included some demonstrations of lighting with gas, and it is likely that Winsor and Hardie formed some association at that time.[68] Hardie's lecturing for Winsor does not seem to have lasted long. He was out of the picture by 1807, probably forced out by the committee of trustees.[69]

With the new house ready and the lectures running, Winsor and his backers launched their drive to accumulate a much larger pool of capital from many investors. His new marketing campaign made his first one of 1804 look minor and subdued. In mid 1806, he published a new 48-page pamphlet titled *To be sanctioned by act of parliament. A National Light and Heat Company for Providing our Streets and Houses with Light and Heat, on Similar Principles, as they are now Supplied with Water, Demonstrated with the Patentee's Authority and Instructions, by Professor Hardie, at the Theater of Sciences, No. 98.*[70] This pamphlet and its subsequent edition contained the most wildly inflated claims that Winsor had yet made. It is striking that merchants and business interests allied with Winsor allowed him to retain such control of the process and to speak in

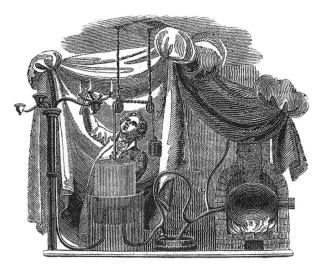

FIGURE 4.3
A display of gas lighting as illustrated in the 1818 edition of Samuel Parkes' book *The Chemical Catechism*. Courtesy of Roy G. Neville Historical Chemical Library, Chemical Heritage Foundation.

such hyperbole. He claimed, for example, that a modest annual profit estimate for his company was £131,583,333, and that the tax the company paid would eliminate the national debt in short order. It is unlikely that any of his backers really believed these figures, but they probably recognized that Winsor was extraordinarily effective at publicity, and many of them had been attracted by the size of Winsor's profit forecasts, all the while realizing that their grand scale was illusory. Many more subscribers were still needed, and the thorny problem of gaining parliamentary approval had not yet been solved. Whatever Winsor could do to generate talk and excitement was going to help their cause. The wild talk was, however, to have its consequences. It allowed Winsor's opponents to portray him plausibly as a quack, and the perception of inflating a short-lived investment bubble, such as the infamous South Sea Company Bubble of 1720, contributed to the defeat of the first attempt to pass an act of incorporation in the House of Commons in 1809.

The title of the pamphlet indicates that Winsor was now thinking of establishing a gas company that would function as network selling gas to customers, much as the water companies did in distributing water. He consciously drew on that analogy, stating in the pamphlet that "the National Light and Heat Company must chiefly be conducted on the principles of the Water Companies."[71] By this he was thinking primarily of their investment model of large initial expenditures of capital and time to establish a supply infrastructure which, once operating, provided reliable service and healthy dividends. Winsor of course thought his National Light and Heat Company had far more benefits than water companies. The idea of selling stoves had receded somewhat, although the manufacturing of stoves and gas apparatus for sale was still present. It was this aspect of their business plan that irked Boulton & Watt in particular and provoked them to reaction and forceful opposition.

Winsor and company proposed in 1806 to raise £1 million by finding 20,000 subscribers at £50 per share. The initial payment was £5, and Winsor promised the rest would not have to be paid as dividends would soon cover the amount owing. The share sale went at a brisk pace. By November, when advertisements began appearing in newspapers, Winsor was reporting that 2,000 of the shares had been taken, meaning he had raised £10,000.[72] Newspapers began to report positively on the venture, the *Morning Post* writing of "the wonderful discovery of Mr. Winsor" that "we rejoice at this happy prospect" and that it must "soon occasion a considerable decrease in our present heavy taxes."[73] The *Morning Chronicle* opined that "a British mine of wealth is on the eve of being worked at home" and "the saving to the country and Government surpass every thing before introduced, and the subscribers must be speedily enriched."[74]

Winsor once again raised the intensity level of his marketing campaign. He advertised in many of the major London papers almost daily from November 1806 through
the first half of 1807.[75] He placed advertisements in society journals and provincial
newspapers.[76] He published more pamphlets, putting at least five into circulation
before April 1807,[77] including *Plain Questions and Answers*[78] and *Analogy between Animal
and Vegetable Life. Demonstrating the Beneficial Application of the Patent Light Stoves to all
Green and Hot Houses*.[79] He sent copies of these pamphlets to influential people, as
he had in 1804. The copies of his pamphlets in the British Library are ones he sent
to Sir Joseph Banks and others and include personal dedications. Winsor tried to get
notables to attend his experiments with some success, as on the occasion of the visit
of Lord Minto, Lord Petty, and others on January 17, 1807, which prompted short
reviews to appear in the *Times* and in the *Morning Post*.[80]

An Irish student who was staying in London visited Winsor's lectures and described
them and his impression of the project in a letter to his father:

I have visited the rooms, and attended the lectures, of Mr. Winsor, the projector of the National
Light and Heat Company. . . . The lights by which his rooms are illuminated are certainly very
vivid; but the inconvenience and disagreeable appearance of the tubes through which the gas
passes, and the lights being from necessity, in general, stationary, will, in my humble opinion,
prevent this philosophical application, for discovery it cannot be called, from ever being used
for domestic purposes. . . . The lecture was delivered in a damp and dreary subterraneous room,
I believe a cellar, rudely fitted up as a little theatre. The lecture unfolded the nature, and origin
of gas light, and the great national uses to which it may be applied. I should think that streets
might be illuminated with the gas to great advantage; as the light is large, beautiful, vivid; but
then the same danger of total darkness . . . presents itself. The amount of the shares in this
speculation are invested in banking-houses, of the first eminence, from which the ingenious
projector is paid the amount of all expences arising from experiments connected with the
project, as a national scheme. Mr. Winsor has injured his philosophical project, by the stupid
puffs which he is continually inserting in the papers; he appears to be a foreigner, and not
without talent. One day in the week is appropriated to an exhibition of his experiments to
the members of both houses and their families: in this he is very politic, as, hereafter, his plan
of lighting the capital must be submitted to the consideration of the senate [meaning Parlia-
ment], and, if approved of, be made the subject of an act of parliament before it can be carried
into effect.[81]

During this intense campaign, Winsor and company attracted much attention, both
friendly and hostile. His rhetoric was bound to make him the butt of many jokes,
such as those in the satirical pamphlet *An Heroic Epistle to Mr. Winsor*, but there were
also more serious critics.[82] William Nicholson, a prominent scientific journalist and

publisher, was asked by a reader of his journal in January of 1807 for his opinion of this new project, and in response he provided a withering commentary: The invention for which Winsor had got a patent was not original, having been voided by the precedents of Cochrane, Diller, Murdoch, and others. Besides, a patent could only be jointly held by five individuals, not thousands as Winsor hoped to do by founding a company. His profit calculations were beneath contempt, but even if they were correct, Parliament would never grant an act for such extortion. Finally, Winsor did not name his major supporters, as any honest entrepreneur would.[83] A month later, perhaps to reinforce the non-originality of Winsor's claims, Nicholson ran an article on how carbonated hydrogen gas had been observed in the seventeenth century.[84]

Winsor responded to this attack, as one might expect, by writing a pamphlet, in which he included a long discussion about patent law and protesting that Nicholson had attacked him unjustly.[85] Nicholson's brief riposte appeared in the April 1807 edition of his journal. Without bothering to refute Winsor's latest pamphlet, Nicholson merely dismissed it and asked again for the names of his committee of supporters.[86]

Another example of the derision and scorn directed at Winsor is found in the May 1807 issue of the *Anti-Jacobin Review* in a review of Samuel Parkes' *Chemical Catechism*. The gaslight scheme was dismissed as "quackery," not because it was not founded on solid scientific principles, but because it was being promoted by ignoramuses who positively rejected science:

This gas . . . is used to produce what is vulgarly called gas-lights, of which the public have heard so much and understand so little. The practicability of using gas for such purposes with some advantage cannot be doubted; but if any considerable saving of expence is designed, it must depend greatly on the profound and extensive scientific knowledge of those who attempt to apply it to general use. Should many persons, however, be disappointed, as is most probable, in their sanguine expectations from the use of this gas, they have only themselves to blame. It is too much to expect an extensive and important application of chemical knowledge from persons avowedly ignorant of the first principles of the science, and who profess to make money, and to despise even a knowledge of the very subject which gave existence to all their delusive speculations. Quackery and avarice have seldom appeared so conspicuous, or so reprehensible.[87]

Winsor had opened himself to charges of being unscientific by running some advertisements in which he scoffed at the persuasiveness and utility of "volumes of Chemistry and Natural Philosophy," hinting that stuffy old "Physicians" were merely hidebound conservatives.[88] The irony of the *Anti-Jacobin Review*'s comment is that, while they were correct in thinking that despite his pretensions Winsor was indeed mostly ignorant of scientific principles, he was nevertheless instrumental in the founding of the Gas Light and Coke Company. Winsor either overreacted to these charges of quackery, as he did in the case of Nicholson's comments, or put on airs of aloofness and condescension, as he did with the *Anti-Jacobin Review*'s.

Winsor was not, however, without defenders. Not long after the *Anti-Jacobin Review*'s comments became available, the *Times* printed a short piece stating that "truth must ever triumph over falsehood," and comparing the invention of gas lighting to the advent of navigation in terms of its importance to Britain.[89] In addition, as the *Times* pointed out, the share subscription was going so well that in January 1807 Winsor allowed subscribers to buy multiple shares.[90] He moved rapidly on to the next stages of the corporate plan, and plans were made to hold a general meeting of subscribers that would pass basic resolutions such as seeking an act of Parliament.[91] Winsor was increasingly targeting the political arena with his marketing. There had been less scope to do this before February 1807, as Parliament was not in session. An election began at the end of 1806, the last members being returned toward the end of December. When Parliament opened its new session on February 13, 1807, Winsor immediately began running advertisements inviting members of Parliament to see his experiments, with free admission.[92] Winsor and his backers also submitted a petition to Parliament, which the Lords agreed to hear on April 9, 1807.[93]

Winsor now called the first general meeting for April 30, 1807, but that meeting never took place, probably because of political instability.[94] Lord Grenville's newly elected government rapidly lost the king's support over the issue of admission of Catholics into the military. After a month of futile efforts to find a replacement government, Parliament was dissolved on April 27, 1807. New elections were called, and the ambitions of the National Light and Heat Company had to be deferred until the new Parliament could be convened. The general meeting of subscribers was rescheduled for June 26, 1807.

While waiting for the general meeting and the new Parliament, Winsor prepared his next exposition of his gaslights.[95] He had always had a eye for finding opportunities of latching onto anything that might win his work prestige or status. He now approached the Prince of Wales, the future Prince Regent and King George IV, one of Winsor's neighbors, who had built himself a palatial residence on Pall Mall, known as Carlton House, in the late eighteenth century. The prince agreed to have the house gardens lit by gas. Winsor had been thinking of an outdoor display such as this, and had advertised an experiment in "how to light a street," which, however, he had canceled because of "the unfavorable state of the weather."[96] The Carlton House gardens spectacle would be an even better opportunity for a public display. The date was set for June 4, George III's birthday.[97] In May 1807, pipes were laid from Winsor's buildings under Pall Mall to the Carlton House gardens. The pipes went under the garden wall and surfaced on the other side. They were then affixed to the wall, running along its length toward the west where some lamps were attached to the wall. The pipes then passed over the wall into Pall Mall again to feed some more lamps on the back gate in the street. Ensconced there was a four-flame burner in "imitation of the prince's feathers." The pipes then went back into the gardens where burners backlit a cut glass plate, referred to as a "transparency," displaying stars crowned with the letters G. R.[98]

Winsor tested the lights successfully on a number of occasions, which he naturally could not resist advertising.[99] The night of June 4, 1807, proved quite successful. Around 8 p.m. the lamps were lit by a lamplighter in front of an assembled crowd. At some point during the night the glass plate was turned around to reveal a poem celebrating the king. The lights continued on until midnight, with a large number of onlookers still present on Pall Mall, "much amused and delighted." The *Monthly Magazine* commented that "from the success of this considerable experiment, in point of number of lights, the distance and length of pipes, hopes may now be entertained, that this long-talked of mode of lighting our streets may at length be realized."[100] Some pranksters decided to annoy Winsor by pouring "liquor of assafetida" along the garden walls and trying to break the glass plate with stones. Winsor responded with an advertisement stating "they may bark at their own scents, but they can never bite my Gas Lights."[101] On the whole, however, Winsor the public demonstration generated positive commentary in the press.

With this success behind them, Winsor and his supporters turned their attention to the upcoming general meeting of subscribers and the freshly elected Parliament. Winsor had many notices printed in the newspapers, and the general meeting was held at the Crown & Anchor Tavern on June 26, 1807. The subscribers resolved to seek an act of Parliament incorporating the company with limited liability for its shareholders, and nominated a provisional committee to prepare formal resolutions for approval by the subscribers at a new meeting called for July 24, 1807.[102] The provisional committee was also asked to verify that Winsor's experiments and apparatus were on solid footing and could indeed provide a basis on which the new company would operate. A special delegation met with Winsor on July 3 and 4, 1807, and he demonstrated his apparatus, the details of which he had in general tried not to reveal in public. Numerical results were later published, and they show that Winsor was working on a very small scale at this point. He carbonized two pecks (36 pounds) of Newcastle coal. At the time, Boulton & Watt's apparatus could carbonize several hundredweight of coal per charge; thus Winsor's apparatus was far behind. However, the special committee decided that his technology was indeed solid and usable enough to pursue the founding of the company.[103] Among the six members of this special committee were James Ludovic Grant and James Hargreaves, who were to serve on the governing committee (called the Court of Directors) of the Gas Light and Coke Company once it was finally constituted, with Grant taking the role of governor. From this point forward, Grant would play an important part in the proceedings of the fledgling company. The Grants were a noble family from Scotland, and James' father, Francis Ludovic Grant, had been a lieutenant-general in the army and an MP. James had entered the Royal Navy at the age of 13 before joining the East India Company in 1788 as a ship's captain. In 1805 the ship he was commanding, the *Sarah*, was wrecked on breakers while being pursued by a Spanish warship off Ceylon. He was looking for a new line of work when Winsor's plans came to his attention.[104]

When the subscribers reconvened at the Crown & Anchor Tavern, the committee presented a number of resolutions. The most important of these were that £20,000 (one-fifth of the fund's capital) be vested in a committee to be given the mandate of lobbying Parliament to get the charter; that the provisional committee be made permanent, with the same members; that the resolutions should be available at the Pall Mall office so that subscribers could sign them; and that the committee be given the power to appoint any honorary members it saw fit.[105] They also asked Winsor to prepare a more extensive demonstration that included lighting many public buildings, the most prominent of which would be the Parliament buildings.[106]

The list of the members of the committee shows how far Winsor's campaign had reached. There were three lords on the committee: the Duke of Athol (John Murray 1755–1830, a member of the privy council in 1797[107]), Lord Viscount Anson (Thomas Anson 1767–1818, MP 1789–1806), and Baron Wolff (probably Sir Jacob Wolff, a baronet and a merchant, son of a wealthy émigré German merchant). Among the remainder, at least three had been or were members of Parliament: Sir Matthew Bloxam (banker), Sir William Paxton (1743/4–1824, banker[108]), William Devaynes (banker and member of the board of the East India Company[109]), and perhaps John Williams. The background of the rest is more difficult to establish, but many were certainly members of London's business community. The Duke of Athol's connection with the project came through his brother, who had married Grant's sister.[110]

While the subscribers were taking care of their internal affairs, the marketing campaign carried on as before. Winsor was inviting "those affected with Coughs, & c." to inhale his gases. He once again offered free admission to his lectures to members of Parliament, and printed another shortened version of his *To be Sanctioned* pamphlet, without, however, tempering his profit calculations.[111] In July 1807 Winsor stepped up his campaign, offering two lectures per day.[112] He railed against possible competitors, which by this point included not only his former assistant Edward Heard, whom he accused of infringing on his patent,[113] but also a new target: the Golden Lane Brewery, which on July 11 had taken Winsor and company by surprise with an extensive display of gas lighting. Although it seemed to be more modest than Winsor's at the Carlton House gardens, the lighting did generate some publicity, with reports in *The Athenaeum* and *The Monthly Magazine*. The brewery's apparatus consisted of approximately 700 feet of pipe in two adjoining streets serving eleven lamps with gas.[114] *The Athenaeum*, never very friendly to Winsor, even suggested that the brewery had beat Winsor at his own game. Winsor's response was to make dark allusions of patent infringement once again, and to warn the public that gas lighting could be quite dangerous if handled by the wrong person since "Gas explodes like gunpowder."[115] He publicly disavowed any connection to the "bungling attempts to light Golden lane and Beech-street," stating that "the insidious report of my having provided the stove, &c. are utterly false."[116]

In August 1807, Winsor reported in the papers that 15,000 shares had been sold, and the remaining 5,000 would be sold at a 50 percent premium, or at £7 10s.[117] So many people were coming to see his demonstrations in Pall Mall that he had to discontinue free admission for subscribers and raise the entrance fees for everyone.[118] The ambitions of Winsor and company, however, were checked once again when Parliament was prorogued in September 1807. The special Tuesday evening demonstrations continued to be reserved for MPs, to which the nobles and gentlemen among Winsor's supporters could invite their contacts as well.[119] September 15, 1807 saw Winsor score another publicity coup when a number of important people came to his house in Pall Mall to see the gaslights. The *Times* reported it the following week:

The first rank and beauty in town met on Friday evening, at Mr Winsor's, in Pall-mall, to see the beauty and brilliancy of Gas Lights. In the elegant circle we noticed the Duchess of St. Alban's, Ladies Beauclerk, Castlereagh, Forster, Dysart, Cooper, and their Lords, who seemed highly entertained, and staid [*sic*] till a later hour, to witness several surprising experiments with Gas flames, which are far superior to any flame produced by the most costly combustible, such as wax, spermaceti, &c.[120]

By October, Winsor and company were moving to close the share subscriptions. They asked all country bankers to send in their subscriptions before December 1,

1807.[121] The last 2,000 shares were being sold at £10, double the original asking price, to "defray all extra costs."[122] The excitement surrounding Winsor and the National Light and Heat Company at this time is captured well in a letter the Countess of Bessborough wrote to Lord Granville on September 7, 1807:

I was interrupted, but must begin again where I left off, and indeed, as there is no other subject thought of or talk'd of, I may as well write of it too, Is it the seizure of Zealand? No! The investing of Copenhagen? No! The Invasion? Oh no! War with Russia? Nothing like it. America? Still less. What can occasion such a ferment in every House, in every street, in every shop, in every Garret about London? It is the Light and Heat Company. It is Mr. Winsor, and his Lecture, and his gas, and his patent, and his shares—these famous shares which are to make the fortune of all who hold them, and probably will involve half England in ruin, me among the rest, and prove a second South Sea Scheme. Yet it promises fair if it did not promise too much—*six thousand a year* for every seven guineas seems more than can be possible but were it hundreds instead of thousands it is immense. 17 *thousand shares* have been sold within these ten days: they were first a guinea, then 3, five, seven; they will be twenty, fifty, a hundred, for there is scarcely means of passing thro' Pall Mall for the crowds of carriages, and people on foot and Horseback. Ld. Anson has 100, the D. of Athol 200, the Royal family 200, Ld. Chol^y 20—every body some, and I five. There is no resisting it and I long for you to have some. At first I dreaded its hurting the holders of Coal mines, but now they say it will raise the coal trade, as, if it passes, all London will be lit from Coal instead of oil. I am not light headed, and only talking to you as every body talks ever where at this moment. I went last night to a very large party; I never saw so off a looking place—something like a cellar with crucibles and strange looking instruments, resembling an Alchymist's Shop in Tenier's picture, and there, mix'd and squeez'd together, were fine Ladies, a few Rabbis, Merchants, Peers, blue Ribbons, and tallow Chandlers—all raving for shares and entreating to sign their names first, lest none should be left. That Shining Lamp which has lit up Pall Mall for this year past has all at once blaz'd up into a comet that bears every thing along with it. I only stipulate to know the amount of the forfeit in case of failure, and not answer for the debt of the Company. He only wants a *million* to begin with, and has almost got it already.[123]

At the beginning of November 1807, with the shares completely subscribed and Parliament not in session, Winsor's attention drifted somewhat. He published a pamphlet on the importance to commerce of having a national deposit bank in the country; it was based on two books he had published in 1799 during his previous stay in England.[124] He even hatched a plan to found such a bank and advertised a share subscription.[125] This scheme panicked some of his supporters in the gas venture and resulted in a testy exchange of public letters in the *Morning Chronicle*, Winsor referring to the subscriber as "a pitiful and designing fellow, below . . . my further notice."[126] Despite Winsor's hostility toward the subscriber, the exchange drew the committee's attention to Winsor's activities, and the deposit bank was never heard from again.[127]

Toward the end of November 1807, the committee of trustees took more serious steps to establish the first gasworks and large-scale demonstrations that had been discussed at the July 1807 general meeting.[128] The committee was hoping to be able to provide gaslights for the immediate area, including the Parliament buildings, the Treasury, the East India House, and other places. Finding that the expense and complication of doing so would be great, they limited their plans to Pall Mall.[129] Winsor finally gave up his public demonstrations to concentrate exclusively on this ambitious new project. The leases at 97 and 98 Pall Mall were purchased so the location could be used as a plant for the new company.[130] "At no inconsiderable personal pains," the committee requested permission from local residents and the paving authorities for St. James' parish to place immediately outside of the Pall Mall offices two decorative lampposts holding three lamps each in glass globes affixed to separate arms. Without such an authorization, the lampposts would have been by law "deemed such public nuisances . . . liable to abatement, by any individual who felt any hostility or objection."[131] The local authorities prevaricated, however, and finally decided to see Winsor's gaslights for themselves.[132] The visit must have been a success, because Winsor ordered lampposts and pipes on November 1, 1807.[133] The supplies were delivered, and by December thirteen lampposts had been erected on the south side of Pall Mall, from the company's office westwards to St. James Street.[134] The committee wanted to extend the pipes to Boodle's, an exclusive gentlemen's club up St. James Street, but the paving authorities denied permission.[135]

This demonstration using true prototype lamps made a deep impression on some observers. Until then, Winsor's demonstrations evidently didn't feature production models. These street lights, on the other hand, gave a real sense that gaslights could provide more brilliance than the old kinds of lamps. *The Monthly Magazine* ran the following review:

We state with much satisfaction that the Gas Lights which have been so ridiculously puffed and vulgarly advertised in the public papers for several years past, have been proved to answer the promised purpose. Part of the south side of Pall-Mall has for a few weeks been lighted with the Gas, and the effect is beyond all dispute infinitely superior to the old method of lighting our streets. One branch of the lamps illuminated with gas affords a greater intensity of light than twenty common lamps lighted with oil. The light itself is beautifully white and brilliant, and the lamp emits neither smoke nor smell. In a word, we can justly say, that every person who has viewed this first public application of the gas lights, has been delighted by the anticipation of seeing our streets and pubic buildings illuminated in this simple, cheap, and brilliant manner.[136]

The Athenaeum was more doubting, as always, but still conceded that Winsor had shown that "at least the whole of Pall-Mall at both sides might be lighted by the Gas produced by one furnace."[137]

As usual, Winsor and company received their share of derision and opposition. When someone put gunpowder into one of the lamps in Pall Mall, causing an explosion when the lamplighter came to ignite the gas, Winsor was forced to hire two night watchmen.[138] More seriously, some people, probably associated with lamplighters who felt their livelihood threatened or related to the oil trade in some way, were questioning the safety of gas lighting. Winsor's chief antagonist was a certain Mr. Hyde, probably a chemist or instrument maker, who had taken it upon himself to give an "Exposition of Gas Lights, in which the insalubrity of the Carbonated Hydrogen Gas, and the fallacy of the *pretended Inventor's assertions* will be proved by the most *unerring* and *conclusive* experiments, together with the Evils that must inevitably result from the Introduction of Gas Lights," followed by a display of philosophical fireworks.[139]

As one can imagine, this provoked Winsor into making vitriolic and hostile comments in the press. Winsor printed a series of notices toward the end of December 1807 accusing the "empirics," "catchpenny philosophers who are without names or abodes," "advertising quacks," "wiseacres," and "the chemical sadler and his learned colleagues" of misrepresenting the situation when they claimed that a malicious person could place "oxigen and gunpowder" into the Pall Mall lamps and cause an explosion. Hyde had also stated (in this case correctly) that "gas was pernicious to inhale," but Winsor retorted that all apothecaries' apprentices knew the contrary, and he himself was curing people of coughs with his gas. Winsor finished the notice by promising, somewhat incoherently, that he would "render them full as notorious, as their attacks on Gas-lights are ignorant; because clowns, the most happy in blundering, will shew by accident all what these quacks exhibit, to frighten, if possible, fools and old women."[140] Hyde's accusations even prompted the normally invisible committee to print a notice—in calmer tones than Winsor's—reassuring the public of the safety of gas lighting, because they had found "the public mind much agitated by alarms which had been excited for the safety of individuals . . . from the apparatus and tubes necessary for making and conducting . . . hydro-carbonic gas."[141] Sensing a wonderful opportunity for an amusing evening, the British Forum arranged a public debate between Hyde and Winsor for January 7, 1808.[142] There is no report as to who won, but it was no doubt entertaining. In any case, Hyde continued his lectures at Winsor's old haunts at the Lyceum Theater and at a house in Piccadilly during early 1808, and even invited members of Parliament to attend.[143]

In January of 1808, perhaps somewhat rattled by the Hyde incident, the committee retained the chemist Friedrich (Frederick) Christian Accum (1769–1838) to evaluate Winsor's apparatus and procedures and to answer a long series of questions regarding the chemistry of the process which the committee put to him.[144] Accum, originally from Germany, had moved to London in 1793 to work as an apothecary. In 1800 he had established himself as a practicing chemist by selling chemicals, offering courses in chemistry and pharmacy, making and selling scientific instruments, and doing research in chemistry. His lessons attracted some prominent students, including the Duke of Bedford and the Duke of Northumberland. After a two-year stint as Humphry Davy's assistant at the Royal Institution, he moved to the Surrey Institution, all the while continuing his popular lectures, becoming one of the best-known lecturers in London. Accum also published a series of articles in the *Journal of Natural Philosophy, Chemistry, and the Arts* and in textbooks on mineralogy, analytic chemistry, and other subjects that would run through many editions over the years, and would be translated into other languages.[145] In terms of finding a chemist with a public reputation in London, the committee had made a good choice. They asked him "whether it is possible to produce ignition or even a single spark in a vessel or tube abundantly supplied with compressed inflammable air" and under what circumstances explosions could occur. Accum was generally reassuring, repeatedly pointing out that explosions could happen only if oxygen and inflammable air were present in the correct proportions, a circumstance not found in sealed pipes or vessels.[146] This marked the beginning of a lengthy association between Accum and the company. Accum became the company's chemistry expert—Winsor not really being credible—and eventually made it onto the court of directors. Not being a good engineer, he had little success with the company once it began operations, but he would go on to make a name for himself by writing the first books on gas technology.

4.4 BOULTON & WATT AND THE ROYAL SOCIETY, 1808

The year 1808 marked a new phase in the effort to establish a gaslight company. Inside the camp of promoters, there was a definite shift in who had the initiative, away from Winsor and toward the business interests who desired a different approach than what Winsor could provide. Winsor was talented in generating attention, and had been spectacularly successful in promoting gas lighting. He had gathered around him a large number of investors with much capital. But with this milestone behind them, Winsor was reined in by the committee of trustees led by James Ludovic Grant, and they took a more active role in promoting the company's interests in the public forum and especially in Parliament. A more gentle and suave touch was needed. Winsor's bombast might prove to be a liability. Winsor's advertisements and pamphlets were generally halted, and the new ones printed for the cause were authored by hard-nosed lawyers and businessmen, not by the voluble and larger-than-life Winsor. He was in fact never to regain the prominent role he had held from 1804 to 1807. His energetic marketing had succeeded in gathering and coalescing an important group of promoters for the company, some of whom now felt he was of little further value. Not all of the investors shared this opinion, however. Disagreements over Winsor's place in the new company would come to a head in 1813, creating the first serious battle over the company's direction.

Outside the company, new players, much more influential than any who had opposed the National Light and Heat Company until now, entered the scene and forced the company into a tactical retreat. Rather than the amateur chemists and satirists poking fun at Winsor's rhetoric, the new opponent was the respected and influential firm of Boulton & Watt, which had much experience in litigation and lobbying Parliament over its steam engine patents. Boulton & Watt marshaled its powerful allies to deal a blow to the perceived threat from the upstart company. The initial weapon of choice was the Royal Society through James Watt's friendship with Sir Joseph Banks, and, at least for 1808, they effectively halted the nascent company in its tracks.

By late 1807, the trustees were making moves to sideline Winsor gently. The process can ultimately be traced back to the general meeting of July 24, 1807, at which the committee to aid Winsor was constituted and took on some of the responsibility for establishing the company. Although as a body they probably did not explicitly want to push Winsor aside, the complexity of the situation called for more people and professional expertise to be involved, and this became more apparent with time, especially after Winsor's abortive foray into banking in November of 1807. The committee had in any case been meeting more and more frequently over the course of the year, and was meeting weekly by November.[147] In that month, the committee hired a lawyer, John Pedder of Middle Temple, to spearhead the legal work needed to get the charter to incorporate the company. Surveyors were also hired to prepare systematic plans of streets and squares for eventual lighting.[148]

The committee of trustees had a pamphlet written in February of 1808—the author is unknown—to make more moderate claims about gas lighting and to make clear that the unpredictable Winsor was no longer heading the affair. This was a necessary first step in convincing Parliament that the gaslight company could be trusted with a charter and would not descend into bankruptcy in short order. The pamphlet openly sought to eschew Winsor's hyperbole. It began by acknowledging Winsor's many precursors, and gave the nod to Murdoch as being "the first who conceived the idea of conducting coal gas through tubes for the purposes of artificial illumination" but noted that "the practice was discontinued at Soho, or exhibited only as a philosophical curiosity." The work at Philips & Lee was also acknowledged, but the "mode of preparing it is still so imperfect, that [Lee] is obliged, from time to time, to take his tubes to pieces to clear them from impurities." In these affairs, the pamphlet claimed, the important point was not that someone "hints obscurely at the truth, discovers some of its bearings, or illustrates it by a few barren experiments," but rather that the experiments are pressed "on a great scale" which "shews that it is capable of extensive applications." Winsor's contribution was not that he made a new discovery; "he did not even invent the mode of conducting it through tubes." Rather, he simplified the apparatus and lessened the associated cost. It is true, the pamphlet went on, that his "publications, indeed, are but ill adapted to promote his cause; and the exaggerated calculations which the sanguine mind of a discoverer is naturally disposed to indulge in, have to superficial observers thrown an air of ridicule and improbability on the whole scheme." In fact, had Winsor not been present, the author claimed, the project would have advanced more rapidly. Now, however, the project was in the hands of "intelligent and respectable gentlemen" who had verified Winsor's experiments. At this point, "more abilities and a greater capital than any one individual can command" were needed "to combat the prejudices and interests combined against it, and to apply it on the extensive scale of which it is capable." For this, an act of Parliament was needed to grant and charter incorporating the company.[149]

Just as this pamphlet was published, Boulton & Watt struck back at the National Light and Heat Company. The countermove had been building for some time. People at Boulton & Watt had been aware of Winsor as early as 1804. Winsor himself had sent the firm promotional material and asked them to buy shares in his first venture.[150] They had taken little note of it, and they probably forgot about him. Once Winsor upped the intensity of his marketing campaign late in 1806, it was only a question of time before his ambitions came to Boulton & Watt's attention. The latest that could have happened was January 1807, when one of their potential customers let James Watt Jr. know that "Mr. Winsor very modestly claims the invention and first application."[151] Watt Jr. did not give it his immediate attention, but when he heard reports from London at the end of February 1807 that Winsor was about to apply for charter from Parliament, he hurried to London to investigate the situation.[152] It was a false report, as it turned out, but Watt Jr. had now experienced the vastness of Winsor's talk and ambitions firsthand, and kept on his guard for specific action on the part of the "quacks" in London, asking his London lawyer, Ambrose Weston, to keep him informed should there be any activity requiring his attention.[153] He returned to Birmingham reassured and went on a "planting expedition."[154]

Once back home, James Watt Jr. cogitated over the situation for some time and concocted a plan to crush Winsor's pretensions to being the inventor of gas lighting: the Royal Society would award the Rumford Medal to Murdoch on the basis of a paper he would write describing his work on gas lighting. The medal had been endowed by Count Rumford in 1796 to be given by the Royal Society to the European man of science who had made significant contributions in research into heat and light in the preceding years. Rumford, who was strongly utilitarian, had been instrumental in founding the Royal Institution, with its program of spreading scientific knowledge, making it useful, and applying it for improvement. Rumford was particularly interested in heating and cooking, and had hoped the Royal Institution would promote research in that area. The medal he had endowed for the Royal Society reflected his earlier interest in the matter. It was to be granted "as a premium to the author of the most important discovery, or useful improvement" on "the theories of fire, of heat, of light and of color" and "chemical discoveries."[155]

James Watt Jr. thought this medal would demonstrate with all the authority of the Royal Society that Winsor was an impostor and a mountebank who was trying to swindle Murdoch out of his just claims to gas lighting. Parliament then would not dare grant a charter to a project based on a fraud's wild claims, and Boulton & Watt's business would be secure. Watt Jr. must have reasoned that the plan had a fair chance of succeeding because of both Matthew Boulton's and James Watt's friendship with the president of the Royal Society, Sir Joseph Banks. The use of their connections to Banks and the Royal Society to gain an edge over their gaslight rivals was part of pattern they followed in their work on steam engines. As David Miller has shown, it was of considerable importance for Boulton & Watt that James Watt be known as a "philosopher" as the firm battled over steam engine patents in late eighteenth century.[156] Presenting Watt in this way had significant commercial importance for the firm because of the prestige and the perceived superiority it accorded him and his work in the context of commercial competition.[157] For example, in their lawsuits in the 1790s, Boulton & Watt partly relied on fellows of the Royal Society they knew as witnesses, while their opponents used people in business with practical experience in steam engines. The court rulings were in favor of Boulton & Watt.[158] In a similar vein, the Lunar Society of Birmingham, of which both Matthew Boulton and James Watt were members, cultivated their relationship with the Royal Society because they thought that the scientific credibility such a relationship could provide may prove helpful in patent disputes.[159] In this case, it would be Murdoch who would be the target of philosophical polishing. In order to do this the work of all the others involved with gas lighting was minimized, and Murdoch became the inventor of gas lighting. The paper presented to the Royal Society, with Murdoch as the author, effectively made this claim. In its opening paragraph, Murdoch stated that "the apparatus for [coal-gas] production and application [has] been prepared by me at the works of Messrs. Boulton, Watt, and Co at Soho."[160] Indeed, this effacing of the group effort extended even to the writing of the paper. Murdoch was to be its sole named author, but the paper was written by James Watt Jr. and was supported mostly by the experiments of Henry Creighton and George Augustus Lee. Murdoch read it only after the work was done.

This emphasizing of Murdoch to the detriment of others working on gas lighting at the firm reflected the growing public prestige accorded inventors. As Christine MacLeod has shown, in the seventeenth century and the early eighteenth century "projectors" or "patentees" had been held in very little regard; they had been considered akin to fraudsters and thieves. By the middle of the eighteenth century, however, inventors had been gaining in regard, a process in which Matthew Boulton and James Watt had been important as they promoted their work by emphasizing Watt. By the beginning of the nineteenth century, although inventors had not yet achieved the status they had enjoyed in France for many decades, they were willing to identify themselves as inventors, and 'patentee' was no longer a term of opprobrium. While the final transformation of Watt into a national hero of invention would come only after his death in 1819, the term 'inventor' had sufficient prestige attached to it that Boulton & Watt clearly thought it was worthwhile to have Murdoch depicted as one. Watt Jr.'s creation of Murdoch as the inventor of gas lighting was then similar to what he did with his father a few years later.[161]

The paper had to be written, and for this Boulton & Watt needed some numbers to put to Murdoch's claims. Since they could not state outright that Murdoch had invented gas lighting, they would need to claim that he was the first to apply it in an industrial context, and to do so they would need information about how the apparatus was functioning at Philips & Lee's mill. Some information was collected by June 1807. Watt Jr. felt they could put something together fairly quickly, so he wrote to his father, who was in London at the time, asking him to broach the question of Murdoch's paper with Sir Joseph Banks.[162] James Watt briefly replied five days later that they would have to try for next year.[163] Watt Jr. then dropped the subject again for a few months.

Gas lighting and the Winsorite threat came into focus once again late in 1807 when Winsor's campaign reached another crescendo in London. This resurrected the Royal Society plan, and Boulton & Watt once again began collecting information. The December 1807 to February 1808 experiments, first at Soho and then at Philips & Lee, by Creighton working with Lee and William Henry, were described in chapter 3. These experiments were coupled with an economic analysis to form the basis of the paper that was submitted to the Royal Society in February. Almost all of the figures in the paper, including the economic analysis, were prepared by Creighton. Watt Jr.'s changes—resisted by Creighton and Lee—were to make the financial picture seem rosier than Creighton's and Lee's reckoning suggested.

Simultaneous with the scientific and technological work, James Watt Jr. wrote to Thomas Wilson, an employee based in Cornwall, asking him to gather what historical evidence he could by interviewing people who might have known Murdoch in 1792–1798, when he had first worked on gas lighting there. This would be used to bolster Murdoch's priority claims over Winsor. Wilson wrote two letters giving details about what he had learned from various people.[164]

James Watt Jr. proceeded to draft a paper based on these results, working with Ambrose Weston and with the blessing of Sir Joseph Banks. Murdoch looked it over and corrected it in a few places.[165] It was then sent off to Banks, who read it before the Royal Society on February 25, 1808.[166] The following day, Watt Jr. wrote a remarkable letter to his father which demonstrates that the intention behind the paper was to claim priority for Murdoch over Winsor, and that Watt Jr. had worked with Banks to come up with a paper well suited to securing the Rumford Medal for Murdoch:

My time has been a good deal occupied in ascertaining what sort of paper would be most likely to answer Sir Joseph Banks's wishes, and secure Mr. Murdock the Rumford Medal. And I have, with some assistance from Mr. A. Weston, drawn up one which fully met Sir Joseph's approbation and by his interest, was last night read at the Royal Society, and I suppose will be published in the next volume of the Transactions, it being the desire of Sir Joseph, that it should appear before the public, as early as possible.

I think however, that although it will undoubtedly affect [sic] its intended purpose of securing to Mr. Murdock the claim to the original idea, and first practical application of the Gas from coal to economical purposes it may be wrested to serve the views of the National Heat & Light Company, by the proof it affords of the gas light being already successfully introduced upon so considerable a scale, and consequently shewing that their pretensions, as far as utility is concerned, are not altogether nugatory.

But this is a danger, which there was no means of avoiding, consistently with Mr. Murdoch's wishes & Sir Joseph's intentions and it will not be a very easy matter for the Gas Committee to reduce their statements to correspond with our matters of fact.[167]

The effect of the paper's publication was deep and pervasive. With the Royal Society's prestige behind it, the Murdoch paper became an object of great interest.[168] The furor conjured up by Winsor had created intense curiosity about gas lighting among the general public, and now the Royal Society had waded in, giving its approbation to the technology. The extent of the interest in Murdoch's paper and in gas lighting more generally is reflected by the number of publications that reprinted the paper in whole or part after it appeared in the *Philosophical Transactions* in July 1808. The following year, it appeared in at least sixteen serial publications.[169]

As James Watt Jr. had planned, the Royal Society, by printing the paper, had acknowledged Murdoch as the true inventor of gas lighting. It had effectively rejected the validity of Winsor's assertions to any patent rights and had thrown into question the value of the National Light and Heat Company, which was based on the premise of working with Winsor and his patents. *The Athenaeum* commented: "This paper of Murdoch proves incontestibly . . . his being the original inventor of the method of using coal gas to produce light, and of the consequent insufficiency of any patents for the invention to others."[170]

The effect on the Winsor camp, however, was not all negative. James Watt Jr. had recognized in his letter to his father even before the paper's publication that it was a two-edged sword. It would refute Winsor's pretensions of priority, but it would also give the Royal Society's stamp of legitimacy to the technology, and perhaps make the National Light and Heat Company's project more acceptable, even as it undermined its patent underpinning. There is no doubt that this indeed occurred, as the number of reprints of the paper shows. It probably also contributed to the relatively quick acceptance of gas lighting in Britain. In any case, from Watt Jr.'s point of view the paper and the medal closed the affair, and he dropped the subject again.

The promoters of the National Light and Heat Company decided to try a different tack. Rather than ask Parliament to pass a bill approving a charter from the king, they would bypass Parliament and apply (via the privy council) to the king for a charter. Perhaps they felt that there was popular pressure on Parliament not to recognize a perceived usurper because of Murdoch's paper. It is more likely, however, that they had begun to think this way even before the paper was published. Even as Watt Jr. was collecting names of MPs who would take their side in a battle in Parliament, he was watching to see when the Winsor camp would submit their petition, but he noted that they missed what he thought was the cutoff date.[171] The Winsorites had probably thought that they could save considerable expense and effort if they could bypass seeking Parliamentary approval, although they had prepared the *Considerations* pamphlet for just that purpose, and had even prepared a draft of the act.[172] Winsor had also restarted demonstrations at Pall Mall exclusively for members of Parliament in anticipation of new session which had begun on January 20, 1808.[173]

In March of 1808, with the assistance of Thomas Myers, a "gentleman with just views on political economy," the committee drew up a report on the financial benefits accruing to the nation's finances from the founding of a large gas company, and submitted it to the Chancellor of the Exchequer, Spencer Perceval, for his consideration. The Chancellor then forwarded it to the privy council. The king asked the Attorney General, Sir Vicary Gibbs, to study the matter and report back as to what steps would be appropriate. He replied on March 19, 1808 that "such charter could not properly be effected from his Majesty, and that an act of Parliament would be required to carry it into execution."[174]

It was now incontrovertible that the only way forward for the National Light and Heat Company was through Parliament. Wary of the expense of having to do this and the lateness of the session, the committee once again wrote to Chancellor of the Exchequer Spencer Perceval to ask his opinion of their chances. Perceval's reply came on March 26, 1808, and it was not favorable. He "was not sufficiently convinced of the public utility or importance of a Light and Heat Company, to induce him to give his parliamentary support . . . particularly in the advanced state of the present session." Grant wrote back asking for the nature of Perceval's doubts, but received no reply.[175] The committee was forced to concede defeat, and on May 11, 1808 it called a general meeting.[176]

Winsor had not been inactive during all this time, even if he was no longer leading the effort. He concentrated on his research work and on extending the lighting of Pall Mall. In January 1808, he built a new furnace in the "front house" at Pall Mall, and installed a condenser for the apparatus.[177] He even took out another patent, granted on March 3, 1808, for improvements to his stove, although, for reasons that are not clear, he renounced it in August.[178] During February of 1808, he hired stokers to keep the furnaces at Pall Mall running during the night and ran some overnight demonstrations and experiments, in one case carbonizing 11 bushels of coal in a

24-hour period during which 200 lamps burned continuously.[179] The committee, however, felt that there was little to be gained from continuing the lighting of Pall Mall, and discontinued it on April 16, 1808.[180]

In preparations for the general meeting of May 26, 1808, Grant and the committee prepared and printed a report for the subscribers. After summarizing everything done to date, they reported that Winsor had worked under various circumstances, such as "the frequent fluctuations of his hopes and fears, and the anxieties occasioned by his pecuniary embarrassments, his personal fatigue and mental sufferings." Making allowances "for this accumulated pressure upon an ardent and irritable mind," the committee suggested that Winsor's expenses, which ran to £6,000 over the last four years, be paid.[181] They then left it up to subscribers to decide if another attempt should be made to obtain a charter, or whether the funds should be returned.

When the subscribers gathered at the Crown & Anchor Tavern on May 26, 1808, they were not sure exactly how to proceed, but they were determined to try again.[182] Winsor proposed that he prepare a new plan, but the subscribers wanted a thorough legal study done to assess his proposal. A report was to be made to the subscribers at a new meeting, scheduled for June 30, 1808. Winsor did not have sufficient time to gather all the information and procure a legal opinion, and the meeting was called off.[183] The committee of trustees was reconstituted by the subscribers to push ahead with the effort in Parliament. It now included Grant, Pedder, Accum, and others. They were mostly inactive until Parliament reconvened in 1809.

Winsor settled into some of his usual activities, albeit with a lower profile than before. He advertised shares for sale, now requiring a £25 payment, five times the original sum demanded.[184] The king's birthday was approaching, and Winsor requested permission from the St. James paving authorities to light Pall Mall once again, which he received, and the lamps came on the night of June 4, 1808.[185] Other than this, he tinkered some more with his stoves, almost burning down the Pall Mall houses at one point,[186] and applied for another patent, which was granted on February 7, 1809.[187] He also worked on a "moveable Telegraphic light house."[188] For the moment, it was mostly a question of waiting for the National Light and Heat Company.

4.5   THE PARLIAMENTARY BATTLE OF 1809

In 1809 the National Light and Heat Company finally made a concerted effort to get an act passed by Parliament. The previous years had seen them accumulate lots of capital and important backers, put together a team to seek the charter, including professionals such as Accum and Pedder, and finally draft an act and sound out the legal and political environment. Their setback in 1808 was due as much to Murdoch's Royal Society paper as to their lack of experience in executing something as difficult as getting an act from Parliament. The new Parliamentary session in 1809 was different. Everything was in place for the effort.

Boulton & Watt, of course, had not left the field. They had assumed that their victory of 1808 had put them in a good position, but when the Winsorites tried again in 1809, they struck back with even greater resources than in 1808. The clash that resulted, in contrast to 1808, was fought almost exclusively in and about Parliament. It was no longer a question of priorities or of the viability of the technology. Those questions had been settled. In 1809 it was purely a political battle. The irony was that a solution was negotiated even as the parties were waging their war. A compromise that was agreeable to both parties was effectively worked out, but the battle had reached such a pitch that it was beyond the two sides to end it at that point. The promoters of the National Light and Heat Company were defeated in 1809, but, after recovering from the shock of their defeat, they came to understand that success was within their grasp in 1810 by resurrecting the compromise they had forged with Boulton & Watt in the previous year.

Parliament opened the new session on January 19, 1809, and the committee retained William Mellish, a wealthy banker and MP for Middlesex, to promote their cause. Mellish might also have been a subscriber to the gas company fund. On February 24, 1809, Mellish put forward a motion for the gas company, but was rejected on a technicality.[189] This attempt caught the attention of the lawyer James Weston, Ambrose Weston's brother, who also worked for Boulton & Watt. He wrote a letter to Watt Jr. on March 9, 1809 including a copy of the petition.[190] Watt Jr. got the letter the next day and immediately sent it on to Boulton, who was then at Bath.[191] Boulton thought that although the "effrontery" of the bill would lead to its defeat, it would be prudent to present a counter-petition arguing that Boulton & Watt had spent large sums on developing the technology, and that it would be unjust for the benefits to fall to a third party.[192]

On March 27, 1809, Mellish tried to present his petition again, and this time it was accepted. The bill was read for the first time.[193] James Weston got a hold of a copy of the proposed bill and sent it to Watt Jr. The bill proposed to create a company that could raise capital up to £1 million and would manufacture and sell gas apparatus, as well as gas, coke, and other byproducts. It also had powers to take up the streets to lay pipes, with the permission of local paving authorities. Weston thought that "it claims no merit of invention and it seems proper that it should be opposed on public grounds rather than be an individual at his own expence," but awaited instructions from James Watt Jr.[194] Boulton had hurried to London to investigate the situation personally, and had come to much the same conclusion as James Weston. Rather than submit a counter-petition, it would be better to allow the bill to be opposed on public grounds, and to come in as witnesses when the time came.[195]

Watt senior, normally unconcerned with gas lighting, also opined on the matter when he replied to Ambrose Weston, in lieu of Watt Jr. (who had gone to Wales on other business), that they should print advertisements declaring that Boulton & Watt already supplied such apparatus.[196] Ambrose Weston agreed and thought that copies of Murdoch's paper should be printed as well.[197] After he read the bill, Watt senior told his son that since it made no "monopolizing claims," it was none of their business, although it was a "shameful bill."[198]

James Watt Jr. agreed with Boulton's proposed course of action of not submitting a counter-petition.[199] Boulton remained in London to monitor the progress of the bill, and wrote to Watt Jr. a week later appraising him of new developments. The Westons also agreed with not presenting a counter-petition in Parliament, and Boulton was getting ready to marshal allies in Parliament to speak against the bill. The first person he would visit was the anti-slavery crusader William Wilberforce, to whom Boulton had given 50 guineas to support his election in 1807.[200] Boulton hoped to make "a convert of so able & zealous an apostle." He was afraid, however, that the bill, "now divested of its former pretensions to the merits of invention," might "be pushed thro' the House by the influence of the gulled subscribers." A pamphlet was needed, Boulton thought. He asked Watt Jr. for "this assistance from your pen as it has already been exerted with so much success in the same cause & can from its better acquaintance with the subject effect the purpose in view with more ease than any other quill."[201] "Although I am flattered," Watt Jr. responded, "I do not feel any inspiration upon the subject and indeed it will take some time to renew my acquaintance with it."[202] He also provided some more names of Parliamentarians who could help Boulton.

The bill, in the meantime, suffered a minor setback in the House when it was rejected on the second reading because it gave the king the power to dissolve the company, effectively granting him authority over Parliament in this case. It had to be withdrawn to be reintroduced a few days later once the "obnoxious clause" had been struck.[203] Members of the committee, catching wind that Boulton & Watt were mounting a resistance, hastily wrote and printed a pamphlet directed at MPs arguing their case and attacking Boulton & Watt for their work. They now openly renounced any claim to invention, but at the same time claimed that Murdoch had not invented anything either. All he had done was "to add to the length of Mr. Diller's pipes, and proposed lighting a manufactory." Even if he had first thought of applying it for "Economical purposes," was his method in fact "safe and œconomical . . . without evident and serious risk?" The committee protested against charges of creating a monopoly, on the basis that it was nowhere to be found in the bill's stipulations. The committee also provided arguments about capital investment in terms that parliamentarians were familiar with because of the various canal, water, and dock companies that had used this legal route in the recent past. The construction of sizable infrastructure required the incorporation of a large company because it was needed to light the metropolis. No individuals, such as "Messrs. Bolton and Watt, respectable as they are," could provide the assurances needed to light a city like London. Only a large company could do that. Besides, Boulton & Watt had only managed to light a few factories in the last few years, which "bear no comparison even to a small parish . . . something much more vigorous and effectual than the isolated efforts of individuals is now requisite."[204]

With the gas company's pamphlet on the street and in the hands of MPs, Boulton felt that the situation was getting out of hand. He addressed Watt Jr. again on April 22, 1809, describing the arguments laid out in the pamphlet. He added:

A gentleman of the name of Grant, a nabob is particularly active & seems to be one of the principal patrons of the scheme. . . . The general opinion of the members whom we have seen is certainly adverse to the granting of the petition upon the grounds that the necessity of required protection is not proved; on the others hand the petitioners possess considerable parliamentary influence & by the activity of their counsels may get their bill through. The arguments used by the advocates . . . [are] that they are possessed of some improvements peculiarly applicable to the lighting of streets, by guarding against the danger of large reservoirs of gas as in the apparatus constructed by us. The benefit of these improvements will be communicated to the public by means of the act of incorporation in which they will own and to give the public the full advantage of them, in a more desirable manner than by a patent; for if they have recourse to this protection they will of course consider themselves at liberty to derive from the public the usual exorbitant profits of patentees. That they claim no exclusive privilege. In short that the public advantage is the only object they have in view & that under such conditions it cannot be promoted so officiously as by the institution of an incorporated company they hope that the opposition of [?] & sordid manufacturers will not be allowed to prevent the attainment of their patriotic views.[205]

Boulton suggested that the committee should be prepared to explain what had been done at Boulton & Watt and why more had not been achieved. He also thought that getting Lee involved would aid their cause, as he would appear more impartial. Finally, they needed to retain counsel before the committee met. Boulton suggested

Henry Brougham, a talented lawyer, a member of the Royal Society, a Whig, and an abolitionist who had launched the radical *Edinburgh Review* and continued to write extensively for it.

James Watt Jr. wrote a note back to Boulton and a letter to Lee asking for his help a few days later.[206] Lee responded that he was "glad you have taken up the subject of the Winsorites which have already advanced too far unnoticed" and pledged his help, writing to MPs he knew and suggesting that Gott could help with those from Yorkshire.[207] Watt Jr. then made preparations to go to London with Murdoch, departing on April 29, 1809 and leaving instructions to Lawson and Southern to prepare figures that could be presented in the Parliamentary committee.[208]

The month of May 1809 was one of frenetic activity for both sides. The bill went to committee, to a third reading, and to a final vote in the House. Watt Jr.'s first point of business was to write a pamphlet with Murdoch responding to the committee's pamphlet. It was printed as a letter from Murdoch to MPs.[209] He prefaced it by saying that, in reference to the gas committee's pamphlet, "had it not been avowed as coming from a committee of gentlemen, I must have considered as the production of an anonymous calumniator." They then proceeded to respond to the various assertions of the first pamphlet. Murdoch had never claimed invention. The "German Diller . . . exhibited for the amusement of the curious and ignorant," but did not "make the application of gas to purposes of public and private *utility* . . . proceeding with all the mercenary caution of a juggler." This "puppet-show entertainment," of which Murdoch was ignorant, had nothing to do with gas lighting as implemented at Soho and at Philips & Lee. The subject has now been "enveloped by Winsor and his friends in mysterious obscurity, and exaggerated and delusive estimates." It was Winsor who was "the successor of Mr. Diller." In regard to the aspersions cast on the safety and economy of the Boulton & Watt gaslights, three year of safe operation contradicted them. The "claim-balancing-committee" was making absurd profit predictions, and threatened to cause another "South-Sea Bubble." Boulton & Watt, as well as Lee, had provided invaluable co-operation and encouragement as gas lighting was developed with "caution and judicious steps . . . which the committee, not very considerately, have stigmatized with the character of *delay*."[210] They finished by attacking the project on the grounds that many of the things the putative company wanted to sell were already available from Boulton & Watt, who did not enjoy the protection of a limited-liability charter.

James Watt Jr.'s use of the South Sea Bubble to attack the Winsor group stemmed in part from the growing resentment toward joint-stock companies. In the last part of the eighteenth century, the East India company had become a lighting rod for attacks over its monopoly on trade to India and the huge fortunes its representatives there could accumulate, and as a result it saw some of its privileges curtailed. Although the company was the most direct target, the joint-stock form of business more generally was facing criticism because of the de facto monopolies it supposedly represented. The argument was that the difficulty in incorporating a company was such that it effectively created an almost insurmountable barrier to new entrants in the market, granting the incumbent a monopoly. It was thought that acts of incorporation should be granted exclusively to a body that would invest in something of great benefit to the public and would not be built otherwise, such as canals or docks. The joint-stock boom that began in 1805 revived fears of a new South Sea Bubble, with opponents of joint-stock companies and the stock market more generally using it to attack these institutions.[211] Although popular sentiment created problems for unincorporated joint-stock companies in the form of lawsuits challenging their existence under the Bubble Act of 1720, the opposition was not strong enough to prevent many companies from winning acts of incorporation.[212] The monopoly objection had enough force in the gas matter that concerns were frequently voiced by MPs in the next month as the gas bill was debated. The bill's defenders claimed in its defense that none of the stipulations of the bill gave it an exclusive right to supply gas, coke, or apparatus, which was true, and that it was better to entrust such a technology to an incorporated public company than to leave it in the hands of private individuals who could easily abandon the business, leaving the public with a great loss.[213]

Both sides now frantically prepared their next steps. Committee hearings were scheduled to begin the next day, May 5, 1809. Watt Jr. sent for Murdoch's Rumford medal, to serve as a visible reminder of whom the Royal Society had supported in this confrontation.[214] The gas company prepared their witnesses. The parliamentary committee was chaired by the geologist and MP for Cornwall Sir James Hall, who was "decidedly Winsorian."[215] The MPs acting as counsel for the bill were Charles Warren and William Harrison. Boulton & Watt initially tried to get their allies in the House to help in committee, but, as Watt Jr. explained, they "found it so difficult to get [their] friends to attend the committee to listen to the nonsense that was going on; and to make those that did attend, put any questions properly"[216] that they were forced to retain Brougham as counsel, as Boulton had suggested earlier. Only Sir Robert Peel would show up with some consistency for Boulton & Watt. Another MP, Isaac Hawkins Browne, whom they had thought was their ally, was there, but never spoke, and they suspected him of siding with Winsor.[217]

The first witness was Frederick Accum, and the strategy that Harrison and Warren pursued with him was to have Accum and some tradesmen they also called claim some sort of superiority for Winsor's stoves and the products they produced. The coke, Accum claimed, was "far superior" to others because it yielded no smoke or dirt, and had "none of the sulpherous smell" common with other coke. It generated more heat. Winsor's tar was as good as any for ship's caulking, and had greater penetrating power. His asphaltum was as good as the best from Russia. His ammonia could be used as a mordant and for fertilizer. Sal ammoniac could be used in making tin and brass, caustic ammonia in making silk and alum. Essential oil could be used in making odorless paints and varnishes.[218] The company's counsels also discussed the lighting of city streets. With Winsor's stoves, the parish of St. James could be lighted for £26,500. In a statement that revealed just how far behind Boulton & Watt the National Light and Heat Company was in terms of experience and technological maturity, Accum stated that no reservoirs or gasometers would be needed in Winsor's plan. Nor did the overall strategy for the company yet include the entire network of mains fed by large gasworks. Their plan to light the parish of St. James envisaged six stations in rented houses, presumably each the size of Winsor's Pall Mall building, supplying 800 lamps and manned by two or three employees.[219]

The talk about the wonderful properties of Winsor's stoves and their products caught Watt Jr. by surprise. The Winsor camp had mentioned the many products of coal distillation, but had not really been using this as a major advantage in arguing for a charter. Now, in the House of Commons committee, they had made it the most important element of their case. Moreover, Boulton & Watt had been ignoring the other products that Accum had praised so highly.

James Watt Jr. wrote to Southern back in Soho to find out anything he could about coke, tar, ammonia, and varnish. When Southern received the letter, he went to see the senior Watt to collect what wisdom he might have to offer on the subject. Watt senior could not recollect perfectly, but he thought that coal tars had been tried for naval use and had been rejected because they destroyed the oakum, were more water-soluble, and rubbed off more easily.[220] Southern performed a feverish series of experiments that ran for a few days trying to determine the heating value of coals and cokes, sending Watt Jr. results as soon as he could get them out.[221] Watt Jr. even had a boy go to Winsor's office in Pall Mall to buy some coke to test it. Although the people at the office seemed to guess the true origin of their visitor's request, they parted with some coke, which was duly analyzed by Murdoch and James Lawson (another Boulton & Watt employee) in London.[222] They found that Winsor's coke was not as spectacular as Accum had claimed. As the information came in from various sources, Watt Jr. felt more confident. He wrote to Southern on May 12, 1809:

I have received your three letters . . . which are very satisfactory, as tending to refute the absurdities which have been advanced here and to confirm the general notions before entertained upon the subject.

Winsor's case is now closed, and a more compleat [septum?] of falsehoods & imposture was perhaps never before brought forward. They have asserted properties equally extraordinary of all the other products of coal except the gas which they do not pretend to be superior to ours & seem willing to admit, at least they have not very boldly denied M[urdoc]ks claims to be the first œconomical application. We are not prepared with evidence to refute their assertions respecting the generalities of their ammoniacal liquor which is likely according to them to prove the grand panaceum. We must therefore simply affirm our ignorance of the facts and indeavour to discredit their witnesses, whose testimony has been in the highest degree suspicious. In short it would appear either that Winsor & Accum have deceived the Gas Committee, or that the latter are themselves indeavouring to impose on Parliament.[223]

Brougham began the questioning with George Augustus Lee for Boulton & Watt on May 12, 1809. Lee gave a history of gas lighting as applied in his mills, insisting that there were no problems with it, and that competition in the manufacturing of gaslight apparatus would be harmed by the emergence of a corporate body with large capital and limited liability. In order to allay the safety fears raised by the opposing side, an insurance agent from the Albion Fire and Life Insurance Company was called to testify that he judged the gas lighting at Philips & Lee to "diminish the risk of fire considerably."[224] Humphry Davy, whom the Winsor camp had tried to recruit to their side, was called in to say that Winsor's coke was not markedly different from any other he had tried, a point Lawson confirmed as well.[225] A manufacturer of coal tar and paint color testified that Winsor's products were nothing new, having been introduced by Cochrane in 1787.[226] The hearings were finished off with lengthy testimony from Watt Jr., who presented the history of the invention, emphasizing that a patent had not been taken out for gas lighting because of prior work from Cochrane and others. He insisted that there had been no unnecessary delays in providing apparatus to those interested. They had needed to experiment with Lee's apparatus before they were ready to sell others. He strongly resisted the idea of incorporating the company because it would create a competitor invested with large capital and shielded from bankruptcy laws, giving it the option of running many more risks, hiring away Boulton & Watt's employees, and probably forcing them to quit the business. During questioning by the opposing counsel, one question was placed to Watt Jr., whose response the promoters of the company evidently noted. It was whether Boulton & Watt would do business with the new company if they did not manufacture apparatus, limiting themselves to lighting the town. Watt Jr. responded that they would then not be interfering in their interests.[227]

The committee voted in favor of passing the bill on May 15, 1809.[228] Boulton & Watt had lobbied to get MPs to attend the committee by distributing invitations, but as usual, not many showed up.[229] A favorable report was then made to the House.[230] The final decision was to be made at the impending third reading, and each side lobbied furiously. Watt Jr. was feeling much more optimistic now, the gas company's claims about wonderful byproducts having been made to look foolish: "We have completely overturned their evidence & established our case, and they are now moving heaven & earth to prevent the evidence from being presented."[231]

Boulton & Watt scored a stroke on May 19, 1809, when William Wilberforce managed to lay the committee evidence before the House—an unusual move that would give more prominence to the testimony, generally favorable to Boulton & Watt.[232] Watt Jr. now wrote and printed another 13-page pamphlet, attacking the company on the basis that it had made no invention and had done nothing in terms of applying the technology and arguing that such a company was hardly necessary considering that the technology was already being sold.[233] In short, "Mr. Winsor's puffs and projects have at last failed."[234] This they sent, together with copies of the evidence collected in the committee, to as many MPs as they could, including Sir Vicary Gibbs (the Attorney General), Archibald Colquhoun (the Lord advocate for Scotland), and many others.[235] By May 27, 1809, Watt Jr. was able to report to his father that they had seen 100 MPs in the intervening period.[236]

The gas company camp began to panic. They realized that, although the vote had gone in their favor in the committee, they had been bested in the actual testimony. They responded to Boulton & Watt's pamphlet with a very brief one of own, two pages in length, affirming that the bill gave them no monopoly.[237] They also tried to seek a compromise with Watt Jr. On May 26, 1809, Harrison made an approach to Brougham and James Weston to propose that they reduce the powers the bill granted to the company.[238] Watt Jr. described the offer to his father:

It is certain they are much alarmed, as their counsel has been endeavouring to get us to agree to modifications of the Bill of his suggesting. They would now be willing to make it a local bill for the lighting of London and Westminster, or a part of these cities only, and perhaps to give up the manufacturing part, and to limit their capital to a less sum than the £300,000 which is fixed in the Bill. But it is now too late for any thing to be done unless by the House itself, as it appears to us we are pledged to go on with our opposition; and should we fail in the Commons, we must follow them to the Lords.[239]

On the whole, James Watt Jr., the two Westons, and Brougham were inclined to accept the offer, but understood that, since they had opposed the bill with such energy to date, it would hardly be credible to drop all opposition suddenly. They sent Harrison a negative answer.[240]

There was now nothing left but to fight to the end. Two more publications were printed, featuring speeches given by Brougham, Harrison, and Warren in the committee.[241] Both of these were over 30 pages in length. It is hard to conceive they had much effect, having become available only a couple of days before the vote. The gas committee decided to try to introduce the proposed clauses—specifically restricting the company's ability to manufacture gas apparatus for sale—and printed a last circular describing the changes.[242] Watt Jr. even drafted a response to the new clauses, but this was mercifully never printed.[243]

The final vote was delayed from June 1, 1809 to the following day. Mellish presented the bill and proposed the new clauses, but these were refused on the basis that the House had not had time to consider them.[244] Davies Giddy (later Davies Gilbert), a Boulton & Watt ally, strongly condemned the bill as monopolizing and as creating another South Sea Bubble. Giddy stated that such acts should only be allowed if the act created a company that would bring the public a great advantage, and that the funds required were beyond the means of private groups, such as the New River Company, a canal, or the West India Docks. William Wilberforce then stood up and opposed the bill as a "gambling project" and presented one of Winsor's pamphlets with its extravagant claims as evidence. The bill was put to a vote and lost by 52 to 38.[245]

Boulton & Watt and their allies were triumphant. Watt Jr. offered the following assessment of their victory to his father:

You will see that we have not treated the promoters of the bill with much respect, and indeed I believe we owe our success as much to the ridicule we have bestowed upon them, as to the intrinsic merits of our opposition for very few indeed of the members have troubled themselves to read the evidence, or to make themselves masters of the general principles of political œconomy, upon which alone, perhaps, this question ought to have been decided. We have however experienced great civility and attendance from all parties in the House, ministers not excepted, who in general took no part in the affair, but I apprehend are very glad to have got rid of the bill. Our canvass has been extensive and there are very few of our acquaintances and friends but what have rendered us some assistance.[246]

James Watt Jr. received congratulatory letters from many people, including Ewart, Lee, and Southern. Lee thought the "gang of schemers & swindlers" were now finished, as did Watt senior.[247] Southern added, however, that in his opinion "it did not deserve so much of your attention as you have I daresay paid to it, and has probably taken more than you calculated upon."[248] Watt Jr. concurred. In his own words, he was "able to think and attend to nothing else while it has lasted," and from this point forward the subject of gas lighting is almost completely absent from his letters.[249] He had had enough of it to last him a lifetime.

4.6 DESPAIR, RECOVERY, AND FINAL SUCCESS, 1809–1812

The gas committee and their supporters were despondent and despairing. In the days immediately following their defeat in the House, the committee, without consulting the supporters, took steps to wind down the entire affair. They placed notices in newspapers calling in their debts and proposed to liquidate all the assets belonging to the putative company.[250] Winsor himself was uncertain as to what to do, although he was still generally optimistic that a company could be founded. On June 9, 1809, he sent out a letter to the subscribers asking if they would support him in a subsequent project, should he try again.[251] Not all subscribers, however, agreed with the plan of killing off the fund, and even saw behind the liquidation a plot on the part of some of the trustees, notably Grant, to dispossess the subscribers of any claim over Winsor's ideas and to reconstitute a new fund with very few subscribers.[252] A prominent London merchant and one of the largest investors in the company, John Van Voorst, took it upon himself to rally the subscribers and to get the decision reversed.[253] He found a few allies who also mistrusted the motives of the committee, and after a visit to the Pall Mall offices to take stock of the situation they called an unsanctioned meeting for all subscribers to take place on July 6, 1809, at the City of London Tavern.[254] The announcement caught the committee and Winsor by surprise. Fearing some sort of coup or worse, they gave notice that they had not called this meeting.[255] Van Voorst met with Winsor and members of the committee whom they trusted to allay their fears, and some of them, including Winsor himself and Sir Matthew Bloxam, went to the meeting at the tavern.[256]

At the meeting, Van Voorst gave a speech which, in keeping with gaslight tradition, he later had printed as a pamphlet.[257] By his own admission, Van Voorst was motivated not by any "romantic ideas of self-devotion or patriotism," but "by a prospect . . . of personal remuneration." He was convinced of the practicability, utility, "and very great pecuniary returns" which "this great discovery" could yield if pursued, and thought it would be a grave error to drop the project at this point. He thought, in fact, that the company did not even require an act of incorporation from Parliament, and in this he was correct, as many unincorporated gas companies were formed and survived after the GLCC was incorporated. The clauses that were proposed at the end of the parliamentary debate by the gas committee were an error, Van Voorst argued, and had the act including these additions passed, the subscribers would have "in an infinitely worse situation than the one" in which they now found themselves. Although feeling "considerable embarrassment," Van Voorst felt it was his duty to call into "question the propriety" of the conduct of Boulton & Watt, although their "general character stands so deservedly high." They were wrong to oppose the bill, there being nothing wrong with the large capital the company was seeking. In addition, the new gas company, once incorporated, would not go to Birmingham to hire away their workmen, there being an ample supply in London (in fact, when the time came, they would go farther afield, to Manchester).[258]

Van Voorst was successful in rallying the "very numerous" shareholders who attended his unsanctioned meeting, and they resolved to push the committee to follow the same line. Winsor and Sir Matthew Bloxam acquitted themselves in a "manly" and "upright" way, and were thanked in the meeting's minutes. A general meeting was demanded of the committee for August 3, 1809, when these resolutions could be made official and binding on the committee.[259] It proved unnecessary, however, because Van Voorst, confident now of the backing of so many shareholders met with the committee of trustees at the Pall Mall offices on July 10, 1809, and they agreed to suspend the decision to liquidate, a feat he jubilantly announced in public notices.[260]

The next general meeting of shareholders was held on August 29, 1809, again at the Crown and Anchor.[261] The printed report of the committee made no recommendation about whether to try again, although it did state that there was "some hope of success."[262] In the event, the subscribers supported Van Voorst and resolved to try again at the next session of Parliament. The committee was reconstituted for this purpose.[263]

All was quiet, without even any advertisement from Winsor, for the next few months (although Winsor's bank, Bloxam & Co., went bankrupt in September 1809).[264] When Parliament opened its new session in January 1810, Mellish was ready with a fresh bill, this time seeking a company with less capital, £200,000; with a sphere of operation limited to Westminster, London, and Southwark; and with a prohibition on the manufacture of apparatus for sale.[265] The bill was introduced on February 20, 1810, and progressed rapidly from there.[266] Once he got wind of the impending bill, Brougham wrote to Watt Jr. encouraging resistance, as did Peter Ewart.[267] Watt Jr. had had enough of the matter, and did nothing, especially after John Pedder, the gas committee's lawyer, sent him a copy of the bill in March to ensure that it contained nothing offensive to Boulton & Watt.[268] There evidently was nothing, because Boulton

FIGURE 4.4

"The Triumph of Gas Lights," by R. Spencer (1810). Courtesy of Science Museum, London.

& Watt were silent as the bill wended its way through Parliament. Even Davies Giddy, the Boulton & Watt ally and strong opponent of the bill in 1809 now supported it with some provisos.[269]

The bill passed through the House quietly, moving to the Lords on April 4, 1810, where it was greeted by the bitter complaining of the Marquess of Buckingham, who had had gas lighting at his home for the past three years, and was unimpressed by the odours.[270] The passage through the House of Lords was shepherded by the Duke of Athol. There was a brief committee examination on May 18, 1810, which consisted of evidence from Accum and a few tradesman regarding the properties of coal gas and various byproducts.[271] A delay was prompted at the third reading over concerns that the bill would create a monopoly, but it passed on June 1, 1810 without fanfare, almost without comment, in such marked contrast to the previous year.[272] One of the changes the Lords introduced was that the group could only apply for a charter once £100,000 was raised, and that a further £100,000 would have to raised within three years of beginning operations.[273] It took two years to meet the first condition, and the king granted the company's charter on April 30, 1812.[274]

4.7   CONCLUSION

The first step in the construction of the gas network in London was the incorporation of the GLCC as a limited-liability joint-stock corporation. This legal, business, and financial model was in the tradition of other large infrastructure projects of late-eighteenth-century and early-nineteenth-century Britain, such as canal, dock, and water supply companies. Parliament, public opinion, and business practice had recognized and standardized a model that gave important privileges to the incorporated company, such as limited liability, strong barriers to entry for competitors, and rights of eminent domain for canals, or permission to take up city streets for water companies. This was allowed because it was recognized that the construction of such large infrastructure was beyond the financial resources of individuals or small private partnerships. Although hostility to the potentially monopolizing nature of these incorporated joint-stock companies existed, opposition to winning acts of incorporation came primarily from vested interests, such as landowners along canal routes. The early years of the nineteenth century saw a growing popularity of the joint-stock form, coupled with an increasing base of investors and popularity of the stock market, leading to a joint-stock boom that Winsor's efforts joined and augmented.[275] As with other joint-stock incorporations, the chief obstacle that the GLCC's backers faced was vested interests, in this case Boulton & Watt, but they negotiated a compromise with them that excluded the GLCC from manufacture of apparatus for sale. Although this compromise was an important step in the consolidation of the network strategy at the GLCC, it would also have important and unintended consequences on the company's relationship with its customers and subcontractors, as described in the next chapter.

The creation of the GLCC as an incorporated joint-stock company provided it with a great deal of stability that proved important in the coming years as it faced technological and organizational challenges. Over the years following 1812, it would encounter some severe financial and organizational difficulties, but managed to survive and eventually thrive, despite the replacement of every person who figured importantly in its foundation. Its survival depended to a great extent on its legal and business form which provided a framework that ensured continuity and sources of more capital and human resources to see it through its moment of crisis and expansion. Without the joint-stock model that had matured in British business and legal practice in the late eighteenth century and the early nineteenth century, gaslight's successful transition from the stand-alone plant to a network strategy would have been impossible.

Entrepreneurship in the Industrial Revolution has not yet received as much attention as has technological innovation. In the case of the GLCC, and gas lighting more broadly, entrepreneurship emerges as distinctly important. Perhaps even more than anything specifically technical, Lebon's contribution to gas lighting was his injection of entrepreneurial imagination to an existing technological tradition, one that prodded both Boulton & Watt and Frederick Winsor to launch their own commercial projects. In the case of Winsor, the entrepreneurial element is even clearer. Winsor had made no contribution to the technical design of gas lighting, but he was indispensable for the initiation of the network strategy, both because of his initial idea of distributing gas throughout the city by means of a network of pipes and because of his ability to attract scores of investors willing to support his vision and the creation of a joint-stock company. Parts of Winsor's vision were unrealistic to the point of absurdity, but he had a tenacity that Boulton & Watt lacked. They were far superior from the technical point of view, with deep experience in manufacturing and innovation, something Winsor was bereft of. Nevertheless, Winsor managed to excite many people to invest and back his plan, including people with business and legal experience, and they carried his vision forward. Like many entrepreneurs, Winsor had an idée fixe that verged on the self-destructive, and he was eventually driven to bankruptcy and flight because of accumulated debt.

The GLCC's founding marked the first steps in the transition to the network strategy for gas. Although gas functioned well as a natural monopoly, meaning that overall costs were reduced when a single institution provides the service, the adoption of the network model was not historically inevitable in the short term. Certainly Boulton & Watt had no sense of a network model. They adopted the same strategy for gas lighting that they were using for steam engine: each mill had its own. The full unfolding of the network model, still not yet achieved in 1812, was a combination of deliberate choice and force of circumstances. The choices began with Winsor's desire to create a huge company, coupled with his suggestion that gas could be distributed like water through a network of mains. The compromise with Boulton & Watt reduced the GLCC's activities to selling products of distillation, giving its business model a more intense focus no longer including the manufacture of apparatus. With initial capital of £100,000, the GLCC had the financial resources to operate on a large scale. Further refinements to the network strategy came once operations began, and the company was limited to only three gasworks to feed all its customers. Simultaneous with the final coalescing of the network strategy, the GLCC had to develop to achieve it; in other words, it had to define and adopt what the specific tactics of the network strategy were going to be, given the specifics of the circumstances in which it had to operate. This is the subject of the next chapter.[276]

Marketing, advertising, and display are other important themes of this chapter. Winsor's advertising, demonstrations, and attempts to associate with prominent people were major elements in his entrepreneurial success in creating the social and political momentum leading to incorporation. His patents, which in the end proved technically useless, and his self-styling as a "patentee," were a cornerstone of his marketing. Boulton & Watt pursued a similar strategy in undermining Winsor's marketing. They attempted to elevate William Murdoch to the status of heroic inventor, a status he has maintained in the historiography of gas lighting. As was shown in chapter 3, however, many people at Boulton & Watt, and also G. A. Lee, were involved in the extensive work of designing and building industrial gas plants. Certainly Murdoch was the first to think of it at the firm, and was important in subsequent work. But just as the Royal Society paper was the product of people besides Murdoch, so too it was the firm as a whole, relying on its suppliers and its customers, that was responsible for establishing gas lighting. James Watt Jr.'s creation of Murdoch as a heroic inventor would be soon repeated with his own father.

5 THE GAS LIGHT AND COKE COMPANY

Since my last visit to London a remarkable undertaking has been launched here. In one of the main streets the shop windows have been lit by gas. This gas—or inflammable air—is conveyed like water in pipes on both sides of the street. Every householder or shopkeeper gets in touch with the gas company and signs a contract for the supply of lamps or wick-holders which he requires and he pays for the gas in accordance with the number and the size of the lamps.

—Hans Caspar Escher, letter from London, November 1, 1814[1]

5.1 INTRODUCTION

Between 1812 to 1820, with remarkable rapidity, the Gas Light and Coke Company built a large urban gas network in London. This was a move away from the stand-alone plants the firm of Boulton & Watt was supplying. The work of that firm and that of its former employee Samuel Clegg was an important base from which the GLCC could expand, but there were many technical, business, and social issues in making the transformation. The new network model required coordination and integration at a level not required in the mill model. These began with the generation of gas. As Boulton & Watt had recognized, retorts, which burned out frequently, were the most capital intensive part of gas plants, and in response Boulton & Watt's customers began to keep them under heat as constantly as possible. For the GLCC, which very soon had 40, 50, and even more retorts in use, whereas most mills had only three or fewer, the question of how to make retorts last was one of pressing economic necessity. The matter was further complicated by the complex interaction between the gasification efficiency (ratio of coal used in the ovens in gasifying a given quantity of coal in the retorts) and the shape and arrangement of retorts. Other technical issues not seen in mills, or at least easily dealt with or ignored, also sprang up. These included the presence of impurities in gas (which clogged mains), leaks in mains, purification of gas on a large scale and related pollution, and supply management. The GLCC was involved in a much more complex web of relationships, particularly in regard to its variety of customers, who, unlike Boulton & Watt's mill owners, did not much care how the gas plant that supplied them ran, or about its economic efficiency. This meant that they were far less concerned with how their use of gas lighting interacted with the whole system, but it definitely did have effects which the GLCC tried to make them feel in other ways. The GLCC's interactions with its customers were further complicated by the fact that it was required by its charter to supply light to interested local lighting authorities at a cheap rate. Often these authorities also controlled access to the pavement under which the GLCC needed to lay its mains. Furthermore, in the case of some users, such as prominent public buildings, the GLCC was more interested in marketing opportunities than in gas sales. The construction of a large, stable gas network over the years required dealing with these many business, technological, and social elements.

How large-scale technological networks were built has been a subject of many studies, particularly since Thomas Hughes' landmark work that investigated how "inventor-entrepreneurs" such as Thomas Edison built systems by combining overall technical design with attention to business and political elements. Edison, keenly aware of how different technological elements of his electrical network interacted, tried to design the network as a whole, an approach that gave him an edge over his competitors.[2] Once the electric network had reached a certain level of maturity, "manager-entrepreneurs" such as Samuel Insull in Chicago ensured its sustained growth by integrating the business and technical dimensions of the electrical network. Relying on acute political and business sense, Insull found new customers with different usage profiles, such as electric trains, in order to maximize the load on the network over the course of the day. He made sure that the generators were functioning as close to capacity as possible during an average day in order to get the highest return on investment.[3] Historians such as François Caron and Hughes himself have pointed out, however, that the building of large networks has not always been marked by such clarity of vision and coordination. In the early days of railways, for example, a coherently integrated approach developed more slowly. Specifically, delays caused by differing visions plagued the French railway system from 1832 to 1859.[4] In another example, the development of the electrical system in London was slow and halting because municipal and national politics together prevented the emergence of standards, which are often crucial for ensuring network stability and growth.[5]

The railways have sometimes been presented as marking the emergence of a new kind of technological network, because the requirements of successfully running an integrated technical and business model were quite different than was the case for canals or roads, where system-wide coordination was not necessary and many separate carriers used a common infrastructure.[6] Railways, in contrast with all networks that came before them, required far more capital investment, were more geographically dispersed, and needed coordination in many more ways, such as train, passenger, and freight scheduling, as well as in accounting. Alfred Chandler in particular argued that it was precisely this need for time-sensitive technical and business coordination, made evident by train disasters, that created the first multi-unit corporations with structured management, including clear lines of communication.[7]

In the years 1812–1820, when the Gas Light and Coke Company built its urban network, its engineers and managers had little by way of example that they could look to in constructing their own integrated network, other than some water companies for distribution. The large-scale development of water provision, sewage disposal, telegraph, and rail transportation, which was to transform cities into more habitable places for the increasing numbers of people moving into them, still lay in the future.[8] In contrast with canals and roads, the GLCC owned its entire network, from gasworks to the burners spread around the city. It was also a tightly coupled dynamic network in which problems and failures propagated rapidly.[9] Even the emergence of the integrated network strategy for gas was a piecemeal process. Frederick Winsor and the founders of the company did not initially conceive of the network as an integrated whole. Winsor's first step toward the network model was to plan to lay piping under streets as the water companies did, but he did not think that there would only be two or three large plants supplying gas for the whole city. Winsor and his supporters thought of many smaller gas plants, each feeding about 100–150 lamps in its vicinity.[10] After operations began, however, and the number of customers grew, the idea of multiplying mill-sized gasworks throughout the city very soon gave way to the large-station model as it became clear that buying or renting many premises was too difficult. The company's managers and engineers then had to deal with a variety of problems in deploying this new model that they had adopted. In doing so, the GLCC was venturing into unknown territory in terms of existing gas technology, but after a Chandlerian reform of its management structure in 1814 it addressed its most pressing problems, such as leaks, gas generation, and network stability, slowly and incrementally.

It was not just inventors and managers but also users and mediators who shaped network technologies, and the gas network was no exception. Users used technologies in ways not originally conceived by the companies that built them, as Claude Fischer showed in his study of the American telephone industry.[11] The first users of gas lighting had a different idea about how to use gas lighting than the GLCC did, but whereas telephone companies were conservative, asserting a telegraph-style model, in the case of gas lighting it was the users who were conservative. Gas users, particularly those in shops and in homes, tended to view gaslights as a replacement for candles and lamps which they could use whenever desired, at any time of day or night. The GLCC, in contrast, wanted to restrict the times at which users could consume gas, so as not to exceed its generating capacity. The GLCC bound its users by contractual arrangements that limited the times at which they could use gas, typically to between 4 and 10 p.m., but the pipes and burners located in homes were out of its direct control.[12] The company's response to off-contract burning, which it considered theft, was to develop a system of inspectors and fines.[13]

The GLCC's efforts to routinize and stabilize its network—normal in the development of large networks—also involved working with a group of mediators who were not initially under the company's control. Because of the compromise with Boulton & Watt during the GLCC's incorporation, it was prevented by law from installing lamps and pipes in homes, and had to rely on outside contractors called "fitters up." Since these fitters had no long-term relationship with customers, their installations were often very poor, leaving the company to placate irate customers and repair their work. As with users, the GLCC introduced a series of measures meant to control the fitters, including the standardization of parts and a certification program.[14]

Section 5.2 describes the first two years of the GLCC's existence (1812–1814), during which it floundered and nearly collapsed before a change of management turned its fortunes around. Section 5.3 explores the many technological changes introduced in the process of building the ever-expanding network, including generating more gas, mitigating leaks, and large-scale purification. Section 5.4 looks at how elements outside the company's immediate control (particularly users and contractors) interacted with and changed the growing network.

5.2   THE NEW COMPANY

LONDON

The London in which the GLCC started its operations in 1812 was the greatest city in Europe, with more than 1.1 million people crammed into a fairly small area, mostly to the north of the Thames, measuring 5 miles across. It was also growing rapidly. Between 1801 and 1831 its population expanded from 959,000 to 1,655,000, and the number of houses it contained jumped from 55,000 to 92,000.[15] Although the majority of the population was poor and uneducated (only about 6 percent had any formal education in 1820[16]), the city's growth between 1810 and 1820 included many middle-class and upper-class areas to the east of Hyde Park and around Oxford Street, in addition to new slums.[17]

London's economic activities were quite varied. The most prominent businesses were in trade, in the finance industry centered on the Bank of England, in the stock exchange, and in a host of companies (including mercantile, commercial banking, and insurance companies). These were supplemented by the ever-busier London docks, which received goods from all over the world, a large brewing industry, a book trade, cabinet making, and a myriad other economic activities. The silk industry, concentrated in Bethnal Green, was run by descendants of the Huguenots who had fled France in the seventeenth century.[18] London's population was varied ethnically, with pockets of immigrants scattered throughout the city, the most numerous of whom were the ten of thousands of Irish, largely concentrated in squalid slums such as the notorious St. Giles. Besides those who had moved permanently, many of the Irish were seasonal workers who labored on farms just outside the city.[19]

The local governance of London was a byzantine labyrinth of a few hundred bodies covering various zones and having different powers, structures, and modes of operating. The most important of these bodies were the City of Westminster around the Parliament buildings, the City of London, which was responsible for the square mile at the commercial center of the metropolis, and the many civil parishes of the counties of Middlesex, Surrey, and Kent, located around the City. The government structure of each varied, but the parishes within the urban agglomeration excluding the City of London proper were usually managed by close vestries, which were groups of around 50 or 100 of the more prominent residents of the parish. There were no elections (these started after reforms in 1831), and the vestrymen usually named new members as needed. The vestries were responsible for the paving, watching, lighting, cleaning, and poor relief in their respective areas, and most contracted out lighting to companies for a fixed term. These companies, typically run by individuals, would be responsible for filling the oil lamps throughout the parish, and hiring lamplighters to tend the lamps. Complaints of corruption, cronyism, and incompetence were not infrequent in this era, and it was not unknown for tradesmen serving as vestrymen to grant themselves overpriced contracts. When the GLCC began operations, it was competing as another lighting contractor to these vestries. There were also eight sewer commissions in the metropolitan area. Sewers at this time were understood as drainage courses for surface water, and not primarily as a means of disposing of waste. The running of waste into sewers was even forbidden in Westminster until 1815.

The City of Westminster and the City of London were also responsible for the lighting and other services in their jurisdictions, but their respective governments had developed very differently. Westminster was largely decentralized, and effectively administered by the civil parishes that constituted it, except that paving was the responsibility of a single committee. Unlike the rest of the metropolis, where parishes ruled, the City of London had developed some central bodies with real authority, and even had jurisdiction in some affairs outside its boundaries, such as over the port of London and taxes on coal, making it quite wealthy. The Common Council of the City of London had taken over responsibility for lighting and paving from the City's constituent wards. Like the parishes, London and Westminster also contracted out the lighting to companies. These various municipal bodies sometimes created specific commissions to which they delegated responsibilities for lighting and other services, as in the City of London.

Besides these important municipal authorities, there was also a host of other special jurisdictions with authority over some aspect of municipal life through parliamentary act, in some cases going back to the Middle Ages. With the passing of time, their existence often outlasted the original logic behind their creation, resulting in sometimes bizarre local divisions. These included liberties, small zones independent of the parishes and originally derived from, for example, the presence of an abbey. In addition, there were many areas which fell under the control of an institution, such as the Parliament buildings, the Artillery grounds, the Tower of London, the Inns of Court, and the Foundlings' Hospital. In some of these cases, they took care of the paving or lighting of a few streets, or even of a part of a single street. There were also turnpike trusts with control of specific streets, typically some of the larger roads outside the City. The result of all this was an administrative morass, leading to situations such as the Strand, which being less than a mile long was under the control of seven paving commissions.[20]

This fractured governance made it practically impossible to coordinate the development of the social infrastructure needed to cope with London's vast and growing population. This would become more apparent as the nineteenth century went on and waves of epidemics swept the city. What construction of infrastructure took place was largely in the hands of private corporations, such as water and gas companies, and later trains. Only local roads and small sewers remained with local authorities. The infrastructure companies were in effect given tasks that the local governments could not do, and they were left to negotiate the thickets of London's local politics.[21] Not until 1855, after the Thames had become an open sewer, did Parliament move to create the Metropolitan Board of Works to coordinate the construction of London's great intercepting sewers.

Where there were moves to coordinate a city-wide service before the 1850s was in policing. The thousands of beggars that filled the streets were not usually threatening, but after 1800, highway robberies on some of the major streets on the edge of the city were worrisome. The Home Office created the Horse Patrole in 1805 for the entire city, and kept the unit under its control. The new police made the highways safe, but street crime within the city increased, and pickpockets were endemic. The policing of the city was as fragmented and disorganized as other services. Each parish hired its own watchmen to patrol the streets at night, but these earned the reputation of being incompetent and useless, and were widely reviled. There was a host of other, often volunteer positions such as beadles to deal with minor offenses. The magistrates ran an office of non-uniformed detectives, whose job was not protection but hunting down known criminals within the city for trial. The detectives were accused of corruption on a number of occasions, sometimes waiting until a reward was sufficiently high before arresting a miscreant, or taking money from criminals. Incredibly, parish officers could not pursue criminals past the boundaries of their jurisdiction. During the war years, there were also around 12,000 regular troops in the city, and 14,000 militiamen, East India troops, and other quasi-military units. They could help in a pinch, such as when looting threatened in the wake of fire, or when political demonstrations turned rowdy, as they did in 1815 during a rally against the Corn Laws.[22]

With this backdrop of largely ineffective policing, 1810–1820 were years of growing concern about criminality and feeble protection. The worries became acute in December 1811 when a couple and their baby were brutally murdered with a hammer in their home on the Radcliffe Highway. The viciousness of the attack shocked the populace, only to be heightened twelve days later when another savage triple murder followed. The murderer was arrested and killed himself in prison, but fears lingered. The assassination of Prime Minister Spencer Perceval in 1812 by a deranged man did nothing to assuage people. The strength of political pressure for reform manifested itself in a number of Parliamentary committees examining the policing of the metropolis. The first was held in 1816, followed by two in 1817, another in 1818, and two more the following decade, eventually leading to the creation of the metropolitan police in 1829.[23] In this context, the brilliance of gaslights as compared to oil lamps on the dark street at night was most welcome, especially as thick coal-smoke fogs of London were beginning. Gaslights were later to be credited with making city streets safer.[24]

There were also other improvement projects in the first two decades of the nine-teenth century, some supported by the government. London's three existing bridges were clogged with traffic. Three acts of Parliament between 1809 and 1813 created companies to build more bridges. The Vauxhall Bridge was opened in 1816, the Waterloo in 1817, and the Southwark in 1819. New approaches were built for these bridges, which altered the aspect of Southwark in particular. An even larger undertak-ing was the Prince Regent's beautification projects for the West End, entrusted to the architect John Nash. Leases on land near Marylebone Park at the north end of the city were set to expire in 1811, creating an opportunity for a large new park, named Regent's Park, together with a grand avenue linking it through Portland Place to Carlton House in Westminster in the south. The road included many new buildings, such as the Quadrant, a terraced quarter-circle building leading into Piccadilly circus. New Street, soon called Regent Street, ran for much of its length along the old Swallow Street, whose poor tenements were cleared away for the splendors of the new road. Construction began in 1814, and finished in 1825.[25]

Other important new buildings were also built in these years. The people of London had a great passion for drama, and crowded to the theaters clustered in the Covent Garden area. The two theaters, Covent Garden and Drury Lane, which held exclusive royal patents for spoken drama in London, burned down in 1808 and 1809 respectively. Reconstruction began immediately afterwards, with Covent Garden com-plete in 1809, and Drury Lane in 1812. Both of these were to adopt gas lighting very soon after it became available, and the GLCC was most happy to have them as cus-tomers because of the public profile it offered gaslights. Other large buildings would also be important customers for the GLCC, including many government edifices, such as Parliament and associated buildings, Somerset House which was an enormous office building on the Thames, Guildhall, East India House, the various Inns of Court for barristers, many churches, and other buildings. All these provided important demand for gas lighting that kept the GLCC alive in its early years.

A DISASTROUS START, 1812–13

The first year and a half of the Gas Light and Coke Company's existence was marked by some of the same potent enthusiasm for gas lighting that was a feature of the heady days of 1806–07, when Winsor had caused a stir in gathering supporters and funds. Enthusiasm served the company's promoters well in that period, allowing them to attract attention and money and win an act of incorporation. Once operating, however, enthusiasm was to serve the company less well: it brought the project to the brink of total failure. The initial corporate period featured frenzied and frenetic activity on the part of the GLCC's directors, who scrambled around seeking customers and marketing opportunities, but with no sense of strategy, no business acumen, and little understanding of the technology or how to deploy it. By the end of 1813, the company was almost insolvent, the directors were all ejected, and the shareholders were feeling cheated and despairing. The GLCC had singularly failed to light anything with gas, except a few demonstration lamps around the Parliament buildings. It was also rent by infighting. The solution to this disastrous state of affairs came after a new slate of directors was elected that retrenched the company and extricated it from this morass into which the first directors had plunged it.

On April 30, 1812, the Gas Light and Coke Company received its charter from the king, though the Prince Regent.[26] Some of the members of the committee that sought the charter, led by James Ludovic Grant, managed to get their names inserted into the charter, ensuring that they would be the directors of the newly formed company. They began the process of formally launching the company by calling a meeting of proprietors for June 1, 1812.[27] The meeting confirmed the committee's transformation into the first directors and the group of nine men took office, with Grant as governor and James Hargreaves as deputy governor.[28] The lawyer Pedder was appointed secretary. The backgrounds of the directors are not easy to determine, but were probably gentry or industrialists. Sir William Paxton and Sir Charles Cockerell were partners in a banking firm. Among the rest, many owned homes in both London and elsewhere in the country, indicating substantial wealth. Frederick Accum was the odd man in this group, being a "practical chemist."[29]

The managerial style of the first group of directors was chaotic and not concerned with the strategic direction of the company. They were both managing the company and involved in day-to-day operations, in many cases personally directing works and negotiating with customers and suppliers. At least three directors were experimenting with apparatus, and on many occasions, the directors did not consult or report back to the Court of directors on their activities. This personal involvement in daily matters distracted the directors from strategic planning, and their failure in this regard caused the firm serious harm. Even the accounting was not strong. Most of the Court of directors probably had never run a large enterprise such as the GLCC, perhaps only being involved with family or small partnerships.

The tasks facing the directors were daunting. They had to negotiate the complicated local political geography of London in various ways. First, the GLCC needed to ask the permission of whatever authority was responsible for paving before streets could be opened and mains laid. Second, it had to provide street lighting before it could enter into a contract with home and shops on a given street. Finally, the company had to offer street lights brighter and cheaper than oil lamps to the authorities.

If the political and business challenges were formidable, the technological ones were also great. Whereas Boulton & Watt had years of experience with machinery, the directors of the GLCC had very little such experience. They now had to transform Winsor's small demonstrations into something that could fulfill their mandate of lighting the streets and buildings of London. Lacking a clear idea of how to proceed, they stumbled along trying many different options simultaneously, causing needless expenditures, by for example leasing properties and ordering goods in a rush and with no tendering process. As a result, they had acquired a large amount of expensive and almost useless equipment by the end of 1813. The uncontrolled purchasing was made all the worse by the prohibition against the sale of apparatus imposed on them by their charter. The GLCC had to buy everything, and costs ran up unchecked until purchasing was carefully managed.

The confluence of these factors led to considerable tensions within the company, and eventually with the shareholders. To make things worse, Winsor's own relationship with the company was uncertain. The committee, reconstituted as the Court of directors, had diminished Winsor's role over the past four years, and now continued to hold him at a distance. Winsor was not made a director nor was he attending the Court's meetings. He was, however, the only person with any experience within the company's circle of acquaintances, and so the directors had little choice but to try to work with him in some capacity.

One of the first Court's initial acts after being constituted was to summon Winsor to present whatever plans he had for lighting a parish and state his salary expectations.[30] After a delay, Winsor gave the Court copies of plans, which, however, were not sufficiently detailed.[31] They asked Winsor whether he had any other plans, as well a series of questions about his proposal. Winsor answered these in some detail, but mostly he referred the directors to his lectures.[32] It is clear from this exchange that Winsor's equipment was still not comparable to Boulton & Watt's—apparently it was small enough to fit in a house.[33]

Not wanting to be tied to Winsor, the Court struck out in other directions, with some of the directors themselves trying their hand at making gas apparatus. Frederick Accum could at least claim experience in chemistry, and he was asked by the Court to buy two of Winsor's stoves for a plant to be built at 96 Pall Mall, the lease of which had been purchased from Winsor.[34] Accum's plan was rejected because of the cost and the "delipidations [sic] it would occasion in the house," which the Court thought would "require the consent of the landlord."[35] Two other directors, James

Hargreaves and James Barlow[36] also tried doing experiments, at the company's considerable expense, but these came to little in the long run, although both men pursued careers in the gaslight industry after they left the GLCC.

Accum soon announced that he had an alternate plan, this one based on the machinery from "a philosopher on the other side of the water."[37] This could have been Winzler, the only person at this time who was making apparatus of any scale on the Continent. To investigate Accum's ideas further, a new committee of chemistry was constituted in November 1812, composed of a few directors. After some discussions with Accum, the committee concluded that his plan was viable and that the company could now advertise for lighting contracts, all without Accum's having built the tiniest part of the apparatus.[38] Confident that they had no need of him, the directors sent a rejection letter.[39]

With the furor of the past years still fresh in people's minds, it did not take long for various parties to express interest in lighting with gas. The parish of St. Pancras and the City of London sent letters, as did the united parish of St. Margaret and St. John in the City of Westminster and the liberty of Norton Folgate (a tiny administrative unit northeast of the City of fairly obscure origins, perhaps relating to a medieval manor).[40] A new parliamentary act had been passed in 1810 requiring the Norton Folgate lighting commissioners to provide lamps all year long, and they were casting about for contractors.[41] There were also discussions with the paving commissioners in the City of Westminster, which the GLCC approached in order to be able to light the Parliament buildings and the nearby house of one of the company's directors.[42] The offer to Parliament was a marketing move.[43] The Court, thinking in Winsorite terms, was hoping to provide Parliament with gaslights at no cost, calculating that the public notice from this would be sufficient to justify the expense.[44]

Among the potentially paying customers, Norton Folgate was the most forward in their application, with the rest hesitating. The contract was for so few street lamps—only fifty or sixty—that there was some doubt that it was worth pursuing. The directors decided that the possibility of lighting the houses within Norton Folgate made it worthwhile, and an offer was made.[45] The directors suggested a 14-year contract (the minimum term required by their charter) whereby the GLCC bound itself to charge 5 percent less than the market cost of oil lamps. It would be at least six months before the contract was finally signed on these terms, stipulating £1 8s 6d per lamp per year.[46]

The contract was to be very much to Norton Folgate's advantage. In addition to the great bargain on lamp fees, they did not have to pay any capital costs. The company came to realize (but only after the current Court was dismantled) that contracts under which customers paid none of the capital cost of the project would bankrupt the GLCC in short order. At this very early stage, they were still thinking of themselves as a direct substitute for oil lamp contractors, who would tender to light and maintain the lamps in a given parish. Gas lighting, because of the infrastructure required to support it, was too different from the oil lamp contractors to be able to function in that way. The GLCC and later other gas companies began to charge the parishes and other customers the capital costs of laying mains and installing gas lamps. The brilliance and convenience of gas lighting and their long-term economy came to be so well recognized that within a year or two there was no shortage of customers willing to pay these capital costs. For the moment, however, such was not the case, and the GLCC's shareholders paid a steep price for being the first gas network and having to establish which pricing model worked. Effectively accounting for and funding of the company's substantial fixed capital investment was, in other words, an important part of the implementation of the network model. The first directors did not, however, realize this.

In December 1812 the Court hired Samuel Clegg for the post of company engineer, paying him a salary of £500.[47] Clegg (1781–1861), born in Manchester, had been taught by John Dalton at the Manchester Academy before beginning an apprenticeship at Boulton & Watt around 1800. He had been involved with gas projects while at Boulton & Watt, and then independently when he installed a plant at Henry Lodge's mill near Halifax in 1805–06. He subsequently built apparatus for other mills and for Stonyhurst College in Lancashire, where he worked on lime purification. In 1809 he was awarded a medal from the Society for the Arts for his design of a gas plant.[48] He went on to install gas plants at other mills before going to London, where he build a gas plant for Rudolph Ackermann's print shop.[49] At this point, Clegg was one the very few gas engineers active outside of Boulton & Watt. Clegg also brought with him some of his workmen.[50] Clegg stayed with the GLCC until 1817 when he left over a pay dispute to work as an independent gas engineer. He later lost all he had after joining an unsuccessful Liverpool engineering company. In 1836 he moved to Lisbon, where he did engineering work for the government on public projects. He later returned to England to work on atmospheric railways with no success. He finished his professional life working for the government vetting bills incorporating new gas companies. The influx of talent and experience gave the company at least some solid footing in the technology. However, Clegg's experience availed the company little, the Court having failed to come up with an effective strategy. Only after the reform of the company's management beginning in 1814 did Clegg's engineering expertise help the company to flourish.

In the meantime, the divisions between the Court and Winsor were growing deeper. The Court's aloof attitude toward Winsor changed into open hostility once Clegg joined the firm. Winsor's own financial situation was becoming precarious: investors who had worked with him directly over the years began making inquiries about where they stood as shareholders in the new company.[51] Winsor was additionally burdened with properties, such as the two other Pall Mall houses next door to the GLCC's offices and a warehouse on Milbank Street, none of which he could afford.[52] When Winsor began making further inquiries with the Court about the agreements the committee of trustees had made with him prior to the charter, the Court responded that the Court bore no relation to the pre-1812 committee. The Court did not recognize these agreements. It added further injury by rescinding payment for some stoves it had recently purchased from Winsor on the grounds that their ownership status was ambiguous.[53] Winsor pestered the Court with letters, which produced only further rejections and payment refusals.[54] The Court of directors then took their hostility toward Winsor to the proprietors. During a general meeting on January 9, 1813, they publicly stated that the GLCC's business was referable to the charter alone, and not to anything that may have happened before it.[55]

It was now clear to Winsor that he would get nowhere with the Court. He appealed directly to the proprietors by holding a meeting on February 6, 1813 at his own home two days before a scheduled meeting of the proprietors.[56] There he presented his version of events, and convinced the proprietors of the justice of his claims. The directors responded at the official meeting by attacking Winsor in a public letter read there,[57] but the proprietors voted to form a special committee, headed by David Pollock (1780–1847) of Middle Temple, to look into his claims.[58] Pollock, a lawyer working in the Insolvent Debtors' Court, had some presence in corporate London.[59] This committee reported back to the proprietors when they reconvened to hear its report on February 25, 1813. The Pollock committee vindicated Winsor in all his claims, clearly calling him the "originator" of the company and thereby repudiating the directors' stance of a rupture between the company and whatever preceded it.[60] They made a series of recommendations that were clearly a blow to the directors: that Winsor be granted an annuity of £600 per year, transferable to his wife or son after his death, to provide advice to the company; that he be a paid director; that, once dividends were paid, he receive 1 percent of the company's profits to a maximum of £5,000 per year. The committee also made clear that they thought Winsor should be elected a director soon, a step which required him to hold sufficient shares, a condition he did not yet meet.

The recommendations were to be put to a vote at a May 1813 meeting of the proprietors.[61] They sided with Winsor and all the recommendations were adopted. These resolutions would have made Winsor quite wealthy in a few years' time, but the intervening lean period ruined him before he could benefit from the shareholders' kindness. Immediately after the meeting, the chastened directors sent a letter to Winsor offering to pay the £600 he had been awarded, but he was still excluded from the company's day-to-day affairs.[62] The next meeting of proprietors, on July 6, 1813, featured an election of two new directors: John Warren, one of the group of shareholders siding with Winsor, and Winsor himself.[63] The proprietors resolved that Winsor's plans be considered,[64] which amounted to a definitive rejection of the Court of directors' view of Winsor's abilities. From then on, in what must have been a strained situation, Winsor began attending the regular meetings of the directors and being involved in the daily affairs of the company. The original directors, led by Grant, still didn't show much interest in Winsor's apparatus, although ordered to do so on a number of occasions by the shareholders.[65]

This year-long conflict was about more than simply friction between the eccentric Winsor and the Court of directors led by the sometimes abrasive Grant. It saw the proprietors publicly doubt the judgment of the directors, eventually overruling them, and laid seeds of further doubts in the proprietors' minds, which became alarm in the last half of 1813. The original committee looking into Winsor's claims in effect became a dissident group of shareholders. By the end of 1813 they had forced out the directors once the extent of the technical and business debacle became clear.

While the battle with Winsor raged, the company continued its mostly inchoate activity. Accum, Barlow, Clegg, and Hargreaves were all working on unrelated apparatus. Clegg was building a fairly small plant at a wharf property which the company had purchased on Cannon Row, while the rest were building or designing various parts or systems on their own. At the beginning of 1813, the company still had little idea where it would start providing gaslights, other than free ones around Parliament. The various attempts at finding customers or even volunteers continued with no geographical coherence among the possible users. The company negotiated intermittently with the Theater Royal on Drury Lane, and looked to lease land next to it for a plant.[66] They offered to light Berkley Square in the West End gratis, which was accepted, but as with the theater, they had no idea how or where the gas could be generated.[67]

The directors did, however, take some steps in 1813 that would bring the company a permanent base. On February 19, 1813, the company purchased from one of its own directors a large plot of land on the south side of Great Peter Street in

FIGURE 5.1
Portrait of Samuel Clegg by Andrew Morton (1840). Courtesy of Science Museum, London.

Westminster, then on the edge of the city.[68] The lot was well located, being close to some of their potential customers such as the Parliament buildings,[69] but farther from the water than their Cannon Row lot, which posed problems for receiving coal and disposing of waste, as they would discover. The plant at the wharf in Cannon Row, purchased a few months previous at the cost of £1,700, was abandoned and put up for sale, and what little apparatus was salvageable was transferred to the Peter Street location.[70] Hargreaves was given direction of the plant at Peter Street, with Clegg working under him. He was asked to prepare plans and estimates for laying pipes in the streets from the plant to Parliament.[71] The company then applied for and received permission in late May 1813 from the parish of St. Margaret and St. John to place their mains under the pavement.[72]

The work at the Westminster site proceeded rapidly enough with Clegg working there that by August 1813 the company was finally ready to lay pipes.[73] Posts for 25 lamps were ordered. Sometime in September, more than a year after it had been incorporated, the company managed to provide its first gaslights.[74] The first paying contract that the company actually fulfilled was with St. John's, a Baroque church completed in 1728 just down the street toward the river from the Peter Street works. As the pipes to Parliament were to pass close by, the directors approached the church and people living on those streets, offering them gaslights.[75] The church vestry was interested, and after some negotiations, the two parties came to an agreement in November 1813 whereby the company would provide interior and exterior lights for £32 per year.[76]

On the east end of the city, the negotiations for the Norton Folgate contract, begun in 1812, dragged on for the first half of 1813 over minor technical points. The contract was finally signed on May 7, binding the company to begin providing gas lighting on September 29 of the same year, a wildly optimistic—not to say fool-hardy in the extreme—move on the part of the directors, who had not yet acquired land in the area for a plant.[77] Accum and others had been looking for some time, as early as December 1812, but it was only in early June 1813 that they found and leased a small lot at 74 Curtain Road, followed by a second adjoining one.[78] This left the company barely over three months to get a plant in operation, which of course created panic. Accum was give charge of directing construction at the new plant. The work was rushed and chaotic, with orders for materials and parts given with no ten-dering. During the construction, Accum complained repeatedly to the Court that people unconnected to the company were entering the grounds for unknown reasons.[79] Despite the rush, Accum still managed to convince the Court to try an experimental set of retorts.[80] The urgency also caused some of the work to be poorly executed. In September, the gasometer cistern collapsed during construction, and Clegg was called in from Peter Street to repair it.[81] As a result of all this, the Curtain Road plant was nowhere near ready when the contract was due to begin, and the company had no choice but to provide oil lamps for Norton Folgate at the company's expense.[82]

The Norton Folgate fiasco was a part of the broader crisis in the company that had been slowly building from the beginning of 1813. By July, the directors were

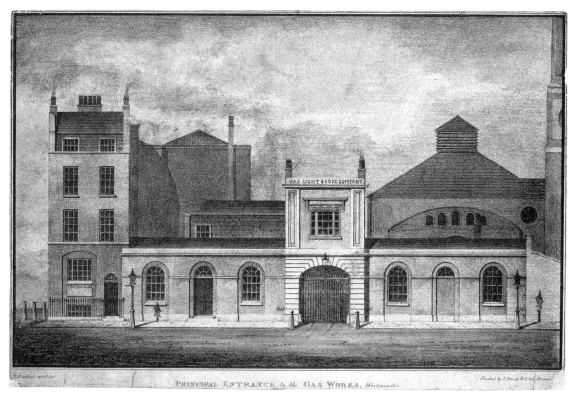

FIGURE 5.2
The Gas Light & Coke Company's buildings in Westminster, London, c. 1820. Courtesy of Science Museum, London.

starting to feel distinctly uneasy about the progress of their two plants. Realizing no one had yet designed a plant or a distribution system with any specifics in mind, they finally resolved to do so. It seems, however, that this was done with no prior work to estimate the potential load on the station, and so the directors used an arbitrary figure for the production capacity of 60,000 cubic feet per day. They asked Winsor to present a plan based on this figure, partly on the insistence of the shareholders, but Winsor was not really capable of producing such a plan.[83] A little over a month later, in September 1813, it had finally become clear even to the directors that they had never understood the basics of gas lighting sufficiently to be able to make decisions about the machinery. They then put together a list of very basic questions, such as how much coal was needed to charge retorts and how much gas was produced, which a new committee was supposed to answer.[84] These were questions Boulton & Watt had investigated repeatedly over several years. Now the directors, having practically no industrial experience, hoped to recreate the results in the midst of accepting contracts that were to begin very soon. In fact, the committee never really managed to produce any meaningful answers, and they never reported back.[85] They were not even certain whether to use lime purification—an invention that their own engineer Clegg had come up with, but which Winsor opposed.[86]

CHANGING OF THE GUARD, 1813

The committee that had looked into Winsor's claims approached the directors in April to talk about the state of the company and in particular about who could be a director as well as about directors' salaries. The directors answered after a month's time by referring the proprietors to the charter and the company's by-laws.[87] In May 1813, these same shareholders requested permission to visit some of the works in progress, which they were granted.[88] What they saw there did little to reassure them that the company was in good hands, and they began to get involved in the company's business. For example, they requested and received permission from one of the local paving authorities around Norton Folgate to lay pipes, something that was useful and accepted by the directors.[89]

In July 1813, they became concerned with the progress of the Curtain Road station, indicating to the directors that the excavations were unstable and had partly collapsed, damaging a neighboring house.[90] Thomas Livesey, an important member of the group, added in a further letter that expenses were out of control, with purchases being made at premiums of 20–50 percent over what could easily have been obtained. The directors looked into the matter further, and on the charge of wasteful expenditure, it turned out that Livesey had made the accusation mostly out of indignation at the perceived inefficiencies, because he could produce only a single example of extravagant expense. The directors then dismissed the complaint and vindicated Accum who was in charge of Curtain Road.[91]

Suffering this small tactical defeat, the concerned proprietors then held back for a few months, but they were by no means satisfied that the directors were competent. Affairs came to a head once again, when, on the night of October 25, 1813, there was an explosion at the Peter Street works. The explosion occurred because a damaged gasometer and purifier began leaking gas when the lime water was being replenished. When the escaping gas came into contact with a flame, it exploded. Clegg was present at the time and was badly burned.[92] The newspapers carried many reports about the explosion, and the directors sent a letter to them explaining what had happened, trying to reassure the public that everything at the works was safe.[93] It was with the shareholders, however, that they had the most problems. They immediately received a letter, signed by 54 shareholders, demanding that a special committee determine if the directors still held the confidence of the proprietors. A general meeting was called for November 1, 1813.[94]

The meeting began with the election of the dissident shareholder David Pollock as a director. After a report on the accident was delivered, James Ludovic Grant offered his resignation, perhaps as a bluff, together with that of the entire Court of directors. These were accepted effective the next general meeting.[95] Grant had done this in a way that typified how the company's affairs had been run: he had not bothered to consult the other directors, and they mostly repudiated the resignations over the succeeding days, eventually including Grant himself.[96] Accum, however, resigned, clearly feeling exhausted by the experience of the last year and a half.[97] When the next meeting took place, on November 30, 1813, the special committee appointed at the previous meeting reported that expenses were out of control, and gave many examples of wasteful practices and incidents. These included the "wholly useless" property at Cannon Row; the use of brick construction when wood would have sufficed, incurring £3,000 in extra expense; not consulting an engineer when items were purchased; not keeping a proper paper record of purchases; the lack of clarity in regard to purchasing authority; and the failure to have construction foremen on site.

The report had been a damning indictment of Grant's tenure as governor, and his position was precarious. He resigned a few days later on December 10, 1813, "on account of indisposition in his family," which was not, however, merely a pretext, as his wife died in January 1814.[98] He and Hargreaves went on to join the GLCC's first competitor, the City of London Gas Light Company in June 1814, but were both forced out there the following year.[99] The gaslight experience had turned into a very bitter one for Grant. He had put enormous time and resources into making the company work, and was finally ejected from the project to which he had devoted himself for six or seven years. He was also financially troubled. His niece recounted many years later in her memoirs that Grant's house in Southampton was auctioned off not long after the incident.[100]

With Grant and Hargreaves (who was disqualified after he sold his shares[101]) out of the picture, the special committee had control of the company, and their first task was to stave off its complete destruction. By the end of November 1813, its funds were almost completely exhausted. It had burned through the first deposit of the subscribers, as well as a £10 capital call on all shares payable on May 5, 1813.[102] A second £5 call was announced for November 1813, but many shareholders were reluctant to pay it, feeling that at this point, they were throwing money away. Pollock pleaded with the shareholders to pay, stating that without it "the welfare of the concern and the charter itself will be hazarded, and the past great expenditure rendered nugatory."[103] He was confident that with a retrenchment in the company's operations and expenses it would not only survive the current crisis but would begin prospering, if only the shareholders supported it.

Despite the GLCC's very public difficulties, interest in gas lighting continued to be strong. In London, for example, prominent places such as Ackermann's shop gave the public a fairly constant example of how it could work. Even in the depths of its troubles, the company received numerous expressions of interest from various businesses and individuals seeking gas lighting for their homes.[104] The parish of St. Margaret and St. John finally signed a contract for some street lamps on November 4, 1813, right after the explosion at the Peter Street works.[105] Work to lay pipes and install lamps continued through the rest of the year.[106] The contract with the St. John's Church was finalized on November 26, and soon the lamps were burning every Sunday morning from 10 o'clock on, making the church the first paying customer the company had.[107] The parishioners suffered from their church's willingness to try such a new technology when a "disagreeable effluvia" issued from the lamps. The company tried to solve the problem by using cannel coal, which Boulton & Watt through William Henry had also identified as a low-sulphur variety, but at least one further complaint about odors was received from the church.[108]

RETRENCHMENT AND RECOVERY, 1814

Once Pollock, Livesey, and the other dissident shareholders had taken control, they pursued a policy of cutting back and stabilization, hoping to expand once current customers were supplied on a consistent basis. The rejuvenation program was divided into two somewhat overlapping stages. The first included putting the company's management on firmer footing by confirming the new directors and then instituting a new management structure. The second stage consisted of devising and implementing a more coherent plan for the technology itself. The general meeting of proprietors of January 7, 1814 marked the beginning of this stage. Grant having left in December, Pollock and the others did a thorough assessment of what had happened over the course of the past year and a half. They reported that the company had entered a "state of apathy and torpor on one hand, and heavy expenditure and unprofitable experiment on the other." The new Court was faced with "very peculiar disadvantages and difficulties," which were "neither few nor small." The Peter Street works were running up large bills, which the directors had tried to rein in by cutting off the free lights they had been providing to Parliament once the session ended in mid December.[109] Of the two contracts to be serviced by that station, only the church of St. John's was being fulfilled. The lighting of parts of the parish of St. Margaret was to have started on December 25, 1813, but the company had missed the deadline, and the Court believed it would not be prudent to start lighting despite having some functioning apparatus.[110] They were afraid, based on Clegg's advice, that with only a single gasometer, the danger of the lights going out was too great. It was better to suspend the contract than to run the risk of a gas failure.

The situation at Curtain Road was even worse. The site had been acquired later than the Peter Street works, and was not ready. As at Peter Street, only one gasometer had been finished, and a recommendation to build a second one was seemingly ignored. Worse still than all the delays at the two stations was the lack of any complete plan for the apparatus.

There were, however, some hopeful signs. Many requests for gas lighting had been received. A petition had also been made to Parliament to amend the company's charter, specifically to remove the requirement to provide street lighting before the houses on a given street could be lit. Removing this would free the company to find many new customers, without having to negotiate with the sometimes prickly local authorities, although permission to lift pavement was still needed.[111]

Changes on every level came quickly after the January 1814 meeting. An experienced accountant named Lonkin was hired to look over the books and bring them up to date.[112] He worked with a special committee of accounts which reported on April 26, 1814. The committee had found little evidence of fraud, but made a series of important recommendations for the company's bookkeeping, including what sorts of accounts should be used, and the adoption of double-entry accounting. It recommended keeping books for each station, and having accounts to reflect the value of buildings and apparatus, or fixed capital.[113]

The company's works also progressed more evenly in the first half of 1814. Its mains were extended from where they ended at the Parliament buildings up Whitehall to Charing Cross, which was to become the most important point of divergence for the company's mains. In a few years' time, important feeder mains would spread from a large cylinder installed at Charing Cross along the Strand, Pall Mall, and other nearby streets. The plants at the two stations were completed with the extra gasometers Clegg had recommended, and by April 1814 the company was able to discontinue the oil lamps and begin supplying gas to Norton Folgate and the parish of St. Margaret's.[114]

The company worked on improving the design of the gasworks. Many of the old retorts were judged "extravagant" and abandoned.[115] In January 1814, Clegg tried a new plan for the retorts, heating four of them at once with a single fire, with two rows of two retorts sitting one on top of the other.[116] This was done to increase the fuel efficiency, or the carbonization ratio, a recently created parameter. It was the ratio of coal burned to the quantity gasified. In the first year, this ratio was measured at 100 percent, meaning the company was burning as much coal as it was gasifying, a sure sign that their retorts were too large. The first tests of the new retort configuration in January 1814 immediately brought a decrease to 45 percent.[117] Further improvements were to come, and the new Court of directors reported in May 1814 that with careful management and the new retort configuration (referred to as setting), they had managed to bring the carbonization ratio down to 20 percent, an improvement of over 70 percent from the previous ones.[118] They thought they could have found even further improvements, but the extreme cold of the winter had hampered their abilities to experiment as the water in the cistern below the small gasometer used for experiments froze regularly.[119] There was in fact much more experimental work to do. The validity of these carbonization figures is suspect, as later work would show, but they had undoubtedly squeezed out greater efficiencies. In addition, the directors admitted they still did not know how much gas was being produced from a given quantity of coal in their production retorts. The gasometers were simply too leaky to be able to come up with a reasonably accurate figure.[120] Until they had some idea, they could not really find out what the gas they were selling to their customers actually cost to generate.[121]

The interest in gas lighting was quite intense, and the company had relatively little difficulties in finding new customers. In January 1814, the company approached the Treasury Board about providing gas lighting for the medieval Westminster Hall in the Parliament buildings, this time for a fee.[122] The Board was interested, and after the company sized the room and sent the estimate to the Board, an agreement was reached. The company would replace the existing 74 oil lamps with 14 gaslights, which would give twice as much light, for a yearly cost of £206.[123] The fittings cost £103.[124] The directors were able to report by mid 1814 that they had received £180 on the GLCC's four contracts (St. John's Church, Norton Folgate, St. Margaret's parish, Westminster Hall).[125] New customers were coming, however, especially for street lighting. In April 1814, the parish of Shoreditch next to Norton Folgate indicated to the company that they wanted to have all their street lamps—about 1,000—replaced with gaslights.[126] St. Luke's parish just to the south of the Curtain Road station likewise entered into negotiations.[127] The company signed a contract to light Cornhill Street well inside the City in May for £130 per year,[128] and tendered for street lighting in St. Margaret's and St. John's for three guineas per lamp.[129] Westminster Bridge asked for an estimate in July 1814.[130]

Nor were local lighting authorities the only parties seeking gas lighting. After the start of gas lighting on the streets of Norton Folgate and Westminster in April 1814, the company received many expressions of interest from homes and shops in the adjoining areas. The company acceded to these as soon as it could, beginning with the house of one of its own directors.[131] The company now began to charge its customers the price of fitting up their homes, an important step toward financial viability.[132] To meet the request of a cluster of ten shops, pipes were expressly laid from Norton Folgate toward Shoreditch church. Many institutions, including the East India House, Lincoln's Inn Fields, and Carlton House, were also interested. The company simply could not keep up, although it was laying pipes rapidly.[133] Some parties were told that they would have to wait until the company could bring their mains closer to their location.

There was a major new contract for gas lighting that ended in disaster.[134] Various leaders of the states allied against Napoleon were to visit London in August 1814, and there were to be illuminations to mark the occasion and to celebrate their victory. The directors of the company suggested to Sir William Congreve that they could provide gas lighting as part of the illuminations. Their idea was to light the front of Carlton House, a perennial favorite for gas lighting.[135] The plan was soon expanded to include a pagoda to be built in St. James Park. The number of gas flames was to be enormous, at least 20,000 at one shilling each for the single night, although the final number was 9,900 lamps.[136] Needing to expand its gas generating capacity to meet the demand, the GLCC purchased an old brewing vat to increase gasometer volume.[137] Pipes were laid to St. James Park, and the lamps were tested successfully two days before the show.[138] Everything functioned marvelously.

Unfortunately for the GLCC, Congreve took a great interest in rockets. On the night of August 1, 1814, with large crowds assembled in the park, he fired off a few even before the pagoda lamps could be lit, and set the pagoda on fire. Its flaming remnants soon collapsed ingloriously into the nearby pool, killing two people.[139] Rumors circulated blaming the fire on gaslights. For the company, this had the potential for being another public relations disaster, coming so soon after the 1813 explosion at the Peter Street works. In the days following the debacle, the company ran notices in the papers that the fire had started even before the gaslights were on. In general, the public's demand for gas lighting did not seem to have been adversely affected either by the pagoda incident or the explosion.

The explosion at the Peter Street plant, did, however, generate official notice, and the Home Office decided to investigate the company. On January 21, 1814, Lord Sidmouth, the Home Secretary, sent a letter to the company announcing that an investigation into the safety of gas lighting was to be carried out by the Royal Society. Sidmouth was reassuring, saying that he thought the inquiry would "establish the company in the good opinion of the public, and to strengthen them in their progress," rather than do them harm.[140] The company arranged the details of the inquiry with the president of the Royal Society, Sir Joseph Banks, who was soon on friendly terms with some of the directors.[141] Shortly after their first contact, Banks sent a letter to the company asking for gaslights to be installed outside the front door of his own home, and the company naturally obliged in short order.[142] As for the inquiry, a delegation came from the Royal Society to visit the works on April 1, 1814, and a report was delivered to the Home Office; it contained a series of recommendations on how to make gasworks safer: gasometers should not be emptied of gas by driving it out with atmospheric air, gasometers should be no larger than 6,000 cubic feet, gasworks should be distant from surrounding buildings, and gasworks should be encircled by earthen embankments.[143] The report was not published at the time, and would not be for another nine years, but the committee was allowed to see some extracts.[144]

When the subscribers had their semi-annual meeting in July 1814, the tone was more optimistic, and the directors' report contained much to cheer the beleaguered shareholders. The bill to amend the company's charter had passed in June 1814, although not without some opposition that forced the directors to weaken the company's proposed powers to break pavement.[145] The most important amendments extended the charter's term and, crucially, gave the company power to light homes without being compelled to provide street lights.[146] The shares, which had been almost worthless six months previous, now had real value.[147] The directors also spoke of how the "increase in interest . . . [was] so considerable," and "the demands . . . so numerous" that there was good cause for congratulations.[148] The directors then suggested a raft of new by-laws for the company to improve governance procedures, which were all adopted.[149] In a optimistic mood, the shareholders held a celebratory dinner a few days later.[150]

A NEW MANAGEMENT STRUCTURE

In early 1815, the new leadership consolidated its direction over the company by instituting a new management structure.[151] The structure was a shift away from the personal form that predominated in British companies at the time. Before 1750 there was no hierarchical management because the complexity of companies could be handled fairly easily by the owners directly.[152] The domestic system of manufacturing, whereby a merchant contracted out manufacturing work to be done at home, required little active management.[153] The rise of the factory system in the late eighteenth century brought with it new demands, since it represented a much greater investment in fixed capital in the form of machines and buildings. It also brought more workers under the same roof who needed active supervision.[154] The scale of companies in other industries, such as canals, iron production, and engineering, also expanded at the end of the eighteenth century.[155] In many of these cases, the management of expanded enterprises was handled by treating the firm as an agglomeration of smaller ones, such as by subcontracting within the firm itself, rather than by developing internal hierarchies.[156] In other cases, however, businesses began to rely on managers who ran the firm more directly, rather than using the subcontracting approach. Some firms, including Boulton & Watt, made their management more effective by adopting accounting methods that better reflected fixed capital investments. Managers were almost always partners in the firm, and there was no separation between management and ownership, except to the degree that the firm had "sleeping partners" who invested but trusted the managing partners to run the business.[157] Although family firms dominated in the eighteenth century, and would continue to be the mainstay of the British economy throughout the nineteenth century, partnerships were becoming more common in the late eighteenth century. Business historians have described this period as the era of personal capitalism, when partners managed almost all aspect of companies, from strategy to function and operations, and delegated almost no responsibility to others. The form typified the British economy to the 1870s.[158]

The rise of joint-stock enterprises with the canal boom of the late eighteenth century, and then in other industries in the early nineteenth century, marked the beginnings of a split between ownership and management, representing proprietorial capitalism. Although the directors of joint-stock companies were almost always shareholders, they owned only a portion of the total shares. This did not, however, produce much change in the management structures of the firms, because the directors were rarely willing to delegate any real responsibilities. Canals, for example, tended to be local affairs, with very few employees doing more than operational work, and the subcontracting approach to management continued to be common.[159] In addition, because financing was a local affair, there was still not much separation between ownership and management. This divorce became stronger as enterprises grew and began to be traded on the stock market. Even before the late eighteenth century, the chartered joint-stock mercantilist companies, such as the East India Company, had in some cases organized their governance into committees. Though in a certain way their structure was similar to what multi-unit corporations adopted in the nineteenth century, historians have argued that the very slow flow of information, with months required to send letters to operational centers around the world, and with weak oversight from London, placed these companies in a different category.[160] In this regard, Christopher Schmitz has claimed that "it is difficult to identify any business concerns in Europe or America prior to the mid-nineteenth century, in which the major proprietors . . . did not exercise effective day to day control of their firms, as well as formulating longer-term corporate strategy."[161]

The rise of the railways in the 1840s and the 1850s has been identified as a cause of the development of new approaches to business organization and management in the British economy and in the United States.[162] Alfred Chandler, among others, associated the railways with the rise of the modern corporation in the late nineteenth century.[163] The first railway, the Stockton and Darlington, was run in the 1820s like a canal, through subcontracting, with almost no hierarchical management. It was able to maintain this structure profitably because it was geographically compact, and it successfully imposed standards on its contractors.[164] The coming of large railways in the 1840s, however, saw the adoption of more hierarchical management structures to cope with a host of challenges, including train, passenger, and freight scheduling, standardization over a wide geographic area, and extensive commercial activity. Large railways instituted geographic divisions run by senior officials, some of them engineers. The local managers reported to the general manager, who worked with the board of directors to establish strategy.[165]

The GLCC's new management structure of early 1815 was similar to what the railways adopted in the 1840s. After studying the matter in a special committee for some weeks, the Court of directors created committees of works for each of the company's three gasworks. The engineers, superintendents, fee collectors, and light inspectors all reported to the works committee. In addition, the Court created two other committees. The accounts committee managed finances, and the light and experiments committee was in charge of exploring new inventions, examining matters related to gasification efficiency, and other technical matters. All of these committees reported to the Court, with two directors sitting on each committee and with the deputy governor sitting on all of them. In addition, any director could attend the meetings of any of the committees. Finally, the authority of directors to make decisions outside of committee was limited.[166] These new committees minuted their regular meetings, and would refer important decisions, including those related to the design of machinery, to the Court of directors.[167]

By the middle of 1814, the GLCC had recovered from its crisis following the explosion at the end of 1813. The company was in new, more competent hands, and it was no longer hemorrhaging cash, although it had to make several more capital calls in the following years as its network expanded. These amounted to another £100,000 through to 1816, and a further £200,000 to 1819, on which dividends of 6–8 percent were paid beginning in 1817.[168] The morale within the company was improving, and shareholders had confidence in the Court of directors. The challenge for the next few years no longer lay in internal politics and company organization. David Pollock, already effectively in charge, would be elected Governor in 1815, and would remain in the post until 1846, when he was appointed Chief Justice for India, dying there shortly after his arrival. He was elected to the Royal Society in 1829.[169] Thomas Livesey likewise stayed with the company as Deputy Governor from 1815 to 1840.[170] The two of them together formed a bastion of stability for the internal affairs of the company over these years, although controversies with some shareholders inevitably occurred. Winsor, however, by now "extremely embarrassed" in his financial situation, had to flee the country in 1815, losing his annuity and eventually his directorship. Always the entrepreneur, he tried to found another gas company in Paris and took out more gas patents there.[171] He was ejected from the new venture, however, and in 1820 his extreme penury forced him to write the GLCC a letter asking for charity. The proprietors voted him a stipend of £200 per year until his death, which came in 1830.[172]

5.3   BUILDING A NETWORK, 1814–1820

With the company's internal travails mostly past, the following years witnessed the maturation of the GLCC with its technology into an urban network. As its mains spread under London's streets in 1814, and at an increasing rate thereafter, the company encountered many new challenges in scaling the technology into a large technological network in an urban environment. These included how to generate gas more constantly and efficiently, how to purify the gas better, and how to distribute it evenly with the least losses due to leakage. None of these challenges was simple, and problems surrounding purification in particular were irksome for the industry and its neighbors. The other challenges were no less significant, and it was with experimentation and innovation over the years 1814–19 that complex and workable solutions were found, solutions that remained in use or were the basis of gaslight machinery for the following decades.

The transformation of the GLCC into an urban infrastructure network was accompanied by the birth of a new industry as many other gas companies were incorporated, together with the development of the new discipline of gas engineering. This process is recorded and reflected in the growing body of literature on gas engineering in the period 1815–1820. These published works reveal a great deal about how the industry matured. The most important were Frederick Accum's *A Practical Treatise on Gas-light* (four editions, 1815–1818) and its successor *Description of the Process of Manufacturing Coal Gas* (two editions, 1819 and 1820) and Thomas Peckston's *The Theory and Practice of Gas-lighting* (1819 and 1823).[173]

The development described in this section was very much institution centered, with individual inventors playing only a small role. Although there was no shortage of people approaching the GLCC with all sorts of inventions that they hoped to sell to the company, the majority were of little use, and even those that were promising took a great deal of refining internally before they were workable. A good example of this is Andrew Rackhouse's oven plan for retorts, described below. Even Samuel Clegg, the company's own head engineer, who had many gas inventions to his name, proved not to be indispensable, despite some important inventions during his tenure with the GLCC. When he left the company over a pay dispute in 1817, the company did not miss a beat in its ongoing expansion. Clegg's own career, in contrast, suffered as he never regained his position in the industry. In the years of almost constant innovation from 1814 to 1819, no individual emerges as a dominant in this process. Many people collaborated, including in-house engineers, most notably Clegg, John Malam, and Thomas Peckston, as well as operational superintendents like Richard Leadbetter at the Peter Street works. The works committees at the stations and the Court of directors coordinated all the work of innovation, largely inside the company, with an occasional idea coming from outside. The GLCC's experience in running its network, its institutional memory, and its resources, both financial and in terms of the availability of production apparatus enabled it to carry out the extensive work described in this section. Many of the design decisions its engineers and directors took over the years was based on experiments that lasted months using their full production gas plant. It was acquiring extensive experience in building and running a large gas network. No individual and no small firm had access to this experience or these resources. If Boulton & Watt's strengths as an institution were essential for the first steps in transforming gas into an industrial technology, this was even more true for the GLCC. Only it was capable of supporting and running the extensive development work that transformed gas into an urban network.

This section looks at how the GLCC scaled up its gas generation, its purification, and its supply network. The other products of distillation, including coke, tar, and ammoniacal liquor are largely left out for two reasons. First, the company itself devoted the vast majority of its time, money, and efforts to gas. The company made some efforts to find uses for them, even setting a tar processing work and entering into discussions with the Navy, but these were minor efforts compared to what it did with gas. The second reason is that coke was the only product other than gas from which the company derived any significant money. In contrast to the gas, however, selling the coke that came from the retorts was simple. The existing demand for coal in London for heating meant that the market for coke was already present. Customers were very willing to buy the company's coke because it was cleaner than coal.

RETORTS AND GAS GENERATION

The scaling issues of gaslight apparatus began with the retorts. The volume of gas the company was generating very quickly surpassed anything the largest of the Boulton & Watt mills supplied. Whereas the mills typically had fewer than ten retorts in operation, and usually fewer than four, the GLCC soon had 40, 50, and even more at its plants. The same issues of extending retort life span that concerned the mill owners was soon pressing on the company, but on a much larger scale.

The question was not, however, simply one of adopting designs and stoking practices that would make retorts last longer. A complicating factor was that the highest carbonization ratio did not occur in the same circumstances as the longest retort life span. Typically, the hottest fires produced the most gas and the shortest retort life. A slow fire lengthened a retort's life span considerably, but decreased efficiency and gas quality, so more retorts were needed, and consequently labor costs increased. The company's directors and engineers then faced the question of balancing these important parameters. They, in turn, depended on retort structure, heating practice, and the type of coal used. Retort cost and life span, labor costs, and coal combined to give the cost of generating gas.[174]

In terms of design, there were two major things that gas engineers could change in the hope of improving the retorts. One was the design (shape and size) of the retorts; the other was how the retorts were arranged relative to each other and relative to the fireplace. Both of these were changed a great deal, by, for example, trying different retort cross-sections, material, and lengths, and many different arrangements or settings for the retorts. Experimentation was, however, very expensive, as resetting a fireplace involved extensive brickwork.

Setting cost was not the only challenge in designing better retorts. Gas engineers soon realized that coming up with reliable figures for all the variables related to gas production was not a question of performing a few experiments on small experimental apparatus. They came to understand that the only way to have any degree of confidence in their figures was to base them on complete production results obtained from full-scale apparatus—preferably over the course of a winter, when production was at its maximum. Peckston wrote in 1819: "In vain may it be attempted, by experiments made in the small way, to come at any thing near the truth; for when those experiments are brought in comparison with the larger and more extended process, experience has taught that no dependence can be placed on them."[175]

Since a retort tended to last months, a specific trial would typically last for six months to a year before the company could come to any conclusion.[176] Even then, the conditions in a full production setting were hardly controllable, and so it is not surprising that in the first decade of large-scale gas production no one form of retort design and setting came to predominate.[177] Another challenge was that ascertaining the efficiency of a given set of retorts when in production was difficult since the gas they generated was mixed with that coming from other retorts. Isolating them completely was the only way to come up with figures, but this was only possible for a day or two, as these test retorts would not be producing gas for sale.[178] This is not to say that the company made no real gains in efficiency and retort life. It did make important improvements, and it acquired an understanding of the various issues involved in retort design, but at a great cost in money and time.

Even before the first Court of directors was dissolved, some experimental work was done on the retorts. Accum recounted that Grant and Hargreaves tested conical and cylindrical retorts to find the best shape, finally settling on the latter.[179] Cylindrical retorts had already been used by Boulton & Watt and Clegg, and so the GLCC's early work was a reproduction of results already known by Boulton & Watt, and even by their own employee Clegg. The experimentation continued with the new Court, under the direction of Clegg and some of its technically minded directors, such as John Warren, although the retrenchment limited the company's financial resources and willingness to take risks.[180] After the first conical retorts were dispensed with, the new retorts had a more efficient circular cross-section, and were about 6½ feet long, made of cast iron.[181]

All retorts from the beginning up to 1817 were set on what was called the "flue plan" (figure 5.3). This meant that the hot gases and flame from the furnace were led

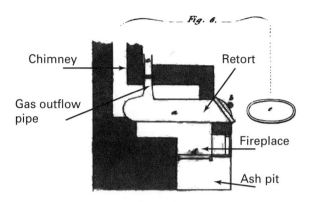

FIGURE 5.3
Retort heated by flues. Creighton "Gas-Lights," *Encyclopaedia Britannica* (1824).

by flues that snaked back and forth under and then above the retort before being exhausted to a chimney. The retorts were not protected by fire brick and came into more or less direct contact with the flame. Company engineers estimated the carbonization efficiency of these new settings to be about 30–36 percent, depending on the number of retorts used.[182] The first settings had one or two retorts heated by a single fire.

In 1816, the company had settled down sufficiently to feel ready for a great number of wide-ranging experiments. The directors asked Clegg to monitor coal usage on a weekly basis, and to determine which retorts were lasting the longest.[183] Perhaps guessing that greater fuel efficiency could be achievable if more retorts were heated by one fire, the company tried setting them in groups of three. A large number of retorts—66 at Peter Street and 30 at Curtain Road—were set that way and tried for four months. The test showed that the carbonization efficiency was no better, and the retorts were destroyed in two thirds the time.[184] The company also tried four retorts to a fire, but this gave even worse results.[185] The fuel efficiency decreased 25 percent, and the retorts deteriorated even more rapidly.[186] Setting a large number of retorts to a single fire had the additional disadvantage that when one retort burned out, the others also went out of service. If the retort was retained but not used, the fire would lose half its efficiency.[187]

In 1816 the company tried modifying the shape and path of the flues, and moving the fireplace to the back side of the retort bench, so that the retorts and fireplace were accessed from different sides. One plan with a shorter flue and the furnace to the front gave the same life span, but with a 5 percent efficiency gain.[188] In another case, in a complete departure from the traditional retort design, Clegg patented a rotary retort, a very complicated scheme that featured moving parts.[189] Within a year, and despite good fuel efficiency, they were judged to be a complete failure because of their complexity and their high cost.[190]

By the end of summer 1816, the directors were disappointed with all the company's experiments, but were still willing to try new designs, such as ones with a square cross-section, originally designed for the Surrey Co., a new gas company south of the Thames.[191] They also tried flat retorts (which supposedly produced less sulphur) as well as earthen and wrought iron retorts, but none of these were heard from again.[192] The company tried to keep a log of all the retorts in use with pertinent information (shape, manufacturer, etc.), so as to glean a firmer notion of which retorts lasted the longest.[193]

In the summer of 1817, the company decided to try a different flue layout that included protecting the retorts against the flames with fire bricks.[194] The design included having the furnace mouths set on the opposite side of the bench as the retort mouths, as in the 1816 tests. The new setting achieved some greatly increased retort life spans, but required a little more fuel. By the time the deepest part of the winter had passed in February 1818, the results were judged to be so positive that some of the other production retorts were reset on this plan. The positive results were confirmed once again a month later.[195] This version of the plan was not, however, to remain in use much longer because it was combined with another idea that came from outside the company after 1818.

In 1817–18, as the company was experimenting with the firebrick flue plan, an inventor, Andrew Rackhouse, approached the company with what he claimed was a revolutionary new retort setting that would save the company £1,000 or £2,000 a year.[196] After some negotiations, the company decided to try his idea.[197] It consisted of using ovens instead of flues to heat the retorts, meaning that the retorts were supported and heated within the oven, and not by hot gases that traveled through flues around the retorts. Rackhouse's plan called for five retorts to be set inside a single oven, heated by three fires, and the company chose to build four of these groups.[198]

There was then a long delay until October 1817, when the company and Rackhouse finally agreed to erect the new ovens at Peter Street.[199] They were still not ready two months later, and the company, which had planned for those retorts to be in service for the winter, grew impatient with Rackhouse and sent him a letter demanding that he finish, which he did fairly soon thereafter.[200] It did not take long for problems to emerge, however. By January 1818, Richard Leadbetter, the superintendent of the Peter Street works, wrote that although Rackhouse's plan was promising, no bricks had been used to protect bottom retorts from the flames, and they were deteriorating rapidly.[201] Two weeks later, the retorts had failed. The company, although frustrated by the experience, judged that the oven plan had some value and was probably better than the flue plan if some modification were made.[202]

In March 1818, the directors decided to replace the recently expired retorts with new ones on the oven plan, but with bricks protecting the bottom retorts from underneath (figure 5.4).[203] This model combined the two new ideas the company had been trying during the winter of 1817–18. The results came quickly. The superintendent reported a month later that these retorts were lasting well, although not as well

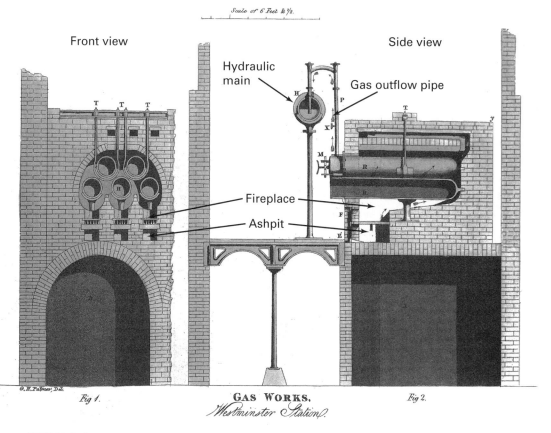

FIGURE 5.4
Side view of elliptical retorts heated in an oven. Accum, *Description of the process of manufacturing coal gas*, (1819), plate iv. Courtesy of Roy G. Neville Historical Chemical Library, Chemical Heritage Foundation.

those with a flue plan setting. There was, however, a major gain in fuel efficiency, with 22 percent less fuel required. Leadbetter thought this advantage could tip the scales in the newest setting's favor, despite the shorter retort life.[204] During the summer season, all the retorts were set on the protected oven plan, and when the results were reported at the end of the winter in March 1819, they were clearly favorable. Although retorts were destroyed more rapidly (320 destroyed per year, versus 210 using the protected flue plan), the fuel efficiency gain (65 percent) meant that it made sense economically.[205] It was then adopted as the standard.[206] The modified oven plan was used for many years.[207] Ironically, Rackhouse was given credit for it, although his actual work for the company proved to be a complete failure. It was only the company's own internal work that included combining it with another test plan that made it viable.

Elliptical-cross-section retorts were a third innovation added to the oven plan and the use of fire bricks in early 1818.[208] The company's minutes do not reveal the source of the idea, but it was hardly original; many different cross-sections had been tried, even at Boulton & Watt. The idea was to spread the coal into a thinner layer and thereby expose it to more even heating. The results were mixed, however, and the company used both circular and elliptical retorts in the coming years.[209]

Many other ideas for retorts came and went. Frederick Accum had set up a novel apparatus at the Royal Mint and offered his ideas to the company. New inventors, including William Congreve and John Outhett, approached the company, but their ideas were not pursued.[210] As Accum wrote in 1819, there had been "much assiduous inquiry . . . and in no branch of the new art of procuring light, has a greater variety of plans of improvement been submitted to the several directing boards of gas works, or more labor and expence been incurred in experiments conducted on a large scale, to ascertain the relative merits of these plans."[211]

PURIFICATION

The purification of coal gas was important for three reasons. The first was that some of the gases, particularly sulpheretted hydrogen (hydrogen sulphide), produced a terrible odor and were very toxic. Other gases, particularly carbonic acid (carbon dioxide), impinged on flame luminosity. The second reason was that some of the impurities, such as suspended liquids, would coalesce in pipes, leading to pressure drops and eventually completely blocking gas flow. Finally, the gas produced from the retorts was a mixture of gases and heavier substances such as tar and ammoniacal liquor; the latter had to be removed before the gas could be sent into the main network.

The basic methods for purification used by the GLCC had been worked out before the company had received its charter, and consisted of a three-stage process: the hydraulic main, the condenser, and the lime purifier. Boulton & Watt had washed the gas with water to remove tar and ammonia, and this method was also used by Clegg in his installations. The hydraulic main evolved very little over the early years. It consisted simply of a large-diameter pipe half-filled with tar and water. The products coming from the retorts were led by pipes into the water. The gases bubbled to the surface, while the condensed tar was drawn off. The hydraulic main also served as a hydraulic valve by isolating the retorts from the rest of the gas plant. (See figure 5.4.)

The second stage of purification was the condenser. The condenser served to cool and separate whatever tar had passed through the hydraulic main, as well as other condensable fluids, including water and ammoniacal liquor. Winsor had used a cooler, which he called a refrigeratory, as did Boulton & Watt, a feature inherited from the Beddoes pneumatic apparatus. The apparatus Clegg built before joining the GLCC included long pipes that circled the gasometer in the cistern that contained water. The condensed liquids were separated from the gases by running them down an incline, while the gas were sent upward to the lime machine for further purification. These pipes were susceptible to clogging, and cleaning them entailed burdensome work.[212]

The company introduced a more complicated self-standing condenser in 1817, after it made tests in which the gas was cooled and condensed more than usual before sending it to the lime purifier.[213] The tests indicated that condensed gas could be more easily purified, and the directors resolved to build better condensers.[214] These were square units that sent the gas back and forth through large vanes immersed in water.[215] The condensed tar and oil was drawn off at the base of the condenser. Although there were slight modifications made to this condenser,[216] it was abandoned in early 1818 in favor of one in which the vanes were replaced with many pipes immersed in water (figure 5.5). This design had the advantage of increasing the distance the gas traveled while exposed to cool water, and it worked very well, increasing the effectiveness of the purifiers.[217] The design had been patented in 1817 by John Perks,[218] an employee of the City of London Gas Company; he claimed the GLCC was infringing on his patent, a charge the company denied.[219] Although a few other designs were tried, the Perks design became the most common one used in gas works.[220] By 1819, condensers were quite large—one was 10 feet by 30 feet by 6 feet.[221]

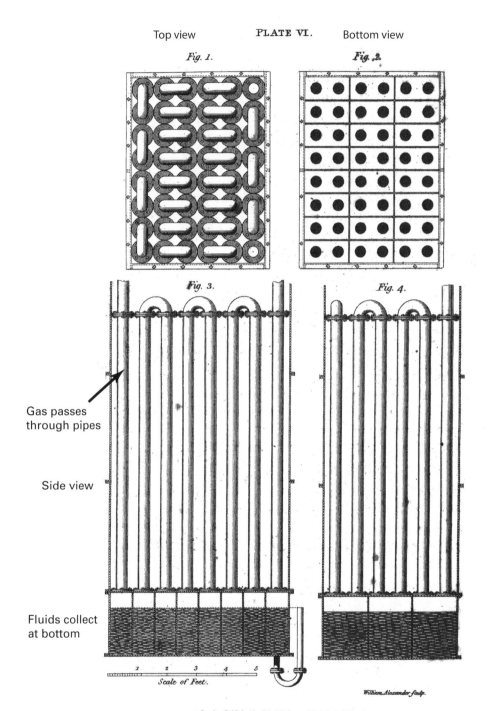

FIGURE 5.5

A Perks condenser. Peckston, *The theory and practice of gas-lighting* (1819, 1823). Courtesy of Roy G. Neville Historical Chemical Library, Chemical Heritage Foundation.

Clegg is often given credit for the invention of lime purification, the third stage in the process, but that is a simplification. In 1806, Edward Heard, Winsor's unfortunate assistant (who, like his former boss, would die in penury), was the first person to take out a patent for removing sulphuretted hydrogen from coal gas with lime.[222] He was never able to make any money from the invention. In his impoverished later years he petitioned gas companies for money for the invention, fundamentally seeking charity. Boulton & Watt had apparently tried lime too,[223] but this could have been no more than a minor attempt, given the absence of any record of it in the archives.

Clegg applied lime purification at one of the first gas plants he built.[224] His first purifiers, which merely bubbled the gas through lime water in various ways, were susceptible to becoming clogged. In 1809, Clegg designed and built an apparatus for a mill in Coventry that included a paddle to stir the lime, keeping it fresh.[225] It was this sort of purifier that Clegg built at Ackerman's shop in 1812 (figure 5.6). These early purifiers had an uneven effect because as the lime became progressively fouled by sulphur, it could absorb less. There was also the danger that clogging would cause increasing back-pressure on the retorts.[226] The 1813 explosion was probably caused by such a problem.

After Clegg was hired by the GLCC, the first purifiers he built were probably much the same as the one he had built at Ackerman's. In its first two years the company seems to have built only two. The orders in 1814 did not show much

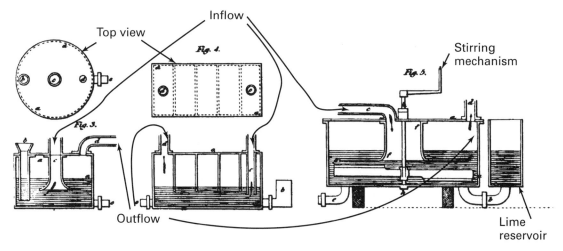

FIGURE 5.6
Lime purifiers. Creighton, "Gas-Lights," *Encyclopaedia Britannica* (1824).

variation in design, although a double machine, presumably with two chambers, was ordered in February 1815.[227]

Clegg was, however, quite active in 1815 in designing new inventions. This was the year that saw him introduce his rotary retorts. He also invented a complicated lime machine, which, like the retorts, he attempted to sell to the company. The company was at this point embroiled in various lawsuits over poisoning of its neighbors, and Clegg, no doubt angling to get the company to build one of his new retorts, suggested that unless the company found a way to reduce the lime used it would be exposed to serious risk of failure.[228] The directors were evidently keen to find a solution to their legal woes, but were not ready to authorize a radical new design for the purifier. Clegg was by now doing consulting work outside the company. He built one of his new purifying machines at another site, probably the Royal Mint, and the company sent Leadbetter to inspect the machine.[229] Although the GLCC had Clegg build a new lime machine during the last months of 1816, it was not very successful, and was cleared away in June 1817, after Clegg had left the company.[230]

In October 1816, simultaneously with Clegg's complicated purifier, the company began using lime in a semi-fluid state, sometimes called dry lime. The results were very good, both in the old and the new purifiers the company had. The directors ordered weekly tests of gas purity with "acetite of lead" and water to monitor the effectiveness of the new method.[231] The tests were positive enough that the company built a dry lime machine at Peter Street.[232] When further tests of gas purity were performed in March 1817, the results were less positive, and the company began looking for another sort of purifier.[233]

The seed of a new plan for the purifier was laid when the company made note in December 1816 that when purifiers were being recharged, gas entered the gasometer directly and therefore unpurified. A second lime machine would be needed to prevent this from happening.[234] Nothing was done immediately, but in April 1817 Clegg's successor as engineer, John Malam, designed a purifier that had three stages instead of only one (figure 5.7). This ensured that the lime machine could be recharged by section, and therefore the gas would always be partly purified. Malam also claimed it would save money and use less lime. The company asked him for an estimate in June of 1817.[235] A few days later, Clegg's purifier was deemed a failure and Malam was ordered to replace it with his.[236] Parts were ordered, and although some were delayed, the machines were ready by 1818.[237]

In June 1813, still not satisfied with the purification, probably due to complaints, the directors wanted to determine just how pure the gas was, and whether

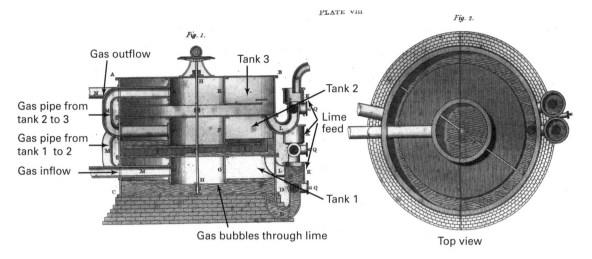

FIGURE 5.7

Malam's three-stage purifier. Peckston, *The theory and practice of gas-lighting* (1819, 1823), plate viii. Courtesy of Roy G. Neville Historical Chemical Library, Chemical Heritage Foundation.

the situation could be improved by using two purifiers instead of just one. They ordered tests for purity to be made at various points during the day for a few days running.[238] The results showed that one machine, with lime changed every 12 hours, was less effective than two used in series, each replenished every 24 hours. The Court then ordered that two purifiers be used, and that Malam's purifier be changed into three separate ones.[239] In fact, Malam's three-stage purifier became the most widespread among gas companies, and the one the GLCC used most often.[240] By the end of the decade, the purifiers were about 9 feet in diameter.[241]

The company also tried other methods of purification, generally with little success. Daniel Wilson, from Ireland, invented or at least thought about a new method that involved mixing the gas with "ammonical gas" and then bubbling it through water.[242] In December 1816, he made the company a fairly complicated offer of a new machine, which was accepted for the Brick Lane station, a third station the GLCC had purchased in 1814.[243] He drew up the details of the design and got to work, but by August 1817 the company ordered it removed as it had "proven useless."[244] Another attempt was made by Reuben Philips in Exeter.[245] He patented a dry lime purifier in 1817, similar to what the GLCC had tried at the end of 1816. In it the lime was mixed with as little water as possible to allow greater freedom of passage for the gas.

It also had the very important advantage of not producing the vile liquid waste that plagued the GLCC's neighbors. In October 1817 Philips approached the company, which sent a delegation to Exeter to see what he had to offer.[246] The delegation was sufficiently impressed that the directors wrote to Philips asking for his terms.[247] After some negotiations and a visit on the part of Philips, the company decided to turn him down.[248] The purifier, was however, adopted at a few gasworks in England, and became much more popular later in the century when it was improved.[249]

One other method for purification was suggested but was not adopted until later. It consisted of using heated iron.[250] The company, interested in this technique (developed by George Palmer in February 1818), ordered experiments.[251] A small machine was subsequently built, but the idea was abandoned.[252] In the middle of the century, iron oxide proved to be a good way of purifying gas, and had the great advantage of producing only gaseous byproducts because the iron could be revivified simply by exposing it to air. It apparently saved the gas industry, as by then regulations against the disposal of spent lime had multiplied.[253] In 1819, however, the company was satisfied with its existing purifiers.[254]

GAS SUPPLY

The expansion of large, tightly coupled networks introduced new problems associated with larger scales, much as the geographic extension of train and electrical networks required new solutions to achieve stability. As the GLCC expanded its network of pipes through London, it found it difficult to get an even supply of gas to every customer. The network, at first small and limited to a few customers, grew constantly from 1814, and supply problems became a persistent issue. The company had to pursue an active approach to managing supply, which included using and improving gasometers, pressure regulators, leak mitigation, valves, main sizing and placement, interconnections, customer management, and area isolation, all techniques and measures that were developed on the fly, and were really only mastered with years of experience. Supply management was made more difficult because the system was not static and predictable. The gas, generated constantly at the retorts, was stored during the day in the gasometers, and the main valve to the supply network was opened at the beginning of the evening. The company soon learned that the notion that the network of mains was mostly airtight during the day, and would begin delivering the gas smoothly once the main valve was opened, was fanciful. Leaks, thieves, and uneven ground all conspired against uniform distribution, and guaranteeing an even supply when gas entered a depleted main network was very difficult. In this regard, Clegg junior wrote in 1841:

There is no branch of science connected with the subject of gas engineering so highly important as that which relates to its conveyance and distribution through pipes. . . . The interests of the Company are not best served simply by increasing the quantity of gas from the same quantity of coal, or improving the lime machine, etc. The laying of street-mains forms the most considerable item in the outlay, and, by a judicious arrangement in the first instance, much may be saved both at first and last.[255]

WELLS, SYPHONS, AND VALVES

The most basic problem was obstructions in the mains, usually from coalescing impurities in the gas. Boulton & Watt's customers were well acquainted with the water, tar, and other substances that would make their way into the distribution pipes, even with careful purification. Water in the pipes was particularly problematic because it was almost always present in the gas. When it condensed and pooled, it could cause the lamp flames to vibrate annoyingly. If enough of it was present, it could occlude the pipes completely. Factories could cope with water by removing sections of pipes as needed, or by installing a single drain at the base of the staircase where the pipes ascended to the upper floors. The GLCC had no such luxury, as its pipes were buried under streets. The implications of this were not, however, realized immediately.

When the company initially laid pipes under the street, it simply left some sections of pipe accessible from the street level, covered by easily removable bricks or wood. A section of the pipe could be unscrewed and water and tar drained from it. Even with just the four customers the company had in June 1814, it became evident that this manner of draining the mains was too cumbersome. A new way had to be found, and Clegg proposed a method, which, while not clear from the minutes, was probably what was called a tar well in the gas industry.[256] It consisted of a cylindrical depression placed at various points along the street mains, into which condensed tar and water could flow (figure 5.8). The street mains had to have a certain incline down toward the wells to ensure that the water could collect there. The tar wells then had to be pumped out regularly, although surprisingly the first wells and syphons could only be accessed by breaking the pavement, and it took a resolution from the committee of works to install them in such a way that this was unnecessary.[257]

Because they were large, tar wells were installed only along major pipes, but water condensation continued to occur in smaller mains, and a somewhat different vessel, called a syphon though not actually functioning as one, was adopted for these cases (figure 5.9).[258] Instead of being integrated with the main as the well was, the syphon drained the water out of the main with a small pipe. Syphons had to be pumped out from the streets regularly, even daily, so the company tried to avoid using them.[259] If

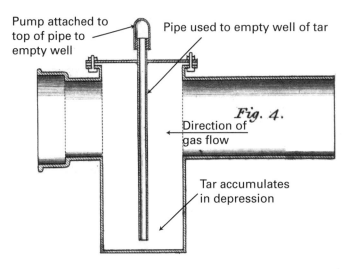

Pump attached to
top of pipe to
empty well

Pipe used to empty well of tar

Fig. 4.

Direction of
gas flow

Tar accumulates
in depression

FIGURE 5.8
Tar well. Peckston, *The theory and practice of gas-lighting* (1819, 1823), plate xi. sCourtesy of Roy
G. Neville Historical Chemical Library, Chemical Heritage Foundation.

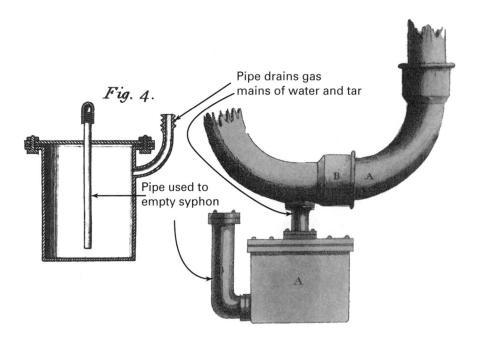

Pipe drains gas
mains of water and tar

Fig. 4.

Pipe used to
empty syphon

FIGURE 5.9
Street syphon. Peckston, *The theory and practice of gas-lighting* (1819, 1823), plate x. Courtesy of
Roy G. Neville Historical Chemical Library, Chemical Heritage Foundation.

the pumping was neglected, the street lights could decrease in strength or fluctuate noticeably, and sometimes would even go out.[260] A smaller version was introduced for use in the service pipes leading to buildings. Unlike the street syphons, gas passed through some versions of the house syphons.[261]

As the company tried to gain better control over how gas flowed in its network, it began using stopcocks and pressure regulators in 1814. The stopcock or valve was attractive because the company discovered not long after it signed up domestic users that being able to cut off customers in arrears or those who used gas outside the scheduled times was an important option to have. In 1814, the company hired inspectors to monitor gas usage, and considered making valves a mandatory part of fittings to provide means to make good on threats to cut off abusers.[262] It did not do so, however, because the first valves it installed proved to be leaky and susceptible to the "meddling of any curious persons." The company redesigned the valves in November 1814 and tried again, but valves were not still consistently installed at every building where gas was provided.[263] In May of 1815, the company adopted a policy of having them at all buildings that had two or more lights, or where service pipes had been laid, but no gas contract was as yet in place.[264] By the end of 1815, the directors decided that all users had to have stopcocks installed, and refused to take new customers until all existing customers had been provided with them.[265] The company then hired more people to monitor and adjust the stopcocks regularly to ensure that no customer was receiving more gas than was required.[266] In terms of actual construction, two valve types were used. House valves were mechanical, and simply obstructed the flow of gas with a mechanical stop.[267] Valves on larger mains, where having the ability to block the flow completely was more important, were hydraulic, meaning that they used water to create a seal (figure 5.10).

PRESSURE REGULATION

The pressure regulator, sometimes called a governor, was adopted in conjunction with the stopcock (figure 5.11).[268] It was introduced in 1814 by Clegg, and he had an important part in its improvement. The regulator was a fairly simple device that consisted of a bell floating on water, encased in a cylinder. Gas entered the bottom of the external cylinder, and communicated with the underside of the bell. The gas continued out one side of the cylinder. As the gas pressure increased, it drove the bell upward, engaging a choke on the gas inflow pipe and causing the outflow pressure to drop. These regulators were very effective at preventing excessive gas flow, and in December 1814 the directors made them mandatory.[269] The order was unevenly

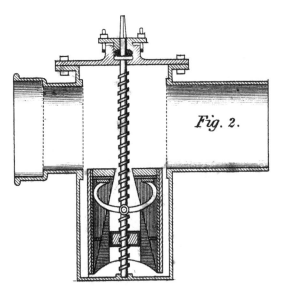

Fig. 2.

FIGURE 5.IO
Mechanical valve. Peckston, *The theory and practice of gas-lighting* (1819, 1823), plate xi. Courtesy of Roy G. Neville Historical Chemical Library, Chemical Heritage Foundation.

applied, but the regulator worked very well, and the company had a campaign in 1817 to have them installed for every customer to save gas in times of supply problems.[270] The company thought regulators so important that any customer refusing one had his gas supply cut off.[271] Regulators were also used on large street mains to control pressure in entire supply areas.[272] The design of the regulators remained almost unchanged for decades, although the material of the plunger was changed to clay from metal to prevent corrosion.[273]

The home regulator could really only prevent local overpressure, and other techniques were needed to control flow and pressure in the entire system. To supply gas at constant pressure, Boulton & Watt's and other early gas apparatus used counterweights on the gasometers, effectively the same method used to regulate pressure on the scientific instrument version of the gasometer.[274] This could work, but required careful calibration of the chains and counterweights. If the counterweight mechanism was not perfectly calibrated, weights had to be loaded on the balance when the gasometer was filled, and removed when the gas was driven into the street mains.[275] As gasometers grew in size, the chains and weights became more expensive and difficult to adjust, and there were fears that if the chain broke, there would be a pressure spike with unpleasant consequences for consumers as their flames became raging torches.

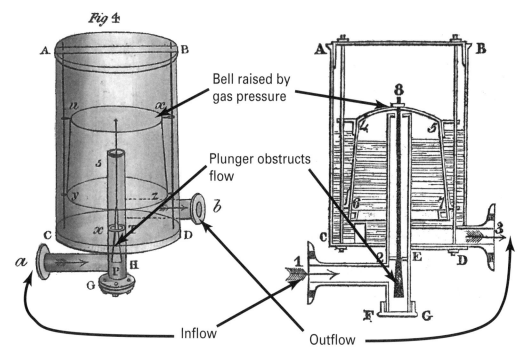

FIGURE 5.11

Small regulator. Peckston, *The theory and practice of gas-lighting* (1819, 1823), plate vi; Accum, *Description of the process of manufacturing coal gas* (1819), plate iii. Courtesy of Roy G. Neville Historical Chemical Library, Chemical Heritage Foundation.

The first gasometers at the GLCC all had counterweights, but in 1815 Clegg developed a version of the pressure regulator that could serve to replace the counterweight system (figure 5.12).[276] It functioned in fundamentally the same way as its smaller relative, but it had an exterior balance with counterweight to dampen the motion of the choke valve and allow for finer adjustment of the pressure. Various versions of this regulator were soon current, but gasometers with counterweights were still in use for many years, probably in small gasworks.[277]

LEAK MITIGATION

Leaks from holes that developed in pipes and from lamps that were left open when not in use were a constant problem for the company, and in the first few years it developed ways of preventing and dealing with them. If the leaks were large enough, air would also enter the pipe network and cause further problems, such as sputtering

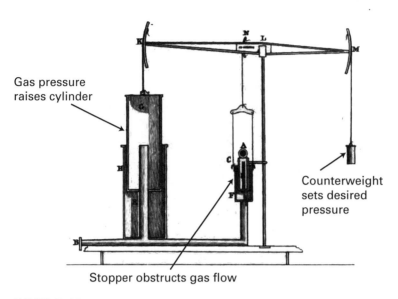

Gas pressure
raises cylinder

Counterweight
sets desired
pressure

Stopper obstructs gas flow

FIGURE 5.12
Large regulator. From "Specification of the Patent granted to Samuel Clegg," *The Repertory of Arts, Manufactures, and Agriculture* second series 30 (1816), reprinted in *Quarterly Journal* (1817).

or dying flames.[278] In homes, leaky fittings were also a source of complaints from customers due to the smells the gas created. Leaks caused the greatest problem right when the time stipulated in contracts for gas use arrived, typically around 4 p.m. The main valve connecting the gasholders to the street network was opened, but if leaks had been emptying the mains under streets during the hours previous, users would turn on their gas only to get air from their burners. It would take some time before the gas from the station would drive the air out and allow for normal use, and by then the customers were quite upset.

One of the first steps the company took to aid in fixing leaks was to introduce easier ways of joining pipes together. The company at first used pipes with flanged ends that could be screwed together and sealed with cement. It replaced this method for most situations with socket joints, also known as spigot and faucet, on Clegg's recommendation (figure 5.13). These were tried in the second half of 1814, and adopted as the standard in October 1814.[279]

The problem with leaks became acute in 1815. Within two years of having been installed, some of the street mains had to be dug up.[280] Clegg reported in May 1815 that there were "many very bad joints" and that the network from the Curtain Road station was losing 200 cubic feet per hour to leaks.[281] As a preventive measure, the

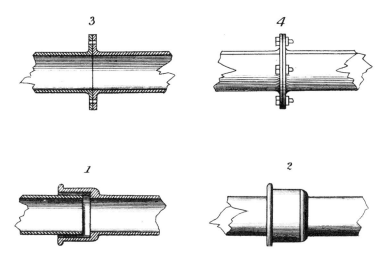

FIGURE 5.13
Pipe connections. Peckston, *The theory and practice of gas-lighting* (1819, 1823), plate xii. Courtesy of Roy G. Neville Historical Chemical Library, Chemical Heritage Foundation.

company adopted pressure testing of pipes before any were installed in the streets, but in the meantime, it was left with little choice but to "feed the leaks" during the day, meaning that gas would be occasionally pumped into the street main network, even when lamps were not in use.[282] The directors ordered Clegg to determine what the leakage in the whole system was, and when it proved to be substantial, he was asked to isolate each main to identify the worst leaks and fix them.[283]

In March 1815, to mitigate leaks inside buildings, the company began to test fittings with an air pump before connecting them to the system, and the Court of directors often reiterated the need to test fittings before new customers were connected.[284] In November 1815, the company started finding people using gas with whom it had never entered into any contract, and who had surreptitiously installed and connected very poor fittings that leaked large amounts of gas. Stopcocks were immediately installed and the gas cut off. The inspectors were ordered to be vigilant for such situations.[285]

With winter and longer nights fast approaching, the supply situation looked stark, especially since the company had failed to finish a new gasholder in November 1815 as planned. A greater number of customers began to complain about low pressure that month.[286] By December 1815, despite all the efforts to find and fix leaks, the company was in a supply crisis. The many remaining leaks were frantically sought, and more inspectors hired to ensure that customers, especially pubs, did not burn allow flames to exceed six inches in length.[287] The opening time for the main street valve was moved up to 1 p.m. to bring the network up to pressure as early as possible.[288] The company limped through the rest of the winter, and when spring and longer days came again in 1816, the directors evaluated what could be done to reduce the leakage. In April 1816, Clegg made a series of recommendations. He suggested that hydraulic valves be placed "at least every 60 yards" along new mains in order to segment them, and then to test these pipe lengths for leaks two weeks after being laid by using a portable gasometer to pressurize the segment.[289] After considering the cost, the directors adopted Clegg's plan.[290]

Clegg's plan worked for preventing some of the worst of the main leaks in new sections, but the committee of works was of the opinion that "the great loss of gas is . . . from some cause or other not yet exactly ascertained."[291] The directors during the rest of 1816 considered and sometimes adopted of a whole series of new measures to reduce the loss of gas to leakage. The company thought about providing large customers such as theaters with their own gasholders to mitigate their impact on the surrounding mains, an idea floated the previous year for individual streets but not adopted.[292] In addition to the mains, pressure testing of new syphons and valves before installation was started.[293] All valves were replaced with hydraulic ones that were less prone to leaks.[294] The "gun barrels" or service pipes joining buildings to the main network were to be tarred once a year, and tar was to be applied to the joints where the service pipes were connected.[295] Finally, since methane and hydrogen are both lighter than air, what air did enter the pipes would collect in the lowest parts of the main network, and the company began installing bleed valves at those points to remove air every day.[296]

These measures and others helped, but the company experienced occasional failures and continued leakage. In one case in 1818, the company's oldest mains, in Westminster, had so much air in them that vandalism was suspected.[297] In another case in September and October 1818, many customers in Oxford and Bond Streets were complaining about low pressure.[298] Tests confirmed that the supply was indeed weak in the area, and the company at first thought it was because they were low-lying. They tried to remedy the problem by planning another supply main to the area.[299] Just before the work was to commence, a large pressure drop was discovered in a 12-inch main in Wardour Street that fed gas to the affected zone. Further investigation revealed that water collecting in the main was obstructing gas flow. The removal of 340 gallons caused a doubling of pressure in the affected areas. Once cleared of obstructions, the main provided such good pressure that a valve had to be inserted to reduce it.[300] With the problem solved, the committee of works investigated why, despite regular tests, the pressure drop had not been detected. The answer was that although the main had been checked, the sample points were not sufficiently close to one another to make the obstructions evident.[301]

The company was also more careful in hunting down people who let air into the mains by leaving the burners open during the day. When a neighbor of the Duke of Marlborough on St. James Street complained that his lights were going out frequently, it was discovered that the duke's servants were neglecting to turn the valves off during the day. The company appointed one of its employees to shut off the street valve to the duke's house every day.[302] This solution may have worked for large consumers, but could not be applied everywhere as such behavior was common. Many consumers simply burned during the day, or left their lamp valves open, depleting the system pressure. The company responded by using inspectors and isolating sections of the network from each other during the day by shutting down valves at major junctions to limit the extent of gas depletion.[303]

SUPPLY MANAGEMENT

In terms of ensuring adequate supply to the whole system and specific supply zones on a larger scale, the company at first simply tried to estimate the best size for a main before installing one, but these estimates could have been little more than guesses.[304] Later, when fresh mains were laid on streets, the company would calculate the maximum demand for gas assuming that all buildings along the street would eventually subscribe.[305] A longer-term strategy for sizing mains was not really present for the first couple of years, and the company was just reacting to customers. Supply problems soon appeared, with individual customers and sometimes entire areas complaining of lacking sufficient supply. The company began to be more judicious in laying mains and sizing the network.

More controlled management started in March and April 1815, when all the existing mains and the locations of syphons were mapped.[306] The directors resolved in December 1815 to keep log books including as many details as possible about their network, such as the locations and sizes of mains, the locations and sizes of all burners, and the numbers, types, and locations of all valves, syphons, and stopcocks.[307] Following through on such an order was essential for supply management, but unfortunately for the company logs were not kept faithfully enough, as later events were to show. The desire for more even supply did produce some results, such as the adoption of pressure regulators described above. Another result was the move to increase the

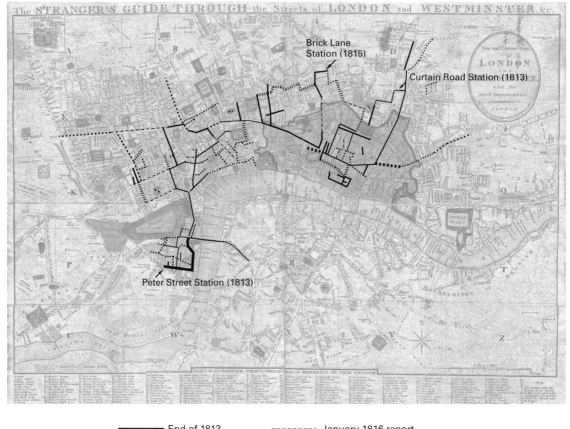

▬▬▬ End of 1813	-------- January 1816 report
─────── January 1815	⁣⁣⁣⁣⁣⁣⁣ To the end of 1816
▪▪▪▪▪▪▪▪ June 1815 report	··········· January 1818 report
············ To the end of 1815	▬▬▬ To the end of 1817

FIGURE 5.14

"Stranger's Guide Through The Streets Of London & Westminster 1814. Pubd. Jany 12th 1814 by Willm. Darton, 58 Holborn Hill, London." Map of London, showing the GLCC's mains in 1818. The location of the mains have been reconstructed from the minutes of the Court of directors, the court of proprietors, and the half-yearly reports to the proprietors up to 1818. After 1818, the reports no longer stated where new mains had been laid. Many smaller mains are not shown. Courtesy of MAPCO: Map and Plan Collection Online. archivemaps.com.

number of junctions between areas to make gas supplies more robust after a problem at one plant caused all the lamps it supplied to go dark.[308] These measures were adequate for the moment, but more serious problems would emerge as the pipes spread to areas with different elevation.

The relative lightness of coal gas as compared to air also caused problems. Part of the reason why gasometers were situated in pits was to use the natural upward flow. For Boulton & Watt and their customers, the final number of burners and their location were planned before the plant was constructed. Locating the gasometer at a sufficient depth below the burners was perfectly adequate to ensure even supply. It was, however, a different question for the gas company in London. Although the system could be pressurized by increasing the gasometer pressure, new supply areas higher or lower than the average would receive particularly strong or weak supply, and have an impact on surrounding areas. One of the first instances of this was on Tothill Street in Westminster in April 1817. Although not very elevated, it was sufficiently so to pose a problem because it drained the surrounding area of gas and lamps there easily produced long flames. The committee of works tried isolating the street from the rest of the network with a regulator, and the idea worked well.[309] This technique was applied at other streets, such as on the Strand and other locations. The use of regulators to isolate the network in elevated zones eventually became the industry standard practice.[310]

The other important factor in ensuring adequate supply was main size, and the company did not have a firm grasp of the capacity of its network until supply problems forced them to look into it. When demand on a street had outgrown the main's capacity, the company began laying a second main parallel to the first, which evidently helped a great deal.[311] This measure was adequate for individual streets, but the strong overall growth in demand for gas meant that the largest supply mains going back to the stations were no longer satisfactory, a situation which came to the fore in the second half of 1818. Supply deficiencies started happening with greater frequency, and the directors ordered a major report of the pressure and cross-sectional area of the mains at Charing Cross, the major originating junction for all the mains supplied by the Peter Street works.[312] Remarkably, the report revealed that the outflow mains from Charing Cross had a total cross-sectional area twice that of the inflow main, meaning that the pressure dropped by a half through the junction. The directors immediately ordered an extra inflow main from the station to Charing Cross.[313]

Further investigations revealed that the weak supply problem was also a dynamic one. One day the company took extra measures to ensure that none of its daytime gas users (or thieves, as the company considered them) were allowed to turn their lamps on. Once the main valve was opened, the gas pressure was good all night. The conclusion was that imbalances were created during the day when pipes were drained too soon, and the system never had a chance to reach a steady state during the night, causing local deficiencies. The company tried to open the main valve early, but this proved to be of little value because the gas thieves turned their lamps on at 2 p.m., too early for the company.[314] The company's next response was to hire another inspector to catch early users, and a month later to segment the supply network with more major valves.[315] Finally, using curved junctions rather than right-angle ones helped.[316] This limited the dynamic effect of daytime depletion to certain zones, and gave greater stability to the whole network.

In the end, these measures were still not enough, and the company understood that the Peter Street works simply did not have sufficient generating capacity.[317] A new solution was to transfer some of its load to another station. The mains from the Peter Street and Brick Lane stations reached each other in 1817, and although they were connected that year, no gas was allowed to flow between the respective systems.[318] Another supply crisis in the winter of 1818–19 prompted the transfer of a large area to the Brick Lane station.[319]

The combination of all these measures, from bleed valves to area isolation to using wells and customer inspectors, meant that the company had to employ a host of people who had daily tasks important in stabilizing the network. The technology functioned with the close and constant help of human hands.

5.4   OUTSIDE THE COMPANY: USERS AND FITTERS

All of these efforts were directed toward elements of the network directly under the control of the GLCC. The company also, however, had to cope with problems stemming from parts of the network it could not completely control—specifically the fitters (who outfitted buildings with the necessary equipment for using gas) and the customers. To bring these under its control, the GLCC introduced standards, certification, and inspectors.

USERS

When Boulton & Watt were selling apparatus to mills, the operators of the gas plant and the user of gas were the same: mill owners. There was no tension between the user and the operator, at least at the level of the person paying the bills. What the workers in mills thought about the smells of the largely unpurified gas was an altogether different question. Once the business model shifted to a network, the provider/ consumer dynamic changed decisively, with important effects on how the technology was used and deployed. The consumers were very separated from the process of gas manufacture, and were little concerned with how their usage of the gas might affect the rest of the network. Mill owners, with their individual plants, had every interest in making sure that whole system work properly and that usage did not unnecessarily strain the system. In the case of the GLCC, the users were interested in gas for light, and the other details of the process, including the company's finances and its share-holders, concerned them very little. This soon meant that issues about control of usage that had never been present for mill owners became very pressing for the GLCC. The orderly behavior the GLCC at first expected from its customers—namely, using gas only during the times stipulated in the contract—did not materialize in many cases. Some customers began acting in ways very different than what the GLCC needed for its original operating model to function. It soon emerged that consumers were not going to be bound by their agreements limiting their gas usage to specific times of the day and excluding Sunday use. Usage patterns, however, had important effects on the network. The problems with users allowing air into the street mains was described in the previous section, but even more pressing was how much gas they burned. If too much gas was consumed over the course of a day, exceeding the company's generating capability, gas supply would fail. To mitigate this danger, the GLCC had to develop ways to maintain control over gas usage or risk repeated failures in supply. Even the threat of cutting off supply was dependent on the company's willingness to carry it out.

The relationship also changed in the other direction. For the GLCC, all its cus-tomers were not simply consumers of gas to be treated on an equal basis. Some consumers were much more profitable than others. This applied most clearly to street lighting. The GLCC maintained throughout the first years of its existence that street lighting was not profitable due to the cheap supply clause in its charter. Nor was profit the only consideration. Consumers were also treated differently because of their public prominence. The GLCC was very happy to provide gas to any building that, because of its public prestige, would somehow share its glory with gas, as the various Carlton House episodes demonstrated. The converse of this attitude was that it was also extremely reluctant to get into any sort of public battle with a prominent but delinquent customer. While it had relatively few qualms in cutting off individual consumers, large public buildings, even if their accounts were notoriously in arrears, were simply not treated with the same harshness.

In the early years of the GLCC, users took to gas lighting with great enthusiasm, despite a series of setbacks and stumbles, such as supply failures and the 1813 explosion. Why did this happen? It was above all the much brighter light that gas gave as compared to oil lamps and candles. Whenever the gas company replaced a number of oil lamps, it would install a quarter as many gas lamps or even less for the same light. References to the brilliance of gas lighting abound in reports of travelers and locals alike. In 1814, Johann Georg May, a Prussian civil servant who had been sent to Britain to produce reports on its industries, observed about gas that "the brightness of the illumination and the variety and elegance of the lamps left nothing to be desired."[320] A French journalist wrote about the new lighting, comparing them to the "little gothic lanterns" that were "inadequate for their purpose" in use in much of the rest of the city.[321] Madame d'Avot spent 1817 and 1818 in Britain, and in her published memoirs wrote: "This most beautiful lighting allows one to read as if in the full light of day. But those neighborhoods that are lighted by oil afford a sad prospect, and one hardly dares to enter streets where these feeble and ancient lanterns look like dim small stars."[322] It was not only foreign visitors who were impressed by the brilliance of gaslights. When Robert Southey went to London in 1820 for the first time in several years, he was struck by the new lighting: "The most impressive sight to me was St Paul's by gas-light. I do not think anything could be more sublime than the effect of that strong light upon the marble statues; and the darkness of the dome which the illumination from below served only to render visible."[323] Likewise, an 1814 guidebook to London stated: "It is scarcely too much to say, that for beauty, brightness, and purity the gas light is as much beyond that of lamps and candles as day is beyond the night."[324] When a gaslight company was being formed in Exeter in 1815, the *Flying Post* wrote that "the brilliancy of the gas light is so superior to that of oil or tallow, as to bear no comparison with it."[325] Gaslights were also easier to use than oil lamps and especially candles, as they did not need to be refilled or replaced regularly. Candles particularly were bothersome. In the early nineteenth century, most candles were made of tallow (animal fat), which ran and needed regular tending. The wicks also needed to be snuffed (trimmed) regularly, perhaps every fifteen or twenty minutes. Otherwise, the flame would gutter (flicker) emitting black smoke, or even grow, possibly becoming dangerous. Gaslights dispensed with all this work.

LOCAL LIGHTING AUTHORITIES

Local lighting authorities were in some ways the easiest for the GLCC to deal with, and the company expended relatively little time and energy on them. All the lights were exterior, with no problems with smells. The company had complete control of street lamps, since providing lamplighters formed part of lighting contracts. The company made very sure lamps were lit and extinguished on time, and that the valves were closed during the day, although it did have to install locked valves to prevent idle hands from playing with them.[326] The most difficult aspect of street lighting for the company was the political power of the local authorities, which was significant and needed some management. This became more important when, after a couple of years of operation, the company largely stopped offering street lighting except where its mains were already in place, or where many domestic customers were requesting gas lighting as well.[327] The company denied it operated this way in Parliament, but its minute books reveal many decisions made on this basis. The company's avoidance of providing street lighting coupled with the desire of lighting authorities to have it led to some political maneuvering, especially when the company and the local authorities were not on the best of terms. Each one tried to use its leverage to achieve their aims. In the case of the GLCC, it encouraged residents to lobby the local authorities to grant the company permission to lay mains in the streets leading to their homes. The parishes, on the other hand, could refuse permission to the GLCC to take up the pavement unless the company would agree to provide street lighting, although this was not common.[328] Usually some agreement was reached to the satisfaction of both parties.

One altercation occurred with the parish of St. Marylebone in 1815. A number of residents of Oxford Street, a wealthy shopping area located in that parish, asked the company for light. The company wanted to accede to the request and met with paving commissioners about laying their mains in May 1815.[329] When the vestry met in July, they refused the company permission to break the pavement, and the company responded by organizing the residents to petition the vestry.[330] The public demand was strong, and the vestry had little choice but to give in, but not before it exacted a little revenge in a fit of pique. The company's new request was rejected on the grounds that it was too informal: it was addressed to paving committee instead of Right Honorable and Honorable Vestrymen.[331] The company duly resubmitted, and permission was given in November 1815.[332]

PUBLIC BUILDINGS

Public buildings were always treated with particular attention by the company. They were the most lucrative customers, usually requiring many lights. The contracts could easily exceed £50 per year, with the largest ones over £100. Unlike street lamps, the responsibility for maintenance lay with the customer. The prestige and exposure of having a well-known public building lit with gas was also important for the company.

Providing free lighting around the Parliament buildings in 1813 was evidently part of this thinking, and the use of gaslights there continued to be an important sign of confidence when they began paying. The commencement of gas lighting at an important building, such as that at Guildhall in November 1815, could be an occasion of some solemnity and could generate notice in the papers.[333] There was, however, a downside involved with some public buildings, although usually not with the government offices (which were well behaved, using gas on schedule and always paying their bills). Many of the rest, however, turned out to be quite demanding on the company in terms of fittings, gas usage, and payments. Fitting up a new building could be an extensive and expensive proposition, running into the hundreds of pounds and taking days or weeks worth of work for the larger ones, although it was the customer, not the company, that paid the charge. Large customers would sometimes modify these fittings, much to the company's chagrin, as this would cause their gas usage to change dramatically. Being the GLCC's largest customers also meant that these public buildings had significant leverage with company.

In what can only be described as a love-hate relationship, the users that brought out these various elements the most strongly were the West End Theaters, and in particular Covent Garden, one of the two theaters with a royal patent for performing spoken drama in London. Most of the theaters became customers in the two or three years after the company's mains reached the parish of St. Paul Covent Garden, where they were located, in early 1815.[334] The Covent Garden Theater was one of the very first, asking for exterior lamps in April 1815. The company complied with their request and extended their mains from the Peter Street station's network expressly for the theater.[335] Gas supply was ready in September 1815, in time for the beginning of the new season, with around 80 lamps installed at the theater.[336] The company, perhaps sensing it would be in their own best interest, suggested that Clegg's newly invented measuring machines be installed so that billing could be by volume rather than by burner as was usual at the time.[337] The meters, however, were nowhere near ready for use in a working environment and were not used.[338]

The Covent Garden experience did not go well for the company. The relationship between the company and the theater ultimately turned around two points. One was the relatively greater importance the company placed on the public perception surrounding the use of gas at the theater over their desire for profit from the lighting contract, and the other was the theater's unwillingness or inability to abide by its contract and its understanding that the company needed the theater for its prestige. The seeds of the conflict were sown over the course of the inaugural season of 1815–16, but were not yet apparent. During this first year, the company was particularly sensitive to how the public received lighting at the theater. A hint of problems arose when a report was printed in the *Morning Chronicle* and other papers in December 1815 that the light would be discontinued at the theater because of the smell.[339] The directors immediately responded by placing an advertisement contradicting the reports. The lights stayed on, but clearly someone was unhappy about them: the fittings were vandalized in January 1816. When some of the directors met about the matter with Henry Harris, the manager and co-proprietor of the theater, he convinced them to rebuild the fittings on a grander scale.[340] Harris was quite a manipulator in his relations with the GLCC, and this was but one of many occasions when the directors allowed themselves to convinced by Harris to act in ways that were not in their best interests. In any case, the interior lights were discontinued in March 1816, probably because the company was having a hard time providing an even supply from the time the theater opened, which was well before the main valve joining the street mains to the gasometers was opened.[341]

With the end of the season in 1816, the GLCC and the theaters made preparations for the coming year. For its part, the company thought about how it could reduce the demands of the theaters and provide better supply. It suggested installing a small gasometer at Covent Garden, and when it quoted for the Drury Lane Theater which asked for gas, the company included a gasometer in its price estimate.[342] For its part, Covent Garden ordered a very large chandelier for its main hall which the company built for it over the summer.[343]

The first of the disputes over billing occurred in summer of 1816, although from the beginning, the company was aware that the theater was burning much more gas than what the original contract had laid out by using burners for many hours a day.[344] In the summer of 1816, when the company's engineers and accountants tried to come up with a figure for the previous season, the theater's management—meaning Harris—disputed it.[345] Unfortunately for the company, it was not in a good position to make demands. The many changes at the theater over the last year had created havoc in its records, and so it could not prove its claims.[346] The committee of works at the Peter Street station clearly felt that they could get nowhere with the theater and the whole matter was referred to the Court of directors to deal with. For their part, the directors decided to bill the theater £550 for the whole year, a sum they were confident was far less than the actual figure, but one which they accepted because of the chaos in the record keeping.[347] This bill was submitted to Harris, but when he rejected it, the Court asked Pollock and Warren to look into it. When Warren went to see Harris on appointment, he managed to be absent.[348] By this point, it was September 1816 and the season was starting again. Harris seems to have evaded the issue altogether and blackmailed the company into keeping the gas on because of the negative effects a cut-off would have on the company's public image. Harris' evasion bought him some time, but the company sent him a letter in November 1816 demanding payment for this year and the £550 owed from the previous year.[349]

The next few months saw little action, and the theater continued its profligate ways, keeping some of its lamps burning for 21 hours per day, a gross violation of some of the basic terms of its agreement with the GLCC.[350] Harris probably relieved some pressure from the company by paying a few smaller installments. In April 1817, with the end of the season approaching, the directors again turned their attention to the theater.[351] A bill was sent, but Harris only paid £100 of the amount owed, and in June 1817 the directors referred the matter to their lawyers.[352] The company now thought it prudent to have a valve installed on the main leading to the theater.[353] As usual, Harris managed to obfuscate sufficiently and pay enough to prevent the gas from being cut off, but nowhere near the amount owed.

September 1817 brought a new season, and another cycle of evasion from Covent Garden, ultimately made possible by the company's indulgence of its behavior. It was not as bad as previous years, however, because the company demanded weekly payments and the debt did not run up as much. Despite all the difficulties the theater had caused the company, the directors clearly thought it was one of their marquee customers and still gave it preferential treatment. A new main was laid from the Brick Lane station (the previous supply was from Peter Street) to ensure that the Covent Garden and Drury Lane theaters had sufficient gas on nights of particularly high demand.[354]

The 1818 season saw the theater request that payments be made monthly as they had been before the weekly regime was instituted, and the company, having a short memory, agreed.[355] By November 1818 the theater was in arrears.[356] Harris, always the negotiator, agreed to pay £100 per week, and the directors gave strict orders to cut off the gas if any payment was missed.[357] The inevitable first missed payment came in February 1819 and the company, instead of stopping supply, was soon demanding payment again.[358] The theater made a payment of £50 to put off the company, but matters came to a head once again toward the end of the season in June 1819.[359] The company sent an ultimatum with June 26, 1819 as the final cut off date, a resolution they reinforced the day before the deadline.[360] Instead of going to see the company personally, Harris sent his secretary to plead, and the directors, incapable as always of being caught in a public disagreement with the popular theater, capitulated and did not cut the gas. The month of July 1819 saw the theater and the company come to terms about a schedule of payment, whereby the theater would discharge its entire debt of £635 in two payments over the next three months.[361]

The directors had fooled themselves into believing the same empty assurances that Covent Garden had offered so often in the past. When the time came in September 1819 to deposit the first check of £200, it was returned by the bank as an unauthorized payment. The directors were shocked, and asked the theater to make good on the bounced check. It also moved payment terms back to weekly.[362] This was an entirely hollow demand, and the theater made no payment.[363] In October 1819, the directors referred the matter to their lawyer.[364] In November they wrote that the "temporizing conduct of their house towards the Gas Light Company has completely exhausted patience of directors," and that unless payment was forthcoming on the ninth, gas would be cut off on the eleventh.[365] Harris responded to this letter by feigning complete ignorance of the communication between the company and his secretary, and by professing surprise that affairs had reached such a point. Touching on the very point that made the directors powerless in their dealings with the theater, Harris also included a warning about the harm it would do the public if the gas was discontinued during the season. Harris even had the temerity to suggest that company should be grateful to the theater for using gas because of the theater's prestige, and that it should be happy to provide gas at no charge. As a final insult, the theater's management had turned to the gas company's old enemy, and installed an oil-gas plant at the theater without informing the company.[366] In a move that beggars belief, the directors suggested new terms of payment, and agreed in December 1819 to negotiate payment with the theater.[367] The theater seems to have made a few payments over the rest of the season.

The company ignored the theater's use of oil until July 1820, when an explosion occurred involving the oil-gas apparatus. The directors made the empty threat of stopping the gas supply unless the theater got rid of its oil-gas plant, feeling that they would be blamed for the theater's negligence.[368] The theater of course ignored the demand, and further letters were similarly ineffectual.[369] When the season began again in September 1820, the oil-gas plant was still in place, and company repeated its threat to cut the theater off.[370] This time, it actually managed to make good on it, probably because the directors knew well that there would be no public scandal.[371] Demands for payment of the £963 owed were blithely ignored by the theater.[372]

SHOPS AND HOUSES

The company's relationship with smaller consumers—houses and shops—was very different than what it had with large public buildings and lighting authorities. In this case, no individual customer was so important that the company could not afford to lose him, and it did cut off many smaller customers for late payments or abusing their gas supply. On the other side, controlling them was no simple affair and small conflicts flared up with some frequency. The source of conflict with smaller consumers was the difference in attitudes toward proper use of gas for light. For customers who had been used to burning oil and candles whenever they pleased, the move to gas with restricted hours of use was not going to be easy. They generally treated gas like candles and lamps and used it whenever they wanted. In addition to such deeply ingrained habits, they had a further disincentive for moderation: the flat rate they paid for their gas. Thus small consumers had little economic incentive to limit their use of gas. For the company, allowing such behavior would have forced a change to its production model, something it was not willing to do. Its response was twofold: it tried to educate its customers in what they considered to be the appropriate use of gas, and to control gas usage better at shops and homes with standardization and the use of inspectors and penalties.

In *Disenchanted Night*, Wolfgang Schivelbusch speaks of "the loss of domestic autonomy" that the introduction of gas lighting entailed. Whereas before gas lighting the home was in control of its light sources, such as candles, lamps, or fireplaces, the new technology transferred some degree of control to an outside influence. "Fire had always remained clearly and physically recognizable as not merely a product, but also the soul of the house." The introduction of gas lighting, he argues, ceded some element of this domestic autonomy.[373] For the GLCC, control of gas lighting was essential, or it risked bankruptcy. The need to control entangled the company with its users, especially it smaller ones, in an intimate way. This section explains how this happened.

The company only gradually became aware of the potential for problems with their smaller customers. Smaller users began to sign on with the company soon after a few shops fitted up gratis for display purposes in April 1814.[374] As requests for gas came in and the company began laying service pipes, they also began a log book of customers to keep track of the contracts.[375] The company soon realized that it would be to its advantage to control gas usage in homes and shops, and in August 1814 standardized the burners it would supply to these locations.[376] This was not adequate, however, and soon the company become aware that there was "a great expenditure of gas in several shop & c." that, unless brought under control, would destroy whatever profit they could have made on the contracts.

The most worrying trend for the GLCC was off-hours burning. It emerged early in 1815 that some of the small customers were not holding to their contracted times, and were using gas either after 10 p.m. or before the time when the gasworks began feeding gas to the street. This was around 4 p.m., but could be earlier, depending on the time of the year. A very few customers paid for extra time. When a number of cases came to the directors' attention, they hired a Mr. Perks in March to be the "inspector of the fittings up and regulating burners."[377] His job included going "along the line of the company's mains one night in every week to observe in what houses the gas is found burning after 10 o'clock at night."[378] The inspector's reports must not have been very cheering, because in May and July 1815 the directors had special meetings to look into the current state of the company's machinery, partly with a view to preventing "any waste or fraudulent use" of the gas.[379] One result was the decision to begin installing stopcocks at customers with more than two lamps, or where the gas had not been ordered but a service pipe had nonetheless been installed.[380] By October 1815, Perks had more work than he could keep up with and another man was hired to go with him to act as inspector and to turn the stopcocks down to the right pressure.[381] A third one soon followed to take over one of the other stations, and even more in December 1815.[382] The inspectors were kept constantly busy. The company supplemented their policing by communicating to their customers what they considered to be reasonable usage, such as a notice that explained that flames should be 1.75 inches long.[383]

The company's minute books are a chronicle of the many problems it had with customers. In one case, the inspectors found a shop where gas was being burned directly from the tubes, meaning no lamps had been installed.[384] In another lamps were used without glasses.[385] In 1816, it was reported that the "Public Houses" were burning ten hours a day and usually with flames greater than six inches.[386] The inspectors even found one case of some people stealing gas through an unauthorized service pipe and burning it in their cellar.[387] One person had enlarged the holes in his burners and when he refused to have them fixed, he was cut off.[388] These were, however, some of the more unusual cases. By far the most frequent problem was the one that had first emerged, which was burning after the stipulated end time or on Sundays. The inspectors at first verbally warned the offenders, but if they proved unrepentant, they were sent notices that they would be charged for their extra usage or even cut off unless the practices were rectified.[389] In 1815, the company reminded its consumers through notices that anyone caught using gas extravagantly would be cut off, and in early 1816 added a clause to its agreement with its customers that anyone found burning after hours or on Sundays would be charged for the whole quarter at a higher rate.[390] This did not deter people, and in 1816 consumers continued to be discovered regularly burning late, even until one or two in the morning. They were duly billed for the whole quarter.[391] The problem of excessive burning was so pressing for the company that some zealous shareholders volunteered in 1817 to act as inspectors, but were politely declined.[392]

Inspectors and threats were one approach, but the company often tried other solutions as well. One was charging by volume, the solution that ultimately solved most of the company's problems in this regard. Clegg had invented in 1815–16 a small measuring machine for use in houses.[393] The company tried them out at a few homes in May 1816, but in June, they were already proving to be problematic and their use was suspended.[394] It took some more years of development work before the gas meter was reliable enough to be useful, and it became the subject of a lawsuit between Malam's and Clegg's respective claims to invention.[395] Gas companies began using large gasworks meters to measure total production in 1823, with deployment of smaller consumer meters beginning slowly after that.[396] Not until the mid 1830s did gas companies make a determined effort to ensure that all their consumers had them installed, and it was the late 1840s before they were universally used for consumers.[397] For the moment, however, the GLCC had to try other options and decided to install stopcocks at all houses and shut off gas every day,[398] but this must have been too labor intensive and the practice seems to have stopped.

Another way to control gas usage was by standardizing fittings as much as possible, and like many other aspects of the network stabilization, the standards were established in the first few years. The company had already chosen its standard lamps in 1814, and in March 1815 asked Clegg to draw up a list of standard tubes based on the number and type of lamps used in shops and homes.[399] To help consumers choose, the company had a room at each station fitted up with a set of sample burners.[400]

FITTERS

The fitters were the group of people building parts of the network with whom the GLCC had the most problems. Like the cheap street lighting clause, this problem also had roots in the Parliamentary battles over the company's incorporation. In the 1808–09 battle, the GLCC had conceded the manufacture and sale of apparatus to Boulton & Watt to buy peace, but the way the clause was framed to prevent it from doing so also stopped the company from selling and installing fittings. The result was that the company had to use outside contractors called "fitters up" to install gas fittings in buildings, and the company had no end of problems with the fitters and their work.

Some of the very first problems that the company had with the extravagant use of gas by some customers was due in part to improper fittings—perhaps installed without proper lamps. In 1814, the directors resolved not to turn the gas on at cus-

FIGURE 5.15

Gas burners. Accum, *Description of the process of manufacturing coal gas* (1819), plate iii. Courtesy of Roy G. Neville Historical Chemical Library, Chemical Heritage Foundation.

tomers until the fittings had been inspected.[401] In January 1815, the directors, "with a view to the safety and œconomy of fitting up," asked their fitters to give the company a list of charges and materials they were using for their work,[402] implying that up to this point the company did not have a clear idea what the fitters were actually installing or what their fees were. This was really just the beginning of the company's problems with fitters. In March 1815, Clegg reported to the Court: "I find very great carelessness in the different shops now fitting up, and even some of the gun barrel ends from our main left open, if these things are not more strictly attended to, the consequence may be of a serious nature. . . . I would likewise give notice to each person employed by the company to fit up houses, that if after repeated instances of carelessness and leaks being detected in their work, that they be immediately dismissed." His proposed solution, adopted by the Court, was to keep another logbook of shops and their fittings, together with a note about who had installed them.[403] Not all fitters complied with the order, and the company had to send threatening letters to fitters who were not reporting their work.[404]

Further problems with fitters followed shortly. In June 1815, Clegg reported that there were "continually complaints coming from the different shops lighted with gas about the smell which arises from leaks, and the fitters up have been repeatedly told of them, but it generally happens that the fitters up have been paid for the work and will not have any thing more to do with it. The joints are usually daubed with paint which serves for a while and then the work is thrown off their hands." As examples he referred to the Attorney General's house and the Ship Tavern, the latter of which had cost the company twice the original cost of fitting to find and fix leaks. The directors' response was to order that no fitter be paid until his work had been inspected.[405] To take some work out of the hands of fitters, the company took over the job of laying service pipes and stopcocks in August 1815.[406]

Poor workmanship was not the only problem with fitters. Sometimes they neglected (or perhaps knowingly omitted to avoid inspection) to report their work to the company. For example, in November 1815 two houses were found to be fitted up without approval, and in a very bad state. The gas was turned off immediately.[407] Throughout 1815 and the first half of 1816, reports of problems with fitters poured in. Some were sanctioned for shoddy work or using poor materials; others had installed extra burners without letting the company know, or had put up burners not supplied by the GLCC.[408] Some charged their customers extravagantly, leaving the customer to complain to the GLCC, which had nothing to with the installation fees.[409] They also continued to ignore requests to repair leaky fittings after they had left a work site.[410] One measure the company tried was to command fitters not to fit up any shop without prior notice to the company, but this availed the company little and complaints continued to pour in.[411]

The situation could not continue, and the company decided in August 1816 to recognize only the work of fitters it had certified and who had agreed to take responsibility for their own work, using only company-approved and standardized components.[412] The standardization of parts had begun in 1815, when Clegg was asked to prepare drawings of all the parts to be used in homes and shops, and a copy of these drawings was given to every fitter to follow.[413] The company even regulated the sorts of seals and solders fitters could use, such as when they were ordered to use leather collars instead of white lead as a sealant.[414] By the end of 1815, the fitters were expected to get all their supplies from the company, especially service pipes and lamps. Gas would not be supplied to homes where the lamps did not bear the company's mark, and fitters were sanctioned for installing non-company burners.[415] In terms of certification, from the end of 1816 fitters were asked to prepare a house, which if judged after inspection to be adequate, would allow the fitter to be added to the approved list.[416] Any house or shop fitted up by a non-certified (or decertified) fitter would have the gas supply refused. The company sent notices advising people of this new policy to all buildings along the company's mains, and similar notes were sent out in areas where it was laying new mains.[417] From then on, when it happened that people were found using lamps installed by unauthorized fitters, they would be cut off.[418]

The new policy of using only approved fitters helped a great deal, but some problems still remained. When fitters were found to have connected homes without proper notification or with faulty fittings, they were threatened with de-certification and prosecution.[419] Leaky pipes were still reported regularly, and the company's superintendents were becoming exasperated with having to deal with endless complaints from consumers. Matters came to a head again at the end of 1818, when the directors, judging that the fitters were doing great harm to the company's reputation, halted the certification of new fitters.[420] They asked Richard Leadbetter, the superintendent of the Peter Street station, to prepare a report on fitters, which he delivered on December, 21, 1818. In the report, Leadbetter wrote:

I consider it a matter of the most vital Importance to the company that some immediate steps should be taken to check the alarming evils accumulating from the very infamous manners in which several of the tradesmen execute their work (for the reception of gas into the houses of the consumers) who make use of every evasion to elude the regulations laid down by the company and if the size of tubing be adhered to consider their instructions sufficiently complied with, and that they are at liberty to affix brass work of any description at the end of it, completely defeating the Intention of carrying an ample supply to the burner, by connecting the tubing to work whose orifices in most instances do not exceed one sixth, and in many not one tenth part of it. and in almost every instance where it is requisite to place cocks, the gas' way through the plugs is constrained in a manner to render it impossible a competent supply can reach the burners under almost any Pressure. . . . Few houses can be found where the interference of the company's servants does not become necessary to remedy some material defect; . . . it is not for me to question the wisdom or policy of issuing regulations to the several tradesmen respecting modes of fitting, by which the company in the opinion of the public, incur all the responsibility of their errors, or the imputation of not furnishing a competent supply.[421]

Leadbetter suggested that the fitters be required to test their work at one-tenth inch of pressure, and that they be liable for repairs one or two years after completion. In fact, this sort of suggestion had been made previously to solve this seemingly intractable problem. The Court of directors created a committee composed of some of the directors, including Pollock, to propose a way forward. The committee took almost a year to investigate, and when it reported on November 12, 1819, it seemed to exculpate the fitters. Its recommendations concerned the types of pipes that were being used. Apparently, it was corrosion, probably caused by the sulphur in the gas, that was eating the pipes and creating the leaks. After trying many different sorts of pipes, the committee suggested a copper pipe lined with lead, and a new form of T junction that was less difficult to drill out and tap.[422]

5.5   CONCLUSION

All the streets of London, as well as most shops, public buildings and many private houses are illuminated [with gaslights], so that the whole of London at night resembles a fairy-tale palace. For this beautiful and beneficent invention, a daughter of chemistry, we have the English to thank.[423]

— Johann Hecke (1820)

Between 1812 and 1820, the managers and engineers of the GLCC built a large urban gas network. Its success is demonstrated by its rapid growth. In mid 1814, it had all of four paying customers and £180 in yearly revenue. By 1816, it supplied about 8,600 lamps to 2,400 customers,[424] with revenue of £35,713.[425] By 1820, its approximately 120 miles of mains were supplying about 30,000 lamps, and its revenues had reached £101,785.[426] The original vision for a system of mains distributing gas to lamps throughout the cities came from Fredrick Winsor, who explicitly referred to water distribution networks as the model for gas. His vision, however, was of a patchwork of independent networks, one where small gasworks fed a hundred or so lamps. Driven by the difficulty and expense of leasing many lots throughout the city to house small gasworks, the managers of the GLCC quickly reformed the model to a larger integrated network supplied by a very few gasworks. Soon after making this shift, the first managers showed themselves to be unprepared for the task for deploying the new model that they had adopted. A new management structure introduced from 1814 helped turn the company's fortunes around, and to 1820, the GLCC's managers and engineers successfully built a gas network in parts of London. While they were proceeding, their work did not seem to be following a single coherent plan, such as Edison had when he first conceived an electrical system. Rather, they were reacting to the circumstances as they developed. By 1819, however, they began to realize they had achieved something significant in developing and operating this network. Thomas Peckston, one of the GLCC's engineers, wrote: "A few years ago, had any one advanced as his opinion, the possibility of lighting, from one gas manufactory, a combination of streets of many miles in length, he would have been looked upon as little better than a madman. Indeed, when the gas was first conveyed to the distance of about half a mile from the manufactory it was considered as a wonderful performance."[427]

From the point of view of stability with increasing scale, the gas network had more in common with the networks that followed it than with the canals, roads, or even water supply networks that preceded it. The transportation networks were not integrated from the business point of view. Gas was also more tightly coupled, with changes in generation, distribution, and consumption patterns having almost immediate effect on the other parts of the network. As the network expanded, the company's engineers and managers developed ways of increasing the gas network's stability by building more robust and efficient means of generating gas, using valves, wells, and pressure regulators to maintain pressure, and developing means to track down leaks and minimize their effects. As with electrical networks, the company also had to keep a watchful eye on load factors, but because gas could be stored, the dynamic worked differently than with electricity. Gas was generated constantly throughout the day and stored in gasholders, but the company had to have a good sense of how much gas was consumed on a given day, or risk supply failures. It tried to match total production and consumption exactly on a given day. As a result, the control of usage patterns became an essential part of network stabilization. The task of stabilization fell to active operational management by the companies' employees, who, besides running the gasworks, would every day bleed air from the mains, test pressure, drain water from wells, adjust regulators and stopcocks, and inspect the network throughout London.

The drive to stabilize the network created tensions between the GLCC and those elements outside of its direct control: its customers and the fitters who installed pipes and burners in homes. Small users, typically shops and homes, tended to treat gaslights like any other light sources, and used it whenever they pleased. Because of the load factor issue, the GLCC could not tolerate unpredictable usage patterns, and at first tried to restrict consumption by contract to between 4 and 10 p.m. for most customers. This soon proved unworkable, however, and the GLCC implemented a further series of measures to control the users, including a system of inspectors to monitor usage by walking the streets at night, fines for off-contract burning, and the standardization of parts, particularly burners, to increase predictability. The fitters were likewise outside the company's direct control, and since they had no long-term relationship with the users, they very often did poor work in installing fittings. In this case, the company implemented a program of certification for the fitters that required them to use only company supplied standardized parts, and inspections were used to verify quality. Any fitter failing a quality test would be decertified, and the company refused to connect customers who had used non-certified fitters.

Not all customers, however, came in for the same treatment. From its earliest days, the company and its promoters were sensitive to public opinion. The introduction of gas into a prominent public building, such as Parliament, was a good opportunity for advertising. The company actively sought and gave preferential treatment to these sorts of customers. The difference in attitude shown to small and large customers came into strongest relief in the case of the Covent Garden Theater, a notoriously delinquent customer that the company never cut off.

The period 1812–1820 was one of almost constant innovation for the gas industry, and the GLCC was at the forefront. Although the GLCC relied a great deal on the foundations laid by Boulton & Watt and on its own head engineer, Samuel Clegg, the implementation of the network model was coordinated and made possible by the company. There were many individual inventors who continued to take out patents and offer their inventions to the GLCC, but the work of innovation was largely internal. Only the GLCC had the experimental scope that a full production apparatus offered. No entity other than the GLCC could work with hundreds of retorts in gasworks, and a large network of mains under city streets to find and identify solutions to all sorts of problems with efficient generation and supply stability. The GLCC, as an incorporated joint-stock company with access to lots of capital and a fairly effective management structure was able to run the company with almost constant innovation in technological and operational affairs.

The history of large integrated technological networks has tended to focus on railways as a turning point because of the nature of the technology involved. Decentralized and distributed management with relatively few standards may have worked for roads and canals, but soon proved disastrous with railways. That Thomas Edison found inspiration in gas networks suggests, however, that the history of integrated networks should also look back to the gas networks of the early nineteenth century and the lessons they might have provided to later network designers. Edison's idea of an integrated production, distribution, and consumption model came from gas networks, whose basic model was first developed by the GLCC. In addition to the production and distribution model, gas, like railways, electricity, telegraphs, and telephones, was technologically demanding, and the creation of a successful gas company and a stabilized network required significant innovations in both management and engineering.

By 1820, gaslight technology was sufficiently mature that it could be transferred to other countries. The GLCC's success generated notice in German publications before 1820, and a few attempts were made to introduce the technology there, but without notable success.[428] However, the picture changed completely after 1820. In Paris, many new companies were founded, including the Compagnie Anglaise in 1821. It was supported by British capital and eventually became the largest gas company in Paris. Two important British promoters of the company were a forge owner, Aaron Manby, and a civil engineer, Daniel Wilson, each of whom played an important role in transferring the British technology to France. Nor was the Compagnie Anglaise the only one to have British supporters with technical knowledge. John Grafton, who held gas patents in England and later in France, was also a shareholder in the Compagnie de l'Ouest.[429] Many of the new French companies were successful, and gas lighting flourished in Paris and France, albeit at a somewhat slower rate than in Britain.

The story is similar in other countries. The first company in Denmark also had British origins.[430] The first German gas companies generally had British backers or were British companies with operations in Germany. William Congreve banded with a large number of investors in 1823 to form the Imperial Continental Gas Association with the mandate of establishing local gas concerns on the Continent, bringing British technology with them. They signed contracts with four cities in Germany in 1825. The Imperial Continental Gas Association was operating in Hanover in the same year, and in Berlin in 1826. Antwerp, Brussels, Rotterdam, Amsterdam, and Vienna followed, and the company became a very important part of the Continental gas industry.[431]

CONCLUSION

A NEW PHASE OF THE INDUSTRIAL REVOLUTION

Although historians have often emphasized the breadth of technological innovation in the Industrial Revolution, the historiography has been dominated by the classic technologies of the textile, coal, iron, and steam engine industries. This is due in part to the fact that economic historians focus on economic and productivity growth. Gas lighting, however, was an important macro-invention, and the gas industry grew rapidly from 1812 on, reaching nearly every town in Britain with a population over 10,000 by 1826. In its creation and its consolidation, gas lighting differed from most contemporary novel technologies in some important ways that indicate the emergence of broader economic and social trends. The first of these was the relationship between science and technology. It is clear that by the late nineteenth century the content of formal science was, in various ways, providing an important impetus to technological innovation, particularly in the chemical and electrical industries. Around 1800, however, formal science itself was inchoate, still more natural and experimental philosophy than the science of later decades. The "scientist" did not yet exist, and scientific disciplines were undefined. They would emerge from a new "Scientific Revolution." Within the more fluid continuum of science and the arts of the eighteenth century, speculative and experimental work on the chemistry of airs created a body of knowledge about materials, techniques, and instruments that provided the foundations of the technology of the gas industry. Thus, although the nature of scientific practice changed a great deal over the course of the nineteenth century, by the end of the eighteenth century it was generating enough useful knowledge to give rise to an entirely new industry. As such, gas lighting represents one of the earliest case of new industrial technologies derived in part from contemporary science, and provides evidence for Mokyr's model of an Industrial Enlightenment.

It was not only in this characteristic that gas lighting represented a new phase of the Industrial Revolution. Gas lighting also was similar to some other nineteenth-century industries in that it required substantial capital to develop and build—far more than an individual or a partnership could provide. Whereas the classic inventions of the Industrial Revolution were deployed by smaller firms, the cost of establishing gas lighting in its network form was far beyond the means such firms had at their disposal. Gas lighting had to be deployed by much larger companies—specifically, joint-stock corporations such as were common in the canal business and in some other types of businesses. Thus, gas lighting also merged two important trends of the Industrial Revolution: technological innovation and the construction of large infrastructures. Canals, roads, ports, and water supply were some of the infrastructure projects of this period that contributed to economic development in various ways, such as decreased transportation costs. Gas lighting followed this trend of infrastructure construction, but relied on new technologies much more than other projects. A few decades later, railways were to be an even more important example of this merger.

QUESTIONS ABOUT THE INDUSTRIAL REVOLUTION

The present study also addresses some of the frequently asked questions about the Industrial Revolution, specifically in regard to why Britain industrialized before other countries. Gas lighting was invented in many countries; however, for social and economic reasons, only in Britain was it developed into an industry. Its multiple simultaneous invention was due in part to the presence of pneumatic chemistry throughout Europe, made accessible to many inventors and engineers through books, lectures, universities, correspondence, and scientific instruments that proliferated in the Enlightenment. The Enlightenment ideal of making knowledge readily available meant that Zachaeus Winzler in Austria, Jan-Pieter Minckelers in the Netherlands, and William Murdoch in Britain were all drawing from the same spring. In this regard, the roots of the Industrial Revolution were found in much of Europe, and existed in the seventeenth century. Another element of the shared background was experimentation in industrial distillation processes prompted by local economic conditions: tar shortages in Britain and wood scarcity on the Continent. The convergence of these two traditions, pneumatic chemistry and industrial distillation, created conditions that made the invention of gas lighting likely.

Local conditions explain the subsequent "failure" of gas lighting on the Continent—more correctly, the bifurcation in the technological tradition of industrial distillation. The presence of a large coal industry in Britain has often figured importantly in explanations of Britain's place in the Industrial Revolution, and it was also crucial in the emergence of gas lighting in Britain. As many contemporary commentators pointed out, coal was a better source material than wood for illuminating gas because its gas was of much higher quality. For this reason, the problems associated with designing and refining distillation technology into gaslight technology were addressed in Britain, where coal was always used in distillation whereas wood was more frequently used on the Continent. Indeed, many of the technical solutions adopted in Britain to make gas lighting practicable—for example, the gasometer and lime purification—were known and discussed by Continental pioneers. They pursued another path, however—one that led to a wood distillates industry. Thus, there was no failure of technical imagination on the Continent.

The use of coal in distillation was not, however, the only local condition that caused this bifurcation. British experience in making pneumatic machinery, especially on an industrial scale, was decisive for the development of gas lighting. Boulton & Watt had acquired this experience by making pneumatic apparatus for Thomas Beddoes, and by making iron steam engines. William Murdoch's invention of gas lighting at Boulton & Watt occurred within a firm that had the resources and the experience to transform it into an industrial-scale technology and foster its commercial success.

BUSINESS AND ENTREPRENEURSHIP

Establishing and consolidating gas lighting as a new industry naturally involved business activities. Important among these activities were display and advertising, which were, in a way, inherent to gas lighting. Among the early forms of gas lighting this was most evident in Diller's philosophical fireworks, but display was also important for Volta as he lectured about his discoveries and displayed his lighter on tours. Display was important to the commercial pioneers too. Zachaeus Winzler held public displays in Vienna, and Philippe Lebon in Paris, as they sought investors. Murdoch did not need to find investors, as he was already funded by Boulton & Watt, but display became important once the firm began looking for customers. The firm, as with its steam engines, gave great importance to demonstrating quantitatively that its lights were better than existing forms of illumination, even publishing dubious results in Murdoch's Royal Society paper. That paper and the Murdoch's Rumford medal also figured in Boulton & Watt's publicity work as the firm battled with Frederick Winsor's group. Boulton & Watt strove to demonstrate that Murdoch was the "inventor" of gas lighting, minimizing the work of the firm as a whole. With Joseph Banks' full complicity, Boulton & Watt received the Royal Society's full approbation.

It was Winsor, however, who was the master of publicity in matters having to do with gas lighting. His early displays at the Lyceum were a first step toward the massive campaign of 1806–1810 in which Winsor gathered a group of important financial and political backers and won an act of incorporation from Parliament. The social and business momentum Winsor created through advertising in newspapers, public lectures, and displays on Pall Mall was so strong that his group overcame the 1809 setback in Parliament, and easily found enough investors to incorporate the Gas Light and Coke Company in 1812. Indeed, the technology was so well known and sought after that the new company had little difficulty finding new customers, and was soon regularly rejecting requests for lights.

Entrepreneurship is beginning to receive more attention from historians of the Industrial Revolution. Specifically, did entrepreneurs of that period seize pre-existing technical opportunities, or did they drive the technical innovations? In the case of gas lighting, they did both. The inventors and engineers who were trying to commercialize gas lighting were already interested in business (industrial distillation) before they saw commercial possibilities in gas lighting. But they also borrowed technical knowledge from pneumatic chemistry. In turn, they provided a good deal more technical innovation to make gas lighting practicable, especially at Boulton & Watt. Winsor, however, was technically incompetent but full of entrepreneurial energy. He seized on existing technologies and was the original impetus behind the establishment of the network model of the technology in its corporate form.

In a related question, was Britain exceptional because it somehow fostered more entrepreneurs than other countries? In the case of gas lighting, there was no shortage of entrepreneurial energy in Germany, if interest in the thermolamp and attempts to make commercial versions were indicative. Indeed, the thermolamp of Hugo zu Salm and Karl Ludwig Reichenbach became a commercial success. Technical details explain the subsequent development of distillation technology in Germany. In France, there was also no shortage of people trying to commercialize gas lighting; however, they failed to get enough support from investors. Philippe Lebon tried hard, but he found no private investors, and only his efforts in tar making were encouraged by the state (because of the possible military applications).

BUILDERS, MEDIATORS, AND USERS

Gas lighting was shaped not only by the engineers and firms who designed and operated the machinery, but also by its users, and by people who mediated between these groups.

Transforming the technology from its early forms (derived from distillation ovens and from scientific instruments) into something usable for industrial purposes was technically challenging. It was the firm of Boulton & Watt that did the most to identify and solve the technical issues surrounding this transformation. Boulton & Watt's work was never encumbered by patents, and was freely borrowed by others in the early gas industry. That firm also trained many of the early gas engineers, the most important of whom was Samuel Clegg, who furthered the technical consolidation of the industry (especially during his years at the Gas Light and Coke Company).

As the technology matured, it was also shaped by its users. Boulton & Watt's customers were industrial mill owners in the north of England. Although at first some of them were very much motivated by enthusiasm for the technology, once a plant was operating they were concerned primarily with keeping it running smoothly and cheaply. The employees of the mill had little say in how the plant was operated. The issues that stemmed from the separation between ownership and operation (on one hand) and users (on the other) did not bother the mill owners who were Boulton & Watt's customers. But once the Gas Light and Coke Company was operating, the preferences and behavior of its customers became very important because they could not be controlled as easily as factory employees. Users in homes and shops treated gaslights like candles and lamps, burning them whenever they pleased with little regard for the destabilizing effect this might have on the gas company's network. In response, the GLCC had to develop measures to mitigate and control the effects of its customers' behavior, including standardization of parts and fitting, inspectors, and a system of fines, warnings, and contracts. Because the company's act of incorporation stipulated that it could not install fittings, it was forced to rely on outside fitters. Because they had no long-term relationship with customers, fitters often did poor work, and the company had to find ways to control the quality of their work. This was done primarily through a certification program, and secondarily by controlling the supply of parts.

TECHNOLOGICAL NETWORKS

The Gas Light and Coke Company implemented a network model for the gas industry, whereas Boulton & Watt provided stand-alone installations. The network model was to dominate the industry for the rest of its history. The origins of the network model can, in part, be traced back to Winsor, who saw the possibility of distributing gas through a network of pipes under city streets, as was done with water. The political battle with Boulton & Watt forced the GLCC to focus exclusively on selling the products of distillation rather than machinery, and that marked another step toward the service model. The fully integrated model, with only a few gasworks supplying gas, was conceived and realized only after the GLCC commenced operations, when it became clear that having many small plants throughout the city would be too expensive and troublesome. As the growing demand for gas led to the network's expansion, the company innovated almost constantly in order to deal with a series of technical issues.

The railways have often been presented as the first tightly integrated technological networks to have proliferated in the second half of the nineteenth century. However, the gas network, dating to the early years of the nineteenth century, was also closely integrated. Problems originating in one part of the network, such as leaks and after-hours consumption, propagated rapidly through the system. The measures the GLCC implemented to stabilize the system included routinization and standardization of components, close monitoring, and inspections. The company's internal stabilization depended on how it was governed and managed, and in 1814 it began to reform its management structure. Committees were given specific responsibilities, such as accounting and oversight of specific gasworks. Decision-making powers outside these committees was limited, and lines of communication between the committees and the Court of directors were established. The creation and implementation of such a structured approach to management was also important for the governance of railway companies, which likewise grappled with an expanding and complex business and technical system.

The stability that allowed the Gas Light and Coke Company to expand and eventually flourish after a precarious start was due in part to the company's legal form as a limited-liability joint-stock company. The GLCC survived the replacement of its entire management, and was able to build its network by drawing on large pools of capital through capital calls to its shareholders. The joint-stock legal form had, in the preceding years, become commonplace among infrastructure companies, such as water-supply corporations and (especially) canal corporations. Likewise, elements of the legal model on which the GLCC relied to gain the power to open city streets to lay its mains came from water companies. Thus, the gas network drew on existing networks for legal, corporate, and technical examples.

NOTES

NOTES TO INTRODUCTION

1. Chandler and Lacey 1949; Everard 1949; Clow and Clow 1952; Elton 1958; Körting 1963; Falkus 1982.

2. Bruland and Mowery 2005, 355.

3. Mokyr 1990, 1999, 2002, 2005, 2007, 2010. See also Jones 2008a,b.

4. Fox 2009.

5. Von Tunzelmann 1993, 298; Mokyr 2002, 31; Berg 2007; Cardwell 1972, 210; Mathias 1972.

6. Mokyr 2002, 51–2, 80–95. Quote on 81; Mokyr 2007, 188.

7. Mokyr 2002, 82.

8. Pomeranz 2000; Vries 2005; Duchesne 2006. For a summary of the debates, see Duchesne 2005.

9. Mokyr 2002, 65; Mokyr 2009; Stewart 1992; Jacob and Stewart 2004; Jacob 1997, 1988.

10. Ashworth 2008; Pomeranz 2000.

11. Landes 2003.

12. Allen 2009.

13. Harris 1976, 1998.

NOTES TO INTRODUCTION TO PART I

1. GW to JWjr, 1801/11/08, BWA-JWP-C2/10. (N.B.: All dates in notes are in the format YYYY/MM/DD.)

2. Chandler and Lacey 1949; Elton 1958; Körting 1963; Veillerette 1987.

3. Berman 1978, 146.

4. Clow and Clow 1952, 426.

5. Mokyr 2005, 295ff.

6. Allen 2009, 240.

7. Mokyr 2009, 141.

NOTES TO CHAPTER I

8. Musson and Robinson 1969; Hall 1974; Clow and Clow 1952; Gillispie 1957a,b.

9. Wengenroth 2003, 229; Shapin 1996, 140–1; Hall 1974; Landes 2003, 61; Mathias 1983, 124; Mokyr 2002, 46–50.

10. Mokyr 2005; Jacob 1997; Jacob and Stewart 2004; Jacob 2007; Jacob and Reid 2001; Stewart 1992. For earlier statements, see Musson and Robinson 1969; Mathias 1972. On public science more broadly, see Lynn 2006; Hochadel 2003; Bensaude-Vincent and Blondel 2008.

11. Golinski 2008, 117–8; Golinski 1992.

12. Klein and Spary 2010; Klein 2005, 248; Klein and Lefèvre 2007. See also Smith 1979; Wengenroth 2003.

13. Peckston 1819, 377. This was also something William Henry understood in 1808; see GAL to JWjr 1808/01/20, BWA-MS-3147/3/478 #18. See also Accum 1815, 131–2; Cooper 1816, 185.

14. For early descriptions of gas plants, see Accum 1815 and subsequent editions; Peckston 1819; Creighton 1824.

15. See, e.g., Franciscus Titelmans, *Compendivm Philosophiae Natvralis* (Lvgdvni Rovillius, 1558), 150–1. (Primary sources cited only once are given in full in these notes and are not listed in the bibliography.)

16. Hunter 2008; Boyle 1665, 767–8, 775–6; Thomas Shirley, "The description of a well, and earth in Lancashire, taking fire by a candle approached to it," *Philosophical Transactions* 2 (1666): 482–4; Boyle 1685, 91; "An extract of a letter of July 28, 1675 by Mr. Lister from York to the Publisher," *Philosophical Transactions* 10 (1675): 391–5; John Clayton, "An experiment concerning the spirit of coals," *Philosophical Transactions* 41 (1739): 59–61.

17. Boyle 1672, 63–6. The process of dissolving iron with spirit of vitriol (sulphuric acid) for the production of medicines had been described in Jean Béguin, *Les elemens de chymie* (Mathieu le Maistre, 1615), 234–9. Béguin's use of iron sulphate as a medicine was common in the early eighteenth century. Fester 1923 [1969], 91; Klein 1994, 145.

18. Principe 2007, 7ff.

19. See, among many others, Isaac Newton, *Opticks* (S. Smith and B. Walford, 1704), 134; Wilhelm Homberg, "Observations sur les matières sulphureuses," *Histoire de l'Académie Royale des Sciences* (1710 [1712]), 225–34, 233; Nicolas Lémery, "Explication physique et chymique des feux souterrains, des tremblemens de terre, des ouragans, des eclairs & du tonnerre," *Histoire de l'Académie royale des sciences* (1700 [1761]), 101–10; Georg Stahl and Ernst Isaac Hollandus, *Chymia Rationalis et Experimentalis*, second edition (Caspar Jacob Eyssel, 1729), 317.

20. Henry Cavendish, "Three papers, containing experiments on factitious air," *Philosophical Transactions* 56 (1766): 141–84.

21. Volta to Priestley, 1772/03/14; Priestley to Volta, 1772/03/14; both in Volta 1949, vol. 1, 59–60.

22. *Gli strumenti di Alessandro Volta* 2002, 56; Home 2000, 121.

23. Volta 1777.

24. For a discussion of Volta's views on the nature of inflammable air, see Holmes 2000, 91.

25. For a fuller account of this story, see Tomory 2009.

26. Kim 2001, 370; Szabadváry 1966.

27. Tiberius Cavallo, *A Treatise on the Nature and Properties of Air, and Other Permanently Elastic Fluids* (London, 1781), 300.

28. Stephen Hales, *Vegetable Staticks* (London, 1727), 183; Parascandola and Ihde 1969.

29. Guerlac 1957a,b; Joseph Black, "Experiments upon magnesia alba, quick-lime, and other alcaline substances," *Essays and Observations, Physical and Literary read before the Philosophical Society of Edinburgh* 2 (1756): 157–225, 198–9.

30. Partington 1961, vol. 3, 76–7, 84.

31. Carl Scheele, *Chemische Abhandlung von der Luft und dem Feuer* (Upsala, 1777), 149–50.

32. Antoine-Laurent Lavoisier and Marie-Anne-Pierrette Lavoisier, *Traité élémentaire de chimie* (Cuchet, 1789), vol. 2, 342.

33. Siegfried 1972; Antoine-Laurent Lavoisier, "De la combinaison de la matière du feu avec les fluides évaporables, et de la formation des fluides élastiques aëriformes," *Histoire de l'Académie royale des sciences* (1777): 420–32.

34. Jean Baptiste Meusnier, "Description d'un appareil propre a manoeuvrer différentes espèces d'airs," *Histoire de l'Académie royale des sciences* (1782 [1785]): 466–75.

35. Levere 2005a.

36. The Dumotiez brothers advertised it in 1795. See also Accum 1804a, 179. For a contemporary history of the gasometer, see Friedrich Parrot, *Ueber Gasometrie nebst einigen Versuchen über die Verschiebbarkeit der Gase* (M. G. Grenzius, 1811).

37. Jean-Antoine Chaptal, *Élémens de chimie* (Jean-François Picot, 1790), vol. 1, 110; "Chymistry, Section II, Vital air" in *The New Encyclopædia; or, Modern Universal Dictionary of Arts and Sciences* (C. Cooke, 1797).

38. Ludwig Gilbert, "Von den veschiedenen Gasmessern, und Beschreibung des von Seguin erfundenen Gazometers," *Annalen der Physik* 2 (1799): 185–93, 186–7.

39. Martinus Van Marum, "Lettre à Berthollet, contentant la description d'un Gazomètre, construit d'une manière différente de celui de Lavoisier et de Meusnier," *Annales de chimie* 12 (1792): 113–40.

40. Volta 1784a, 331–2, footnote. Brenni 2003.

41. Volta 1777, 53–4.

42. Volta 1778, 235.

43. See concluding note to article "Aria infiammabile" in Pierre Joseph Macquer and Giovanni-Antonio Scopoli, *Dizionario di chimica* (R.I. Monastero di S. Salvatore, 1783), vol. 2, 286; Volta 1918, vol. 6, 409.

44. Krünitz 1793, vol. 59, 291.

45. Volta to Jean Senebier 1779/07/10, in Volta 1949, vol. 1, 355.

46. The first text describing a lighter was Joseph Weber, *Beschreibung des Luftelektrophors nebst angehängten neuen Erfahrungen . . . Neueste mit der Beschreibung der elektrischen Lampe* second ed. (1779 Eberhard Kletts), which also had a 1778 edition. This work was followed soon thereafter by Ehrmann 1780 and Ingenhousz 1782. Krünitz 1793 contains quite an extended discussion. For a recent article on the lighter, see Brenni 2003.

47. Volta 1784b; Volta 1918, vol. 7,118.

48. Johann Gottfried Ebel, *Instructions pour un voyageur qui se propose de parcourir la Suisse* (J. J. Tourneisen, 1795), vol. 2, 26.

49. "Weber, Josef," in *Allgemeine deutsche Biographie*, vol. 41 (Historische Commission bei der Königliche Akademie der Wissenschaften, 1875); "Ehrmann, Frédéric-Louis," in *Nouvelle biographie générale*, vol. 15, ed. M. Hoefer (Didot, 1858). For a list of people who worked on the lighter and a bibliography see Poppe 1811, 16–18, and "Lampe, elektrische," in *Physikalisches Wörterbuch; oder, Erklärung der vornehmsten zur Physik gehörigen Begriffe und Kunstwörter*, vol. 3, ed. J. C. Fischer (J. C. Dieterich, 1800).

50. Volta 1783, 10, footnote; Volta 1784a, 331–2, footnote.

51. Barbier de Tinan to Volta, 1779/04/29 in Volta 1949, vol. 1, 344; Ingenhousz 1782, 214; Barbier de Tinan to Volta, 1779/09/18 in Volta 1949, vol. 1, 374–5.

52. Ehrmann 1780, 26–7.

53. Ingenhousz 1782, 214–5. Ingehousz visited with Barbier on his way to Vienna, and they conducted experiments with Volta's inflammable air pistol. Barbier to Volta, 1780/08/02 in Volta 1949, vol. 1, 417–8.

54. Ingenhousz 1782, 220–1.

55. Cowper to Volta, 1778/07/09 in Volta 1949, vol. 1, 261.

56. Cowper to Volta, 1778/09/08 in Volta 1949, vol. 1, 281; Brenni 2003; Taylor 1966, 214.

57. Volta 1784b; Volta 1918, vol. 7, 118; Volta to Cowper, 1779/07/27 in Volta 1949, vol. 1, 357; Volta 1784a, 331–2, footnote.

58. Nairne to Cowper, 1778/11/13 in Volta 1949, vol. 1, 300.

59. *London Evening Post*, 1778/05/14.

60. Krünitz 1793; "Lampe, elektrische," in *Physikalisches Wörterbuch; oder, Erklärung der vornehmsten zur Physik gehörigen Begriffe und Kunstwörter*, vol. 3, ed. J. C. Fischer (J. C. Dieterich, 1800); "Electricity: XLVI. The Inflammable Air-Lamp," in *Encyclopaedia; or, A Dictionary of Arts, Sciences, and Miscellaneous Literature*, ed. J. Akin (Thomas Dobson, 1798); "Lamp (Inflammable air)," in *The New Encyclopaedia, or, Universal Dictionary of Arts and Sciences* (Vernor, Hood, and Dharpe, 1807); Edme-Gilles Guyot, *Nouvelles récréations physiques et mathématiques*, second edition (Paris, 1799), vol. 2, 330–1; "Lampe à gaz inflammable," in *Encyclopédie méthodique: chimie et métallurgie*, vol. 4, ed. A.-F. Fourcroy (Pancoucke, 1805), 616; George Adams, *An Essay on Electricity* (London, 1784), 286–8. See also Millburn 2000; Poppe 1811, 16–8.

61. Winzler (1803b, 169) mentions criticisms of the "elektrischen feuerzeuge" in regard to explosive accidents caused by the mixing of hydrogen and atmospheric air. See also "Lampe à gaz inflammable," in *Encyclopédie méthodique: chimie et métallurgie*, vol. 4, ed. A.-F. Fourcroy (Pancoucke, 1805), 616.

62. Krünitz 1793, 283. See also "Thermolampe," *Der Verkündiger* 6 (1802): 273–6, 273, footnote. Krünitz mentions that Georg Pickel lighted his laboratory with inflammable gas in 1786. For a description of lighters derived from Volta's, see Brenni 2003. Richard Lorentz took out a patent in England on a lamp that was fundamentally based on Volta's design; see "Specification of the patent granted to Richard Lorentz," *The Repertory of Arts, Manufactures, and Agriculture*, second series, 111 (1807): 250–3.

63. Jaspers 1983, 25ff.; Jaspers and Roegiers 1983.

64. Kim 2006, 299.

65. Ibid., 305.

66. "Lettre de M. Lapostolle aux Auteurs du Journal," *Journal de Paris*, 1784/01/24, 106–107. See also "Lettre écrite le 8 du mois dernier a M. ***, membre de plusieurs académies, par M. Carra," *Journal encyclopédique ou universel* 2 (1784): 503–9.

67. The name is often misspelled "Minkelers". This is due to the printer's omission of the "c" when his paper on inflammable airs was printed. See Jaspers and Roegiers 1983, 219. For more on Minckelers in the context of the ballooning craze, see Austerfield 1981, 138–60; Jaspers and Roegiers 1983.

68. Minckelers 1784.

69. Jaspers 1983, 432ff.

70. Jaspers 1983, 47–8; Charles Morren, "Invention de l'éclairage au gas," *Bulletins de l'Académie royale des sciences, des lettres et des beaux-arts de Belgique* 2 (1835): 162–4; Morren, "Notice sur la

vie et les travaux de Jean Pierre Minkelers," *Annuaire de l'Académie royale des sciences et belles-letres de Bruxelles* 5 (1839): 79–91, 82.

71. Austerfield 1981, 395, citing Adry 1925.

72. Jaspers and Roegiers 1983, 251, note 45. The eudiometer is mentioned in paragraph 16 of the *Mémoire*.

73. Barthélemy Faujas-de-Saint-Fond, *Description des expériences de la machine aérostatique de MM. de Montgolfier*, vol. 2 (Cuchet, 1783–4), 244; Minckelers 1784, 12.

74. Werrett 2007.

75. Ibid., 332. Brisson 1797, 62; De Clercq 1988, 124.

76. De Clercq 1988, 124–5, 138.

77. *General Evening Post*, 1783/12/23.

78. "Philosophical Fire," *The Scots Magazine* 50 (1788): 164.

79. Diller sold at least two versions in the Netherlands. The one sold to the Stadholder cost seven times as much as the one sold to the Renswoude Foundation (De Clercq 1988, 138).

80. Werrett 2007, 328–32.

81. "Divers articles d'inventions dans les arts, & de découvertes dans les sciences, &c.," *Journal encyclopédique ou universel* 6 (1787): 153.

82. De Clercq 1988, 124.

83. *Journal de Paris*, 1787/06/28, 433–4. See also Franklin 1904, vol. 11, 338–9.

84. "Divers articles d'inventions dans les arts, & de découvertes dans les sciences, &c.," *Journal encyclopédique ou universel* 6 (1787): 153; *Times*, 1788/04/07.

85. "Extrait des registres de l'Académie Royales des sciences," *Observations sur la physique* 31 (1787): 188–95; "Mercredy 27 Juin 1787," "Mercredy 4 Juillet 1787," and "Samedy 24 Mai 1788," in *Procès-verbaux* (Académie royale des sciences), vol. 106–7.

86. For a more detailed description see "Physique," *Journal de Paris*, 1787/06/28, 433–4. On Diller see Brisson 1797. On Bienvenu see Bret 2004.

87. "Des feux d'air inflammable," *Observations sur la physique* 33 (1788): 72–3; Louis-Joseph and Pierre-François Dumotiez, "Les feux d'air inflammable de M. Diller," *Observations sur la physique* 34 (1789): 18; "Foreign literary intelligence," *The Critical Review* 66 (1788): 408; "Vermischte Notizen: Chemische Augenbelustigung," *Allgemeines Journal der Chemie* 2 (1800): 734–6; Winzler 1803b, 163.

88. *Times*, 1805/01/01.

89. *World*, 1788/07/14; *Journal de Paris*, 1788/07/04, 814.

90. On the Netherlands, see Antoine-François Fourcroy, "Extract of a memoir concerning three different species of carbonated hydrogenous gas," *A Journal of Natural Philosophy, Chemistry and the Arts* 1 (1797): 44–5, 49–55. The Dutch chemists state that ethylene was preferred in "lamps supported by inflammable air." On Germany, see "Nachrichten von einer Lampe für brennbare Luft des Hrn. Bienvenu," *Magazin für das Neueste aus der Physik und Naturgeschichte* 5 (1788): 89. On Britain, see *World*, 1788/07/14; *Morning Post and Daily Advertiser*, 1788/07/15; *A Catalog of Optical, Mathematical, and Philosophical Instruments, Made and Sold by Willm. and Saml. Jones, at Their Shop, No. 135, Next Furnival's Inn, Holborn, London* (London, 1793), 8. In the aforementioned catalog, an instrument described as "a new perpetual inflammable air lamp, lighted by the electrophorus" is offered for £4 4s.

91. "Herr Diller," *Magazin für das Neueste aus der Physik und Naturgeschichte* 5 (1788): 171–2; "Report from the Royal Society on Diller's invention," *The Critical Review, or, Annals of Literature* 64 (1787): 380–1; "Among the Nouvelles Literaires of this month, is a Notice des feux d'Air inflammable," *The Analytical Review* 2 (1788): 503–4; *Public Advertiser*, 1787/07/24; *St. James's Chronicle or the British Evening Post*, 1787/07/14; *Whitehall Evening Post*, 1787/07/17; *World and Fashionable Advertiser*, 1787/07/19.

92. *Times*, 1788/04/07. Advertisements for his show appeared in the *Times* during May, June, and July. For an example see *Times*, 1788/07/07.

93. "Philosophical fire," *The Scots Magazine* 50 (1788): 164.

94. "Foreign literary intelligence," *The Critical Review* 66 (1788): 408.

95. Gilbert Elliot, *Life and Letters of Sir Gilbert Elliot, First Earl of Minto, from 1751 to 1806*, vol. 1 (Longmans, 1874), 217–8. Another description can be found in Caroline Powys, *Passages from the Diaries of Mrs. Philip Lybbe Powys of Hardwick House, Oxon. A.D. 1756–1808* (Longmans, 1899), 231–2.

96. *World*, 1789/03/11.

97. George Adams, *Lectures on Natural and Experimental Philosophy*, vol. 1 (R. Hindmarsh, 1794), 496. G. S. Klügel (*Encyclopädie, oder, zusammenhängender Vortrag der gemeinnützigsten, insbesondere aus der Betrachtung der Natur und des Menschen gesammelten Kenntnisse*, vol. 2, (Friedrich Nicolai, 1806), 460) also mentions that colored flames were produced by mixing gases in fireworks.

98. William Jardine Proudfoot and James Dinwiddie, *Biographical Memoir of James Dinwiddie, L.L.D.* (E. Howell, 1868), 25.

99. See, e.g., *Times*, 1788/04/07.

100. "Cartwright (John)," in *The Thespian Dictionary, or, Dramatic Biography of the Eighteenth Century* (J. Cundee, 1802).

101. For advertisements of displays in theaters see *Morning Chronicle*, 1801/06/21; *Times*, 1800/05/17 and 1809/04/04; *Aberdeen Journal*, 1811/11/27. Cartwright subsequently moved to the United States with his show; see R. S. Coffin, *Oriental Harp* (Smith & Parmenter, 1826), 16–7.

102. *Times*, 1788/10/06.

103. Many references to the fireworks can be found in journals of the early nineteenth century. See Werrett 2007, 334; "Full and interesting account of the various ascents of Messrs. Garnerin, Sowden, Locker, &c. in balloons," *Sporting Magazine* 20 (1802): 196; "Obituary: Mr. William Clarke," *Gentlemen's Magazine* 86 (1830): 469. In Italy, the inventory of Volta's instrument cabinet included a Diller device; see *Gli strumenti di Alessandro Volta* 2002, 354. Diller's fireworks were also known in Austria; see Winzler 1803b, 163.

104. Antoine-François Fourcroy, *Leçons élémentaires d'histoire naturelle et de chimie*, vol. 1 (Société Royale de Medicine, 1782), 407.

105. As was mentioned above in connection with the Dumotiez brothers, instrument makers began making simple versions of Diller's apparatus. See "Vermischte Notizen: Chemische Augenbelustigung," *Allgemeines Journal der Chemie* 2 (1800): 734–6; Neret, "Nouveau réchaud d'un amateur de physique," *Observations sur la physique* 9 (1777): 57–8.

106. Antoine-François Fourcroy, *Elémens d'histoire naturelle et de chimie*, fifth edition, vol. 2 (Cuchet, 1793), 332.

107. Langins 1983.

108. For an extensive bibliography summarizing contemporary knowledge of inflammable gases, see Johann Friedrich Gmelin, *Geschichte der Chemie seit dem Wiederaufleben der Wissen-*

schaften bis an das Ende des achtzehnden Jahrhunderts, vol. 3 (J. G. Rosenbusch, 1799), 367–80. For a summary of ways of producing inflammable gases, see "Ballon aérostatique" in *Encyclopédie méthodique: dictionnaire de physique*, vol. 1, ed. Gaspard Monge (Pancoucke, 1793).

NOTES TO CHAPTER 2

1. Kim 2001, 370; Szabadváry 1966.

2. Winsor 1804a,b; "Specification of the patent granted to Mr. William Murdock, of Redruth," *The Repertory of Arts and Manufactures*, first series, 9 (1798): 97–9.

3. Wrigley 1988; MacLeod 2004.

4. Mokyr 1990, 13; Crafts 1994, 59; Crafts 1995, 757. For an opposite view, see O'Brien, Griffiths, and Hunt 1996, 175.

5. Beaver 1951.

6. Williams 1935, 169–174; Clow and Clow 1952, 390–393; Outland 2004, 8ff.

7. This story of the trade in tar ties in with trade-related accounts of the Industrial Revolution such as that of Kenneth Pomeranz (2000), who argues that intercontinental trade allowed Britain to grow by finding sources for wood and other materials it had exhausted at home. It also supports Robert Allen's view that intra-European trade had an important contribution in the seventeenth century, before American trade became important (Allen 2009, 128).

8. See, e.g., Becher 1682, 64–5.

9. "Stone dike which crosses a fell and colliery," *Gentlemen's Magazine* 60 (1790): 20–1; Henry Ecroyd Smith, *Annals of Smith of Cantley, Balby, and Doncaster, County York* (Hills and Co., 1878), 197–8; Granger and Appleby 1794, 15–9.

10. Granger and Appleby 1794; Bailey 1810, 290–2; Macfarlan 1925.

11. "A succedaneum for tar," *The Scots Magazine* 41 (1779): 706; Richard Watson, *Chemical Essays*, vol. 2 (J. Archdeacon, 1782), 347–8; John Latimer, *The Annals of Bristol in the Eighteenth Century* (1893), 441; Watt and Muirhead 1854, vol. 3, 283. See *DNB* entry on Champion. See also Granger and Appleby 1794, 18.

12. Hugh Owen, *Two Centuries of Ceramic Art in Bristol* (Bell and Daldy, 1873), 36–7.

13. John Champion to Matthew Boulton, 1790/06/15 and 1790/07/02, Matthew Boulton Papers 226/39, 66, both in Boulton & Watt archives in Birmingham Central Library.

14. Ambrose Weston to JW, 1809/04/06, JWP 4/86, Boulton & Watt archives.

15. *Memorandum Concerning the Progress and the Uses of the Discovery of Extracting Tar from Coal, 28th May 1785*, cited in Clow and Clow 1952, 393.

16. Granger and Appleby (1794, 16) state that a tar works should be located at a coal pit close to a navigable river. This shows the importance of a navigable river and canal system for technological development in the Industrial Revolution. See Szostak 1991.

17. Granger and Appleby 1794, 16–7.

18. "Specification of the patent granted to Earl of Dundonald," *The Repertory of Arts And Manufactures*, first series, 1 (1794): 145–8.

19. Archibald Cochrane (Lord Dundonald), *Account of the Qualities And Uses Of Coal Tar And Coal Varnish* (W. Smellie, 1784), 3; Clow and Clow 1952, 412.

20. Aiton 1811, 605; Harris 1966; Rees 1971.

21. William Nicholson, "Remarks on Mr. Winsor's projected Heat and Light Company," *A Journal of Natural Philosophy, Chemistry and the Arts* 16 (1807): 73–5; Thomas Thomson, "On coal gas," *Proceedings of the Philosophical Society of Glasgow* 1 (1843): 165–75, 173–4; John Hart, "Contribution to the history of gas lighting," *The Mechanics' Magazine* 40 (1844): 410–2.

22. Mather 1992, 369; Mather, Fairbairn, and Needle 1999, 67. The situation was similar in Sweden; see Lindqvist 1984, 34–61; Lindqvist 1983.

23. See, e.g., Gaspard de Courtivron, "Discours sur la nécessité de perfectionner la métallurgie des forge, pour diminuer la consommation des bois," *Histoire de l'Académie royale des sciences* (1747 [1752]): 287–304. The price of wood quadrupled between 1773 and 1811 (Woronoff 1984, 245; Woronoff 1994, 112–120).

24. Allen 1984; Mather 1992; Mather, Fairbairn, and Needle 1999; Whited 2005, 79–84. For movements in the price of wood, see also Benoit 1990, 109–112.

25. Sieferle and Osman 2001, chapter 4; Gericke 1998; Radkau 1983. See also Radkau 1986; Strittmatter 1986. For a link between the Holzsparkunst literature and gas lighting, see Johann C. Hoffmann, *Holz-Spar-Kunst: Ein Buch für Jedermann*, vol. 2 (Bruder und Hofmann, 1806), 139ff.

26. Genssane 1770; Barthélemy Faujas-de-Saint-Fond, *Essai sur le goudron du charbon de terre, sur la manière de l'employer pour caréner les Vaisseux, & celle d'un faire usage dans plusieurs arts* (Imprimerie Royale, 1790), 31ff.

27. Storrs 1966, 261.

28. Brose 1985.

29. "Nouvelles littéraires. Programme de la société royale d'agriculture de Lyon," *Observations sur la physique* 25 (1784): 154–5; "Société royale d'agriculture de Lyon," *L'esprit des journaux, françois et étrangers par une société de gens-de-lettres* 12 (1784): 258–9; "Prix," *Journal de Paris*, 1784/09/22, 1124–5; Jean-Baptiste Lanoix, *Mémoires sur les fours de boulanger, chauffrés avec du charbon de terre, & plans des mêmes fours, couronnés par la Sociéte royale d'agriculture de Lyon, en l'année 1784* (Bernuset, 1784).

30. J.-P. Pointe, "Jean-Baptiste Lanoix," *Revue du Lyonnais* 22 (1845): 397–419.

31. Veillerette 1987, 95.

32. Werrett 2007, 327–8; Lebon 1801c, 7–8. Compare "Philosophical fire," *The Scots Magazine* 50 (1788): 164.

33. Veillerette 1987, 95.

34. Ibid., 103–4.

35. Ibid., 136, quoting a Lebon paper on "Disposition favorable à donner aux machine à feu" (1791; in archives of École nationale supérieure des Ponts et Chaussés). Caution should be exercised in regard to Veilleret, as his quotes from Lebon's patent indicate some unacknowledged editing and errors. He also misdates the diagram from Lebon's patent.

36. Barthélemy Faujas-de-Saint-Fond, *Essai sur le goudron du charbon de terre, sur la manière de l'employer pour caréner les Vaisseux, & celle d'un faire usage dans plusieurs arts* (Imprimerie Royale, 1790). See also Genssane 1770. Cochrane's treatise was also translated into French.

37. Williot 1999, 15.

38. Veillerette 1987, 134.

39. Williot 1999, 15; Veillerette 1987, 135.

40. Lebon 1801c, 1. Title of paper given in Veillerette 1987, 138: "Moyens nouveaux d'employer les combustibles plus utilement & à la chaleur & à la lumière, & d'en recueillir leur sous-produits"; Lebon 1799.

41. Lebon 1799, 234.

42. Marchais 1801, 384. Lampadius got hold of a thermolamp in 1799 and performed experiments with it in Leipzig. See Lampadius 1816, vol. 1, 118.

43. Lebon 1801a, 240, 244.

44. Ibid., 243, 246.

45. Veillerette 1987, 164–5.

46. "Description d'un thermolampe, perfectionné par M. Wentzler," *Bulletin de la Société d'encouragement pour l'industrie nationale* 1 (1802): 241–3; Marchais 1801; "Lebons Thermolampen," 1801; "Nützliche Entdeckungen," *Der Genius des neunzehnten Jahrhunderts* 3 (1801): 261–2; "Neue Erfindungen," *Oekonomische Hefte* 17 (1801): 471–4; "Thermolampe," *Der Verkündiger* 6 (1802): 273–6; "Auszüge aus Briefen und ein Paar Zeitungsartikel," *Annalen der Physik* 10 (1802): 488–510; "Thermolampe und Phlogoscop," *Der Genius des neunzehnten Jahrhunderts* 5 (1802): 298–305; "Thermolampen oder Oefen, die mit Oekonomie erwärmen und zugleich erleuchten," *Französische Annalen für die allgemeine Naturgeschichte, Physik, Chemie, Physiologie und ihre gemeinnützigen Anwendungen* 1 (1802): 47–59; "Notizen: Lebons thermolampe," *Allgemeines Journal der Chemie* 9 (1802): 349–55, 582–6; "Ausführliche Beschreibung der Thermo-Lampe," *Oekonomische Journal* 1 (1802): 466–9; Carl Gottfried Bünger, *Abbildung und Beschreibung einer Thermolampe* (Friedrich G. Pinther, 1802); Johann Michael Daisenberger, *Beschreibung der Daisenberger'schen Thermolampe* (Stadtamhof, 1802); F. Scherer, "Thermolampe," *Allgemeines Journal der Chemie* 9 (1802): 582–6; Wagner 1802; Johannes Baptista Wenzler, *Beschreibung einer Thermo-Lampe, oder eines Leucht- und Spar-Ofens* (Ambrosi, 1802). These works refer to many others.

47. "Neue Erfindungen," *Oekonomische Hefte* 17 (1801): 471–4, 473; Williot 1999, 18.

48. "Lebons Thermolampen," 1801. The footnote in Wagner 1802 quotes at some length from German journals describing both Lebon's apparatus and his demonstrations. See also Williot 1999, 21.

49. Lebon 1801b.

50. Blagdon 1803, 23.

51. "Lebons Thermolampen," 1801, 211. See also Wagner 1802, 497; "Versuch mit der Thermo-Lampe," *Magazin der Handels- und Gewerbskunde* 1 (1803): 514–5; A. J. B. Defauconpret, *Six mois a Londres en 1816* (Eymery and Delaunay, 1817), 35.

52. Veillerette 1987, 182. Veillerette says that it was never even opened.

53. Winzler 1803b, 25, 34.

54. Veillerette 1987, 191.

55. "Gilbert, Ludwig," in *Allgemeine deutsche Biographie*, vol. 9 (Historische Commission bei der Königlich Akademie der Wissenschaften, 1875).

56. Wagner 1802, 492, footnote; Ludwig Gilbert, "Ueber die sogenannten Thermolampen und den ersten Erfinder derselben," *Annalen der Physik* 22 (1806): 51–7, 52; Karl Maria Ehrenbert von Moll ("Aus einem Schreiben aus Paris vom 29. Dec. 1803," *Annalen der Berg- und Hüttenkunde* 3, 1805: 309) suggests that there was official interest in the thermolamp, but that it was judged to be dangerous.

57. "Sur la carbonisation du bois, et sur les produits de sa distillation en grand," in *Mémorial forestier, ou recueil complet des lois, arrêtés et instructions relatifs à l'Administration forestière*, ed. Goujon (Arthus-Bertrand, 1810), 234; Antoine-François Fourcroy, Claude-Louis Berthollet, and Louis-Nicolas Vauquelin ("Verkohlung und Erleuchtung im Grossen, mit Thermolampen-Oefen," *Annalen der Physik* 30, 1808: 393–414, 402) make similar remarks about how well known Lebon's thermolamp was.

58. Marchais 1801.

59. Antoine Thillaye-Platel, "Carbonisation de la tourbe," *Annales de chimie* 58 (1806): 128–48, 128–9.

60. "Sociétés savantes. Sociétés de Liége. Nouvel éclairage," *Bulletin de pharmacie* 12 (1811): 573–5.

61. For an exception, see "Lebon's thermolampe in Paris," *The Anti-Jacobin Review and Magazine* 16 (1803): 389–90.

62. "Chauffage au moyen de la vapeur de l'eau chaude," *Bulletin de la Société d'encouragement pour l'industrie nationale* 7 (1808): 89; "Eclairage par le gaz hydrogène," *Bibliothèque physico-économique, instructive et amusante* 8 (1810): 110–1; *Extrait du rapport général sur les travaux du conseil de salubrité pendant l'année 1822, avec des notes et observations pour servir de réponse aux critiques publiées contre l'éclairage par le gas hydrogène* (Ladvocat, 1823), 7.

63. Wagner 1802, 492–4, footnote.

64. Veillerette 1987, 206–9.

65. *Mercure de France*, 1811/08/3, 240.

66. For discussions of the heroic view, see Rosenberg 1982, 48ff.; Miller 2004, 11–26; MacLeod 2007.

67. Ludwig Gilbert, "Ueber die sogenannten Thermolampen und den ersten Erfinder der-selben," *Annalen der Physik* 22 (1806): 51–7, 51. Like Gilbert, others also commented that Lebon's work was not original and relied on widely availble chemical knowledge. Poppe 1811, 21 and "Thermolampe," in *Handwörterbuch der Naturlehre*, ed. K. P. Funke, vol. 2 (C. F. E. Richter, 1805), 289.

68. Friedrich Kretschmar, "Zusatz des Herausbegers. Einige Erfahrungem enthaltend, über die Thermolampe, und deren Answendungen," *Annalen der Physik* 22 (1806): 83–95, 83–4. Others who worked on the thermolamp include Daisenberger, Bünger, Kretschmar, Griesch, Wenzler.

69. A longstanding source of confusion in the historiography of the early gas industry is the existence of three people with similar surnames: Zachäus (or Zachaeus) Andreas Winzler, Johannes Wenzler (who built a small thermolamp in 1802), and Friedrich Winzer (who went on to promote the first successful gas company on London and anglicized his name to Fred-erick Winsor). Winzler was from Ulingen in Swabia; Winsor was from Braunschweig.

70. Much of the following biographical information is from "Winzler, Zachäus Andr." 1835 and from HalvaDenk 1991.

71. "Winzler, Zachäus Andr." 1835, 165; Georg Christoph Hamberger, "Winzler (Zachäus Andreas)," in *Das gelehrte Teutschland oder Lexikon der jetzt lebenden teutschen Schriftsteller*, ed. J. S. Ersch (Meyer, 1835), 633; HalvaDenk 1991, 4.

72. Winzler 1803b, 13. For a description, see "Auszüge aus Briefen und ein Paar Zeitungsar-tikel," *Annalen der Physik* 10 (1802): 488–510, 496–7, footnote. "Lebons Thermolampen" (1801) has a plate with simple drawings. See also "Ausführliche Beschreibung der Thermo-Lampe," *Oekonomische Journal* 1 (1802): 466–9, but note that it is not clear whether the thermolamp depicted is Lebon's.

73. Winzler 1803b, 13.

74. He cites Lavoisier, Fourcroy, Guyton, Chaptal, Nicholson, Scheele, Priestley, Hales, and Volta, as well as secondary figures (Minckelers, Leutmann, Göttling, Lampadius, Nordenschöld, et al.).

75. Winzler 1803b, 122–3, 131–2

76. Winzler 1803b, 84–5, 145.

77. An ypfer was a graphite crucible made in Passau. See Berzelius 1841, vol. 10, 521.

78. Winzler 1803b, 15–16.

79. Peter Woulfe, "Experiments on the distillation of acids, volatile alkalies, &c.," *Philosophical Transactions* 57 (1767): 517–36.

80. Winzler 1803b, 38, 77–82, 149.

81. Ibid., 20.

82. "Winzler, Zachäus Andr." 1835.

83. Winzler 1803b, 21.

84. Ibid., 38.

85. Ibid., 39; Hatwagner 2008, 75–6. See also J. M. Zwicker, *Der patriotische Forstmann* (Polt, 1805), 208.

86. "Winzler, Zachäus Andr." 1835, 165.

87. Winzler 1803b, 23.

88. Winzler 1803a.

89. Winzler 1803b, 177–180.

90. Ibid., 41.

91. "Art. IX. Gemeinnüzige Anzeigen," *Gnädigst-privilegirtes Leipziger Intelligenz-Blatt* (1803), 304–5; HalvaDenk 1991, 4; *National-Zeitung der Teutschen*, 1803/11/22, 855–6; "Économie forestière: Améliorations, économie du combustible," in *Annales forestières, faisant suite au Mémorial forestier* (Arthus-Bertrand, 1808), 231.

92. "Correspondance. Suite de la correspondance de M. Boudet, Pharmacien en chef," *Bulletin de pharmacie* 2 (1810): 283–8, 286.

93. zu Salm-Reiferscheid 1808. *Vaterländische Blätter für den österreichischen Kaiserstaat* (1812/12/02, 581–3) describes the Blansko thermolamp as the first large one. For many details, see Hollunder 1824, 5–25.

94. Wankel 1882, 112–3. Wankel's chronology is hard to reconcile with Hugo zu Salm's article in the *Annalen*. I assume that the account in the *Annalen* is correct.

95. Wankel 1882, 75, 112.

96. "Salm, Hugo Franz Altgraf" 1836; Gustav Trautenberger, *Die Chronik der Landeshauptstadt Brünn,* volume 4 (Josef Alder., 1897), 230.

97. Kennelly 1928, 9; Hollunder 1824, 12-3, 457. The Klafter was first a unit of length. The Austrian Klafter is defined as 1.896 meter, while the Viennese Klafter was 1.791 meter. Secondarily, the Klafter was also a unit of volume used to measure firewood. One Klafter of firewood was a square 6-Fuss block (where 6 Fuss was a linear Klafter) and, variously 2, 2½, or 3 Fuss high.

98. zu Salm-Reiferscheid 1808; Wankel 1882, 117.

99. *Vaterländische Blätter für den österreichischen Kaiserstaat,* 1810/08/31, 291–2; *Allgemeine Literatur-Zeitung,* 1811/02/19, 155; Hollunder 1824, 12–3; Wankel 1882, 113–4.

100. "Hugo Franz Altgraf zu Salm-Reifferscheid-Krautheim, Herr der Mährische herrschaften Raitz, Jedownitz und des Ollmützer Lehens Blansko," 1840, 580.

101. *Vaterländische Blätter für den österreichischen Kaiserstaat,* 1810/08/31, 291–2; "Hugo Franz Altgraf zu Salm-Reifferscheid-Krautheim, Herr der Mährische herrschaften Raitz, Jedownitz und des Ollmützer Lehens Blansko," 1840.

102. "Winzler, Zachäus Andr." 1835, 165; *Allgemeine Literatur-Zeitung,* 1817/09, 814; "Thermolampe," in *Oekonomisch-technologische Encyklopädie,* vol. 183, ed. W. D. Korth (Leopold Wilhelm Krause, 1844), 255; *Vaterländische Blätter für den österreichischen Kaiserstaat,* 1812/12/02, 581–3; "Dritte Classe der Kohlen. Die thermolampen Kohlen," in *Darstellung des fabriks- und gewerbswesens in seinem gegenwärtigen zustande,* ed. S. v. Keess (Mörschner und Jasper, 1824), 61; A. L. Millin, "Nouvelles: Aurtiche," *Magasin encyclopédique* 3 (1812): 411–2; "Winzlers Thermolampe bei Kloster-Neuburg". *Hesperus* (1812), 417–9. He is known to have moved on to Slovenia, but no further information is available. See HalvaDenk 1991, 4.

103. Hollunder 1824, 14ff.

104. Luxbacher 1991, 248; Liebig 1849, vol. 4, 441–2; Knapp 1847, vol. 2, 502–3.

105. "Hugo Franz Altgraf zu Salm-Reifferscheid-Krautheim, Herr der Mährische herrschaften Raitz, Jedownitz und des Ollmützer Lehens Blansko," 1840, 581–3; Luxbacher 1991, 244 ff.; Farrar 2008.

106. *Vaterländische Blätter für den österreichischen Kaiserstaat,* 1812/12/02, 581–3, is quite clear on this point.

107. "Thermolampe," in *Oekonomisch-technologische Encyklopädie*, vol. 183, ed. W. D. Korth (Leopold Wilhelm Krause, 1844), 255; Knapp 1847, 501–3; Liebig 1849, 441–2.

108. "Sur la carbonisation du bois, et sur les produits de sa distillation en grand," *Mémorial forestier, ou recueil complet des lois, arrêtés et instructions relatifs à l'Administration forestière*, ed. Goujon (Arthus-Bertrand, 1809), 234–5; Antoine-François Fourcroy, Claude-Louis Berthollet, and Louis-Nicolas Vauquelin, "Rapport d'un mémoire de MM. Mollerat, concernant la carbonisation du bois en vaisseaux clos, et l'emploi de différens produits quelle fournit," *Annales de chimie* 58 (1808): 128–48; Jobert 1991; Benoit 1990, 104–5.

109. Fourcroy, Berthollet, and Vauquelin 1808a, 176.

110. Murdoch 1808; "Sur l'application du gaz tiré de la houille aux usages économiques," *Annales des Arts et Manufactures* 33 (1809): 66–78; J.-N. Barbier-Vémars, "Sur l'application du gaz tiré de la houille à divers usages économiques, et sur-tout à l'éclairage," *Annales des arts et manufactures* 35 (1810): 292–309; "Notice sur l'application du gaz tiré de la houille aux usages économiques par Mr W Murdoch," *Bibliothèque britannique* 41 (1809): 68–79; "Eclairage économique par le gaz inflammable tiré de la houille par M Murdoch," *Archives des découvertes et des inventions nouvelles* 2 (1810): 312–3.

111. Accum 1815.

112. Accum and Winsor 1816; Accum and Lampadius 1816.

113. See Lampadius' comments in Accum and Lampadius 1816, 68, 105. See also J. Prechtl, *Anleitung zur zweckmässigsten einrichtung der apparate zur beleuchtung mit steinkohlen-gas* (C. Gerold, 1817); Williot 1999; Körting 1963.

114. Williot 1999; Körting 1963.

115. "An account of Improvements made in Gas Light. By Messrs. Sobolewsky and Horrer, at Petersburgh," *The Repertory of Arts, Manufactures, and Agriculture*, second series, 24 (1814): 315–8.

116. Winzler 1803b, 145.

117. Ibid., 38, 81, 149.

118. *Vaterländische Blätter für den österreichischen Kaiserstaat*, 1812/12/02, 582, footnote; "Mr. Winsor's Thermolampen-Ofen in London," *London und Paris* 15 (1805): 109–12, 109.

119. "Thermolampe" 1805, 70.

120. "Account of the Method of Carbonising Turf used by M. A. Thillaye Platel, from the Annales de Chemie [*sic*]," *The Athenaeum* 2 (1807): 187.

121. Maxime Ryss-Poncelet, "Sur l'éclairage par le gaz hydrogène de la houille," *Annales des arts et manufactures* 41 (1811): 53–81, 58.

122. Jean-Antoine Chaptal, *De l'industrie française*, vol. 2 (Renouard, 1819), 52–3.

123. The first German town where wood-gas lighting was used was Munich in 1850. The process was developed by Max Pettenkofer in 1848–9 and implemented in various German cities by L. A. Riedinger. By 1860, however, almost all wood-gas plants had been converted to coal. See Körting 1963, 122–4, 129–30. After that time, wood was used only in gas plants remote from railway lines (ibid., 229).

124. Millward 2005, 15.

125. Mokyr 1999, 45.

126. "Salm, Hugo Franz Altgraf," 1836.

127. Mokyr 2005. Mokyr 2002.

128. O'Brien, Griffiths, and Hunt 1996, 175.

NOTE TO INTRODUCTION TO PART II

1. For a list of nearly 500 mills that adopted gas lighting in the nineteenth century, see West 2008, appendix 3, 235–48.

NOTES TO CHAPTER 3

2. Berg 1994, 177–8; Mokyr 2002, 74; Mokyr 1994, 41; Inkster 1991, 8ff.

3. Mokyr 1999, 36–9; Mokyr 1990, 103–5, 240–1; MacLeod 2004, 124; Mokyr 2009, 107, 113; Mathias 1983, 121–7; Harris 1976; Harris 1998, 554–560; Landes 2003, 61–3; Allen 2009, 54–5, 204.

4. Jacob and Stewart 2004; Stewart 1992; Jacob 1997; Mokyr 1999, 76–81; Mokyr 2002, 65; Cohen 2004.

5. Landes 2003, 84–7, 92; Mokyr 2002, 80–85.

6. Nuvolari 2004, 135; Mathias 1983, 127; Allen 2009, 204. Mokyr (2009, 114–5) lists skills in instruments, naval work, and mining.

7. Mathias 1983, 122, 127; Hills 2002–2006.

8. Cartwright 1967; Miller and Levere 2008; Stansfield and Stansfield 1986; Jay 2009.

9. Lindqvist 1990, 313; Cardwell 1972, 111–2. See also O'Brien, Griffiths, and Hunt 1996, 171ff. Inkster (1991, 39ff.) argues that from 1780 there was an acceleration in technological development based on science. On changes in the nature of innovation at this time see Inkster 2003, 30ff; Bruland and Mowery 2005.

10. Mokyr 2010, 192.

11. Mokyr 2009, 347; Stewart 2007.

12. Morus 1996b; Morus 1998; Marsden and Smith 2005, 56.

13. Miller 2000; Miller 2004; MacLeod 2007.

14. Chandler and Lacey 1949, chapter 3; Elton 1958, 262–8; Griffiths 1992; Clow and Clow 1952.

15. Falkus 1982, 233.

16. Griffiths 1992.

17. Wilson 1995, 28; Wilson and Thomson 2006, 52.

18. Uglow 2002; Schofield 1963; Musson and Robinson 1969.

19. Jones 2008b, chapter 3; Jones 2008a. Wedgwood was only an affiliate, not a member.

20. Levere 2007, 157–9.

21. Barbara M. D. Smith, "Keir, James (1735–1820)," in *DNB*.

22. BWA-MS-3147/3/2 #11; Griffiths 1992, 102.

23. For details, see Griffiths 1992 and John C. Griffiths, "Murdoch, William (1754–1839)," in *DNB*.

24. Griffiths 1992, 221–2.

25. Iron(II) sulfate, used for making dyes and as a mordant.

26. "Specification of the Patent granted to Mr. William Murdock, of Redruth," *The Repertory of Arts and Manufactures*, first series 9 (1798): 97–9.

27. Cornish word for pyrite.

28. Henry 1805, 73, corroborates this. On Murdoch's activities at this time see Griffiths 1992, 224–5.

29. The chronology of the invention in Cornwall was investigated in great detail by Griffiths (1992, 224–249). Griffiths does not completely cite the letters he uses. They are Thomas Wilson to B&W, 1808/01/27 and 1808/01/29, BWA-MS-3147/3/363 #10 and #11.

30. Murdoch 1808, 131.

31. Hall 1809, 51.

32. BWA-MS-3147/3/479 #1.

33. Henry 1805, 74.

34. Hall 1809, 51.

35. JWjr to MRB, 1799/04/04, BWA-MS-3147/47/3/48.

36. On Watt see Hills 2002–2006.

37. Robinson 1954; Jones 1999.

38. Tann 2004b; Tann 2004a; Dickinson 1937, 169–70. There were in fact even more companies. See Dickinson 1937, 209.

39. The Soho Manufactory and the Soho Foundry were both involved in receiving orders for gas plants. The orders seem to go first to the Manufactory (Boulton & Watt), which then ordered parts from the Foundry (Boulton, Watt, & Co.). See Manufactory order book (BWA-MS–3147/4/105 and 106) and Foundry order book (BWA-MS–3147/4/115 and 116).

40. For more general details, see Tann 2004b; Tann 2004a; Robinson 2004; Dickinson 1937; Dickinson 1936; Jones 2008b, 48ff, esp. 54; Miller 2004, 83–9; Marsden and Smith 2005, 45–65.

41. GW to JWjr, 1801/11/08, BWA-JWP-C2/10.

42. JWjr to GW, date unclear but located in the letter book before a letter dated 1801/12/14, BWA-JWP-LB/7.

43. The order was from local merchant Shakespeare, Johnson, & Berry. BWA, Matthew Boulton Correspondence, Soho House MS 1682.

44. Creighton 1824, 449. Coal is about 70 lb/ft^3, and 15 lb has a volume of 0.2147 ft^3, or 7 in^3.

45. Creighton 1824, 449; BWA-MS-3147/3/479 #1.

46. Clegg 1841, 6.

47. Beddoes and Watt 1795, part 2, 27–35.

48. Cartwright 1967; Miller and Levere 2008; Jay 2009; Stansfield and Stansfield 1986.

49. Miller 2004, 50.

50. Stewart 2007, 160. See also Jacob and Stewart 2004, 104–7.

51. Musson and Robinson 1969; Hall 1974, 145ff.; Mokyr 2002, 38; Jacob 2007; Jacob and Stewart 2004, chapter 4.

52. Letter from James Watt to David Brewster, May 1814, in "History of the origin of Mr Watt's improvements on the steam-engine. Contained in a letter from the late James Watt, LL.D. F. R. S. &c. &c. to Dr Brewster," *Edinburgh Philosophical Journal* 2 (1820) 1–7 on 6.

53. According to Creighton (1824, 449), coal gives off 330–360 ft^3 of gas per cwt, or 3 ft^3/lb. A single candle brightness lamp consumed about 0.35–0.45 cubic feet of gas per hour.

54. Levere 2005b, 234–5; Levere 2007, 165–170.

55. See the various financial reports in BWA-MII/7/3.

56. Knight 1996, 53.

57. Clegg 1841, 6.

58. Banks 1823, 3.

59. J. J. Mason, "Lee, George Augustus (1761–1826)," in *DNB*.

60. Philips & Lee to B&W, 1803/07/19, BWA-MS-3147/5/804.

61. Matthews 1827a, 51; Lee testimony on 1809/05/12 in Hall 1809, 38.

62. GAL to JWjr, 1805/03/27, BWA-MIV/L6.

63. For evidence of new work on gas lighting at this time, see BWA-MS-3147/3/479 #1.

64. BWA-MS-3147/3/539 #38.

65. J. Northern, "Experiments on pit coal," *The Monthly Magazine* 19 (1805): 235; William Henry, "Response to Mr. Northern," *The Monthly Magazine* 19 (1805): 313. See also G. J. Wright, "Response to Mr. Northern and Mr. Henry," *The Monthly Magazine* 19 (1805): 427–8.

66. Ewart to JWjr, 1805/06/01, BWA-MIV/E2.

67. JWjr to MRB, 1805/08/17 (BWA-MS-3147/3/54 #5); Philipp Andreas Nemnich, *Neueste Reise durch England, Schottland, und Ireland* (J. G. Cotta, 1807), 127.

68. JWjr to MRB, 1805/11/12, cited in Watt and Muirhead 1854, vol. 2, 303. The original letter is not known. Andrew Ure had also set up a gas lighting in the Andersonian Institution. See Clow and Clow 1952, 400 (figure 88).

69. JW to JWjr, 1805/11/27, BWA-MIV Box 16/14.

70. John Sinclair, *Appendix to the General Report of the Agricultural State and Political Circumstances of Scotland*, vol. 2 (Constable, 1814), 305; Falkus 1982, 230.

71. Fisher 1866, vol. 1, 145–6; "Comments on Mr. Wm. Henry's experiments on the gases obtained from the distillation of wood, peat, and pit-coal," *Retrospect of Philosophical, Mechanical, Chemical and Agricultural Discoveries* 1 (1805): 129–34, 133; *Morning Post,* 1805/06/15.

72. Eidingtoun Hutton to B&W, 1805/08/24, BWA-MS-3147/3/263 #10.

73. JWjr to B&W, 1805/08/30, BWA-MS-3147/3/54 #7; Wood Daintry & Wood, 1805/07/28, BWA-MS-3147/3/478 #46.

74. The storeys were called K, Twist, L, M, N and O. Notes 1806/06/28, BWA-MS-3147/3/478 #13; Eidingtoun Hutton to B&W, 1805/09/09, BWA-MS-3147/3/263 #12; Eidingtoun Hutton to B&W, 1805/09/12, BWA-MS-3147/3/263 #13; Ground plans, 1805/09/14, BWA-MS-3147/5/804/11.

75. JWjr to John Southern, 1806/03/01 (BWA-MS-3147/3/60 #9) gives 7 feet as the diameter, as does JWjr notebook, 1806/03 (BWA-MS-3147/4/5, 54 and 52). Southern to JWjr, 1806/03/09 (BWA-MS-3147/3/479 #1) gives the surface area as 140 ft^2, which corresponds to a height of 5.25 ft. See Memo, 1806/03, BWA-MS-3147/3/479 #4.

76. If Creighton's *Britannica* diagram represents the correct proportions. Corroborated by Soho Foundry order book, order for Philips & Lee, 1805/10/27, BWA-MS-3147/4/115.

77. Creighton (1824, 449) states about 15 cwt. See also JWjr to John Southern, 1806/03/01, BWA-MS-3147/3/60 #9.

78. Foundry order book, 1805/09/27, 1805/10/27 and 1805/10/30, BWA-MS-3147/4/115; Manufactory order book, 1805/09/27, 1805/10/27 and 1805/10/31, BWA-MS-3147/4/105, 88, 94, 96.

79. Murdoch to B&W, 1805/12/20, BWA-MS-3147/3/289 #16. See also Foundry order book, 1805/12/09, BWA-MS-3147/4/115.

80. Murdoch to B&W, 1805/12/23, BWA-MS-3147/3/289 #17.

81. Clegg 1841, 13; "Materials for a memoir of Mr. Samuel Clegg," *Mechanics' Magazine* 22 (1835): 470–2. See also Awty 1974.

82. Murdoch to B&W, 1806/01/01, BWA-MS-3147/3/289 #18.

83. Other firms that expressed interest in acquiring a gas plant in 1805 include Daintry & Wood and Greg & Ewart. See Eidingtoun Hutton to B&W, 1805/11/03, BWA-MS-3147/3/263 #19. Radcliffe & Ross of Stockport inquired in early November with Hutton: Eidingtoun Hutton to B&W, 1805/11/03, MS 3147/3/263 #19. John Grieve in Edinburgh and Messrs. G. Dunlop & Co. of Glasgow also expressed interest: Eidingtoun Hutton to B&W, 1805/12/05, BWA-MS-3147/3/263 #26.

84. Jay 2009.

85. Jones 2008b; Mason 2009.

86. Marsden and Smith 2005, 57–60. See also Tann 1978.

87. Jacob and Stewart 2004, 100–1; Stewart 2008, 375–9; Stewart 2007, 170.

88. Hills and Pacey 1972.

89. MRB to JWjr, 1806/01/14, BWA-MS-3147/3/42 #16.

90. JWjr to Southern, 1806/03/01, BWA-MS-3147/3/60 #9. This may be the third gasometer, according to the order books (Foundry order book, 1806/01/07, BWA-MS-3147/4/115; Manufactory order book, 1806/01/07, BWA-MS-3147/4/105, 117). Lee ordered a further 60 cockspur lamps (Murdoch to B&W, 1806/02/05, BWA-MS-3147/3/289 #19).

91. Murdoch to B&W, 1806/02/07, BWA-MS-3147/3/289 #20.

92. JWjr to MRB, 1806/02/18, BWA-MS-3147/3/55 #2.

93. JWjr to MRB, 1806/02/21, BWA-MS-3147/3/55 #3.

94. JWjr to MRB, 1806/02/26, BWA-MS-3147/3/55 #4; Greg & Ewart, Radcliffe & Ross, James Kennedy, A & G Murray, Atkinson.

95. JWjr to MRB, 1806/02/26, BWA-MS-3147/3/55 #4.

96. JWjr to MRB, 1806/02/26, BWA-MS-3147/3/55 #4.

97. JWjr to JW, 1806/02/27, BWA-MII/13/1.

98. JW to JWjr, 1806/02/28, BWA-JWP-LB/4, 134.

99. JWjr to MRB, 1806/03/03, BWA-MS-3147/3/55 #5.

100. JWjr to MRB, 1806/03/12, BWA-MS-3147/3/55 #8. Hutton was sent to collect payment.

101. JWjr to MRB, 1806/03/03, BWA-MS-3147/3/55 #5; J. Douglas & Co. (Holywell): JWjr to MRB, 1806/03/08, BWA-MS-3147/3/55 #7; Daintry & Co.: JWjr to MRB, 1806/03/12, BWA-MS-3147/3/55 #8. See also JWjr notebook, BWA-MS-3147/4/5, which has sizing calculations from late February and early March 1806 for Oldham, Birley & Marsland, James Kennedy, Greg & Ewart, McConnel & Kennedy, Pooley, Wood & Daintry, Douglas (Pendleton and Holywell), Garside & Butterfield, Horrocks, Strutt, Peel, Wormald & Gott (several mills), Radcliffe & Ross, and a few illegible ones.

102. MRB to JWjr, 1806/03/09, BWA-MIV/B6.

103. MRB to JWjr, 1806/03/09, BWA-MIV/B6.

104. These were Wormald, Gott, & Wormald (Leeds), Watson Ainsworth & Co. (Preston), and S. Horrocks (Preston): JWjr to JW, 1806/03/19, BWA-MII/13/1. The firm of William Strutt & Co. (Derby) also placed an order at this time; see drawing, 1806/04/26, BWA-MS-3147/3/478 #34. One of the Strutts lit his house with gas. See John Farey, *General View of the Agriculture*

of the County of Derbyshire Drawn Up for the Consideration of the Board of Agriculture and Internal Improvement, vol. 3 (Sherwood, Neely and Jones, 1817), 197.

105. JWjr to JW, 1806/03/19, BWA-MII/13/1.

106. Henry Creighton to B&W, 1806/03/25, BWA-MS-3147/3/247 #7.

107. JWjr to MRB, 1806/02/26, BWA-MS-3147/3/55 #4; Peter Marsland (Stockport): 1806/03, BWA-MS-3147/3/478 #26; JWjr to MRB, 1806/03/03, BWA-MS-3147/3/55 #5; Radcliffe & Ross (Stockport): JWjr to MRB, 1806/03/03, BWA-MS-3147/3/55 #5; McConnel & Kennedy (Manchester): 1806/02, BWA-MS-3147/3/478 #19; 1806/03/06, BWA-MS-3147/5/821/34; Greg & Ewart (Manchester): 1806/02, BWA-MS-3147/3/478 #3; JWjr to MRB, 1806/02/26, BWA-MS-3147/3/55 #4; Daintry & Co. (Macclesfield): JWjr to MRB, 1806/03/12, BWA-MS-3147/3/55 #8; J. Douglas (Pendleton): 1806/03, BWA-MS-3147/3/478 #1. For drawings and notes see 1806/04, MS 3147/5/812/14 and 16; notes, 1806/04, BWA-MS-3147/3/478 #47 and 48a; Henry Creighton to B&W, 1806/04/07, BWA-MS-3147/3/247 #8; Henry Creighton to B&W, 1806/04/12, BWA-MS-3147/3/247 #9.

108. Henry Creighton to B&W, 1806/04/14, BWA-MS-3147/3/247 #11; Eidingtoun Hutton to B&W, 1806/04/14, BWA-MS-3147/3/264 #1; Eidingtoun Hutton to B&W, 1806/06/27, BWA-MS-3147/3/264 #22; Samuel Oldknow's Mellor Mill: 1806/04/26, BWA-MS-3147/3/478 #30; W. G. J. Strutt: 1806/04/26, BWA-MS-3147/3/478 #34; Douglas & Co.: 1806/05/24, BWA-MS-3147/3/478 #2; Penson at Wigan: Henry Creighton to B&W, 1806/05/09, BWA-MS-3147/3/247 #12; 1806/05/24, BWA-MS-3147/3/478 #31; Strutt's mills in Derby: 1806/06/02, BWA-MS-3147/3/478 #35; Horrocks: 1806/06/27, BWA-MS-3147/3/478 #5; Watson Ainsworth: Henry Creighton to B&W, 1806/06/29, BWA-MS-3147/3/247 #17; Drawings, 1806/06, BWA-MS-3147/3/478 #44; For J. Douglas & Co see 1806/05/24–7, BWA-MS-3147/3/478 #1–2; Notes 1806/06/28, BWA-MS-3147/3/478 #13. This document refers to Douglas & Co.'s Old and New Mills at Pendleton.

109. Eidingtoun Hutton to B&W, 1806/07/12 and 1806/07/26, BWA-MS-3147/3/264 #27 and 29.

110. A small gaslight apparatus was built for at least one customer (Ridgeway). It was an experimental model rather than a production model. Ridgeway does not turn up anywhere in Boulton & Watt's accounts as a gaslight customer. Henry Creighton to B&W, 1806/06/14, BWA-MS-3147/3/247 #16; Henry Creighton to B&W, 1806/09/18, BWA-MS-3147/3/247 #24.

111. Henry Creighton to B&W, 1806/06/29 and 1806/07/22, BWA-MS-3147/3/247 #17 and #21.

112. Drawing for Joyce Cooper (Staverton) 1807/05/02, BWA-MS-3147/3/478 #6.

113. Benjamin Gott to JWjr, 1807/07/06, BWA-MIV/G2.

114. Eidingtoun Hutton to B&W, 1807/10/08, BWA-MS-3147/3/265 #26.

115. James Kennedy (Manchester), 1808/05/14, BWA-MS-3147/3/478 #8; Extras per Agreement 30 Sep 1806–30 Sep 1807, BWA-MII/7/4.

116. Murdoch to B&W, 1806/02/05, BWA-MS-3147/3/289 #19.

117. JWjr to JW, 1806/02/27, BWA-MII/13/1.

118. Eidingtoun Hutton to B&W, 1806/02/11, BWA-MS-3147/3/264 #8. He asked again a month later (Eidingtoun Hutton to B&W, 1806/03/15, BWA-MS-3147/3/264 #13).

119. James Watt jr to Southern, 1806/03/01, BWA-MS-3147/3/60 #9.

120. Watt Jr. mentions the two letters in JWjr to MRB, 1806/03/12, BWA-MS-3147/3/55 #8.

121. Southern to JWjr, 1806/03/09, BWA-MS-3147/3/479 #2.

122. Clegg 1841, 70.

123. Henry Creighton to B&W, 1808/01/25, BWA MS 3147/3/247 #43, Answer #1, 2, 10, 20. See also Lee's testimony on 1809/05/12, "Select Committee on Gas-Light," 44, and Watt Jr. on 1809/05/13, 53.

124. The drawings are: BWA-MS-3147/5/804/5–7. The drawings are not dated, but Lee refers to the eight gasometers represented there in another letter: GAL to John Southern, 1806/08/11, BWA-MS-3147/3/478 #14.

125. GAL to Southern, 1806/08/11, BWA-MS-3147/3/478 #14.

126. Memo about McConnel & Kennedy's apparatus, 1806/09/29, BWA-MS-3147/3/478 #20.

127. Murdoch memo, 1806/06, BWA-MS-3147/3/478 #12. Orders in the Foundry and Manufactory order books: 1806/03/01–04, BWA-MS-3147/4/105, 129–36.

128. Murdoch to B&W, 1806/01/01, BWA-MS-3147/3/289 #18.

129. Henry Creighton to B&W, 1807/12/28, BWA-MS-3147/3/247 #40a.

130. GAL to JWjr, 1808/12/07, BWA-MIV/L6.

131. Henry Creighton to B&W, 1807/12/28, BWA-MS-3147/3/247 #40a.

132. Henry Creighton to B&W, 1808/01/25, BWA-MS-3147/3/247 #43 [hereafter CA] Answer #11.

133. JWjr to Henry Creighton, 1808/01/19, BWA-MS-3147/3/478 #17.

134. CA Answer #11.

135. CA Answer #1 and 2, #20. See also Lee's testimony 1809/05/12, Hall 1809, 44, and Watt Jr. 1809/05/13, 53.

136. CA Answer #10. About £100 today.

137. CA Answer #1 and 2.

138. CA Answer #6 and 7.

139. GAL to B&W, 1807/12/28, BWA-MS-3147/3/247 #40b; GAL to JWjr, 1808/01/20, BWA-MS-3147/3/478 #18.

140. CA Answers #8 and #9. It is frequently stated that Samuel Clegg invented the hydraulic main, but these traps were hydraulic mains.

141. CA Answer #8 and #9.

142. Henry Creighton to B&W, 1808/01/10, BWA-MS-3147/3/247 #41.

143. CA Answer #12 and #13.

144. CA Answer #12, #13, and #14.

145. CA Answer #15.

146. GAL to JWjr, 1808/12/07, BWA-MIV/L6.

147. For a summary of these arguments, see West 2008, 171–83.

148. For a description of the method see Accum 1815, 22–7. For the confusion that resulted, see, e.g., M. Ricardo, *Observations on the advantages of oil gas establishments* (C. Baldwin, 1823), 7. Doubts were also expressed by the GLCC. See GLCC MCD-b1, 319, 1814/05/06.

149. Eidingtoun Hutton to B&W, 1805/07/01, BWA-MS-3147/3/263 #7.

150. Henry 1805; Henry 1808.

151. For these experiments, the coal gas yield was calculated at 100 pounds per hundredweight, rather than the 112 lb/cwt used in all other experiments. I have changed these values to 112/cwt for consistency. The sources of data in table 3.1 are as follows. Lines 1–2: Retort capacity 15 lb of coal 1808/01/28, BWA-MS-3147/3/479 #12, 1; JWjr notebook 1805/07/13–15, BWA-MS-3147/4/5, 55; Line 3: Retort capacity 8 cwt: JWjr notebook, 1805/07/13–15, BWA-MS-3147/4/5, 54–55 and 1808/01/28, BWA-MS-3147/3/479 #12; Line 4: 1808/01/28, BWA-MS-3147/3/479 #12.

152. JWjr to GAL, 1808/01/23, BWA-MS-3147/3/478 #17.

153. JWjr to John Southern, 1806/03/01, BWA-MS-3147/3/60 #9; JWjr to GAL, 1808/01/23, BWA-MS-3147/3/478 #17; 1808/01/28, BWA-MS-3147/3/479 #12; JWjr notebook, 1806/03, BWA-MS-3147/4/5, 52, 54–55; See also Austerfield 1981, 414–5.

154. Notebook, 1808/01/28, BWA-MS-3147/3/479 #12, 1.

155. Memorandum respecting Mr. Lee's Photogenous Apparatus, 1807/06/02, BWA-MS-3147/3/478 #16.

156. JW to JWjr, 1807/06/09, BWA-MIV Box 16/15.

157. All data in table 3.3 are from Notebook, 1808/01/28, BWA-MS-3147/3/479 #12. These findings were obviously very promising for Wigan cannel coal, and Boulton & Watt must have communicated the result to Lee because James Lawson informed Watt Jr. at the end of March 1807 that Lee had ordered "Wygan" coal and was going to try it out. See James Lawson to JWjr, 1807/03/30, BWA-MIV/L3.

158. The small retort held 14 pounds of coal. JWjr to Henry Creighton, 1808/01/19 (BWA-MS-3147/3/478 #17) mentions that the small oven he used at Soho had already burned out. See also 1807, BWA-MS-3147/5/821/19. The notes here state it "burnt out soon." It is filed with McConnel & Kennedy, but this is clearly the experimental retort for the Soho Foundry. The Soho Foundry account sheet lists 50% of the cost of an experimental retort expense on 1807/12/31 for £48.12.6. See Soho Foundry Amounts of Sales of Goods manufactured 30 Sep 1807 to 30 Sep 1808, BWA-MII/7/4. Source of table 3.4: data listed in Synopsis of Photogenous Experiments with the Small Retort 1807, BWA-MS-3147/3/479 #10.

159. Henry Creighton to B&W, 1808/01/25, BWA-MS-3147/3/247 #43, 3; Experiments on the new light, 1808/01, BWA-MS-3147/3/479 #11. These are press copies of notes taken during the experiments. They are almost completely illegible. Henry Creighton to B&W, 1807/12/28, BWA-MS-3147/3/247 #40a.

160. Source of data in table 3.5: Henry Creighton to B&W, 1808/01/25, BWA-MS-3147/3/247 #43, 3; Experiments on the new light 1808/01, BWA-MS-3147/3/479 #11.

161. GAL to B&W, 1807/12/28, BWA-MS-3147/3/247 #40b.

162. Sources of data in table 3.6 are as follows. Lines 1–2: GAL to JWjr, 1808/02/04, BWA-MS-3147/3/478 #18; Lines 3–7: Henry Creighton to B&W, 1808/01/10 and 25, BWA-MS-3147/3/247 #41 and #43, 3.

163. Henry Creighton to B&W, 1808/01/10, BWA-MS-3147/3/247 #41; Experiments on the new light 1808/01, BWA-MS-3147/3/479 #11.

164. JWjr to Henry Creighton, 1808/01/19, BWA-MS-3147/3/478 #17; Henry Creighton to B&W, 1808/01/25, BWA-MS-3147/3/247 #43 [hereafter CA Answer] #11.

165. CA Answer #17.

166. JWjr to Henry Creighton, 1808/01/19, BWA-MS-3147/3/478 #17: question #19.

167. CA Answer #19.

168. Memorandum respecting Mr. Lee's Photogenous Apparatus, 1807/06/02, BWA-MS-3147/3/478 #16. Creighton's new figures are in Henry Creighton to B&W, 1808/01/25, BWA-MS-3147/3/478 #18, 4–5.

169. Memorandum respecting Mr. Lee's Photogenous Apparatus, 1808/01/19, BWA-MS-3147/3/478 #18. This letter gives 629 cockspur and 275 Argand, but Creighton corrected himself in Henry Creighton to B&W, 1808/01/25, BWA-MS-3147/3/478 #18, 4–5. The correct values were used in Murdoch's 1808 Royal Society paper.

170. Murdoch 1808, 127–8. Creighton had actually recommended using £600 instead of £677 as yearly depreciation for this reason. Memorandum respecting Mr. Lee's Photogenous Apparatus, 1808/01/19, BWA-MS-3147/3/478 #18.

171. Edwards 1989, 83–5.

172. GAL to JWjr, 1808/01/20, BWA-MS-3147/3/478 #18.

173. JWjr to GAL, 1808/01/23, BWA-MS-3147/3/478 #17.

174. Henry Creighton to B&W, 1808/01/25, BWA-MS-3147/3/247 #43. 3750 ft^3 of gas per day, requiring 10.75 cwt. The total yearly cost of coal was £78, and the depreciation on £5,000 at 12.5% was £625, for a total yearly cost of £703. Tallow cost was £3,000, oil £1800. Net gain £2297.

175. JWjr to GAL, 1808/01/23, BWA-MS-3147/3/478 #17.

176. GAL to JWjr, 1808/01/25, BWA-MS-3147/3/478 #18.

177. JWjr to Henry Creighton, 1808/01/27, BWA-MS-3147/3/478 #18.

178. Henry Creighton to B&W, 1808/01/29, BWA-MS-3147/3/247 #44, also mentioned in BWA-MS-3147/3/478 #18.

179. GAL to JWjr, 1808/02/04, BWA-MS-3147/3/478 #18.

180. Murdoch 1808, 126.

181. Watt Jr.'s testimony on 1809/05/13 in Hall 1809, 52.

182. Henry Creighton to B&W, 1809/02/23, BWA-MS-3147/3/247 #66.

183. Henry Creighton to B&W, 1808/01/10, BWA-MS-3147/3/247 #41.

184. William Balston to B&W, 1807/01/23, BWA-MS-3147/3/42 #24.

185. Josiah Wedgwood to JWjr, 1807/06/03, BWA-MS-3147/3/529 #64.

186. Henry Creighton to B&W, 1808/01/10, BWA-MS-3147/3/247 #41.

187. Henry Creighton to B&W, 1808/01/29, BWA-MS-3147/3/247 #44.

188. Gasometer pit 1808/10/17, BWA-MS-3147/5/817/2.

189. Henry Creighton to B&W, 1808/01/29, BWA-MS-3147/3/247 #44.

190. Henry Creighton to B&W, 1808/04/16, BWA-MS-3147/3/247 #49.

191. Kennedy: Gasometer pit, Movable water traps, Retort section 1808/10/17, BWA-MS-3147/5/817/2, 4 and 6 and 1808, BWA-MS-3147/5/817/10; Birley: Retort and gasometer pit, 1808/09/19, BWA-MS-3147/5/817/5; Gasometer 1808/10/03, BWA-MS-3147/5/817/2, 9; 1808, BWA-MS-3147/5/817/1, 9; McConnel & Kennedy: Gasometer suspending apparatus, 1808/09/28, BWA-MS-3147/5/821/12; Gasometer suspending apparatus, 1808/10/05, BWA-MS-3147/5/821/10; JWjr to MRB, 1808/09/20, BWA-MS-3147/3/57 #2.

192. Lighting apparatuses per agreement 30 Sep 1808 to 30 Sep 1809, BWA-MII/7/4; This records a sale to Birley on 1808/11/28 for £700 and 1808/12/31 for £79; Another to Kennedy on 1808/12/30.

193. MRB to JWjr, 1808/11/21, BWA-MIV/B6; JWjr to John Southern, 1808/12/09, BWA-MS-3147/3/60 #15; Retorts, 1808/12/16, BWA-MS-3147/5/813/9; Gasometer and plans, 1808/12/20, BWA-MS-3147/5/812/1 and 6; Plan of Burley mill, 1808/12/21, BWA-MS-3147/3/478 #48.

194. Henry Creighton to B&W, 1809/02/10, BWA-MS-3147/3/247 #64; Henry Creighton to B&W, 1809/02/14, BWA-MS-3147/3/247 #65.

195. Henry Creighton to B&W, 1809/02/23, BWA-MS-3147/3/247 #66.

196. Lighting apparatuses per agreement 30 Sep 1809 to 30 Sep 1810, BWA-MII/7/4: addition 1809/12/30 for £93; Bill for three retorts, 1810/08/11, BWA-MS-3147/5/817/14; Lighting apparatuses per agreement 30 Sep 1809 to 30 Sep 1810, BWA-MII/7/4 shows two retorts on 1810/08/17 for £25.17, and retort and additional apparatus on 1810/09/18 for £61.2.

197. "The imperial tourists," *The Literary Gazette and Journal of Belles Lettres, Arts, Sciences, etc for the Year 1818* 2 (1818): 113–4. See also Johann Conrad Fischer, *Tagebuch einer im Jahr 1814 gemachten Reise über Paris nach London und einigen Fabrikstädten Englands vorzüglich in technologischer Hinsicht* (Heinrich Remigius Sauerländer, 1816), 124–35.

198. Lighting apparatuses per agreement 30 Sep 1808 to 30 Sep 1809, 1809/03/02, BWA-MII/7/4.

199. Henry Creighton to B&W, 1809/07/07, BWA-MS-3147/3/247 #70; Cross section, 1809/07/08, BWA-MS-3147/5/820/1; Wormald, Gott & c. to B&W, 1809/03/31, BWA-MS-3147/3/439 #124; Soho Foundry Amounts of Sales of Goods manufactured 30 Sep 1808 to 30 Sep 1809, 1809/08/31, BWA-MII/7/4; Gott to JWjr, 1809/09/01, BWA-MIV/G2.

200. Gasometer, 1810/07/02, BWA-MS-3147/5/812/3; Bill, 1810/07/20, BWA-MS-3147/5/812/3; Lighting apparatuses per agreement 30 Sep 1809 to 30 Sep 1810, 1810/09/29, BWA-MII/7/4.

201. Lighting apparatuses per agreement 30 Sep 1810 to 30 Sep 1811, 1811/03/09, BWA-MII/7/5; Four retorts: 1811/10/22, BWA-MS-3147/5/812/5; Gasometer: 1811/12/02, BWA-MS-3147/5/812/5.

202. Lighting apparatuses per agreement 30 Sep 1809 to 30 Sep 1810, 1809/08/16, BWA-MII/7/4; Lighting apparatuses per agreement 30 Sep 1810 to 30 Sep 1811, 1811/01/21, BWA-MII/7/5; "Gas lights used in the manufactories in Scotland," *The Tradesman, or, Commercial Magazine* 4 (1810): 547; Retort: 1809/05/04, BWA-MS-3147/5/809/1; Southern to JWjr, 1809/05/05, BWA-MS-3147/3/332 #9; Eidingtoun Hutton to B&W, 1809/08/06, BWA-MS-3147/3/267 #17; Eidingtoun Hutton to B&W, 1809/12/19, BWA-MS-3147/3/267 #26.

203. Drawing 1809/07/29, BWA-MS-3147/3/478 #25; Henry Creighton to B&W, 1809/08/02, BWA-MS-3147/3/247 #72; 1809/08/29, BWA-MS-3147/5/811/9; Lighting apparatuses per agreement 30 Sep 1809 to 30 Sep 1810, 1809/12/14, 1810/02/09, and 1810/02/19, BWA-MII/7/4.

204. 1809/09/07, BWA-MS-3147/5/821/50; Lighting apparatuses per agreement 30 Sep 1809 to 30 Sep 1810, 1810/04/27, BWA-MII/7/4 for £690; Lighting apparatuses per agreement 30 Sep 1810 to 30 Sep 1811, 1811/08/29, BWA-MII/7/5, for £316.

205. Bill 1810/04/06 and drawings 1810/04/24, BWA-MS-3147/5/813/10, 1, and 7; Lighting apparatuses per agreement 30 Sep 1809 to 30 Sep 1810, 1810/07/26, 1810/09/29, BWA-MII/7/4; Lighting apparatuses per agreement 30 Sep 1810 to 30 Sep 1811, 1811/01/16, BWA-MII/7/5.

206. Gasometer: 1810/06/01, BWA-MS-3147/5/814/1, 1, and 7; Lighting apparatuses per agreement 30 Sep 1810 to 30 Sep 1811, 1810/10/21, 1810/10/30, and 1810/11/26, BWA-MII/7/5.

207. Gasometer: 1810/06/20, BWA-MS-3147/5/815/1, 3, and 4; Order summary 1810/08, BWA-MS-3147/5/815/6; Lighting apparatuses per agreement 30 Sep 1809 to 30 Sep 1810, 1810/08/24, 1810/09/29, 1810/11/30, and 1810/12/24 BWA-MII/7/4.

208. Ground plans, 1811/04/07, BWA-MS-3147/5/805/3; Bill for retort, 1811/09/09, BWA-MS-3147/5/805/3; Lighting apparatuses per agreement 30 Sep 1810 to 30 Sep 1811, 1811/09/24, BWA-MII/7/5. There was a fire in Benyon and Bage's factory in 1814. See *Times*, 1814/12/29.

209. Pipe design 1811/07/09, BWA-MS-3147/5/805/3. Lighting apparatuses per agreement 30 Sep 1810 to 30 Sep 1811, 1811/07/26, BWA-MII/7/5.

210. JWjr to B&W, 1812/12/30, BWA-MS-3147/3/61 #18; Order details, 1811/03/01, BWA-MS-3147/5/806/3; Ground plan, 1812/03/19, BWA-MS-3147/5/806/3; Lighting apparatuses per agreement 30 Sep 1810 to 30 Sep 1811, 1811/09/10, BWA-MII/7/5 for £195; Lighting apparatuses per agreement 30 Sep 1811 to 30 Sep 1812, 1812/02/13, BWA-MII/7/5 for £292.

211. Goods sold from 30 Sep 1813 to 30 Sep 1814, BWA-MII/7/5; Drawing, 1814/08/14, BWA-MS-3147/3/478 #45; Pipes, BWA-MS-3147/5/807/4.

212. Notes, 1815/02/14, BWA-MS-3147/5/808/1; Ground plans, 1814/11/18, BWA-MS-3147/5/808/2 and 3.

213. BWA-MII/7/6.

214. Ground plans, 1815/03/15, BWA-MS-3147/5/808/2 and 3.

215. JWjr to MRB, 1809/04/20, BWA-Lunar Society #118.

216. "Specification of the patent granted to Edward Heard, of London, chemist," *The Repertory of Arts, Manufactures, and Agriculture*, second series 10 (1807): 31–2; "Mr. Edward Heard's discovery," *The Monthly Magazine* 23 (1807): 67.

217. Clegg (1841, 13) claimed that they never used it, but he is partial to his father. Matthews (1827a, 25) claimed that Boulton & Watt used quick-lime at some point, but "very imperfectly." I have found no mention of lime purification in the archives.

218. GAL to JWjr, 1813/10/28, BWA-MIV/L6. Lee would go on to implement lime purification at his mill without Boulton & Watt's collaboration. William Henry, "Experiments on the gas from coal, chiefly with a view to its practical application," *Memoirs of the Literary and Philosophical Society of Manchester* second series, 3 (1819): 391–429, 413.

219. Divall and Johnston 2000, 80; Rosenberg 1994, 191–200; Rosenberg, Landau, and Mowery 1992, 76–116.

220. Wright 1992.

NOTES TO INTRODUCTION TO PART III

1. August Niemeyer, *Beobachtungen und Erfahrungen auf einer Reise nach England im Jahr 1819*, second edition (Hallische Waisenhaus, 1822), 349–50.

2. Buchanan 1986.

3. Ward 1974; Albert, Freeman, and Aldcroft 1983; Szostak 1991.

NOTES TO CHAPTER 4

4. Mokyr 2010, 194–5.

5. Winsor 1804a, 14; Winsor 1806.

6. Ball and Sunderland 2001, 266–8; Millward 2005, 15.

7. Falkus 1977, 140; Millward 1991, 99; Ball and Sunderland 2001, 265.

8. Millward 1991, 98–9; Millward 2005, 43–4; Ball and Sunderland 2001, 33.

9. *An act [50 Geo.III cap. clxiii] for granting certain Powers and Authorities to a Company* 1810, Clause XXV, XXVII.

10. Pollard 1964; Neal 1994, 151–2; Hudson 1986, 262; Quinn 2004, 160; Mathias 1983, 145; Von Tunzelmann 1993; Deane 1979, 180–1; Cottrell 1980, 10–11, 34–35; Harris 2000, 5–6, 8–11; Wilson 1995, 46. For a survey by sector, see Harris 2000, 173–183.

11. See Mokyr 2009, 232 and most of the sources in the preceding note.

12. Inkster 2003, 36.

13. Harris 2000, 85–100, 182–3; Mathias 1983, 145; Ward 1974; Trew 2010.

14. Harris 2000, 9–10, 99; Wilson 1995, 48.

15. Trew 2010; Pearson and Richardson 2001, 659–660; Harris 2000, 85–100. On the 1807 mania see Harris 2000, 100 and 216.

16. Harris 2000, 124–6.

17. MacLeod 1988, 78–88.

18. Miller 2000; Miller 2004; MacLeod 2007.

19. Winsor 1816, 132.

20. Danby Pickering, *The statutes at large from the Magna Charta, to the end of the eleventh Parliament of Great Britain, anno 1761 [continued to 1806] Anno tricesimo quinto Georgii III. Regis, being the fifth session of the Seventeenth Parliament of Great Britain* (J. Bentham, 1795), vol. 40, part 1, citing "An Act for naturalizing Frederick Albert Winzer", listed under Private Acts, #10; "Thermolampe," 1805, 69–70.

21. Winsor 1799, Dedication.

22. Winsor 1804a, 3; Poppe 1811, 20.

23. Winsor 1804a, 5.

24. Winsor 1804a, 11; Accum and Winsor 1816, 63.

25. Winsor 1802.

26. Winsor 1816, 133. He may have gone to Vienna: "Thermolampe." 1805, 69.

27. Ibid., 70.

28. Daniel Lysons, *The Environs of London: being an historical account of the towns, villages, and hamlets, within twelve miles of that capital*, second edition (T. Cadell and W. Davies, 1811). When Winsor later filed a patent, he listed himself as being from Shooter's Hill ("List of patents for 1812," *The Edinburgh Annual Register* 5 (1814), part 2: 341–2), as did his son in his wedding notice (*New Monthly Magazine* 12, 1819, 119). For Winsor's connection to Shrewsbury House, see Chandler and Lacey 1949, 25–27, citing William Thomas Vincent, *The Records of the Woolwich District* (J. P. Jackson, 1888).

29. Matthews 1827a, 29.

30. "Patent granted to Frederick Albert Winsor for an improved oven," *The Repertory of Arts, Manufactures, and Agriculture*, second series, 10 (1807): 320. The specification is in the National Archives (UK) (reference number C210/101).

31. William Nicholson, "Remarks on Mr. Winsor's projected Heat and Light Company," *The Athenaeum* 1 (1807): 186–7.

32. Winsor 1804a, 14, 16, 17, 23–6, 21, 47.

33. Winsor 1807b [reprint], 264.

34. "Comments on Mr. Wm. Henry's experiments on the gases obtained from the distillation of wood, peat, and pit-coal," *Retrospect of Philosophical, Mechanical, Chemical and Agricultural Discoveries* 1 (1805): 129–34, 133; *Morning Post*, 1805/06/15; Winsor 1807b, 42. Winsor claimed that another person working for him, John Humphries, was responsible.

35. Winsor 1807b [reprint], 262 ff. See also *Times*, 1806/12/27.

36. Edward Heard, *Summary account of an improved method for illuminating theaters, assembly-rooms, public gardens, streets, light-houses, public offices, dwelling-houses, &c, &c, &c with the philosophical lights* (R. Jenkins, 1805).

37. "Specification of the patent granted to Edward Heard, of London, chemist," *The Repertory of Arts, Manufactures, and Agriculture*, second series, 10 (1807): 31–2; "Mr. Edward Heard's discovery." *The Monthly Magazine* 23 (1807): 67.

38. "Patent granted to Frederick Albert Winsor for an improved oven," *The Repertory of Arts, Manufactures, and Agriculture*, second series, 10 (1807): 320; Winsor 1807b, 45.

39. Brereton 1903, 28.

40. Winsor 1804b, 3.

41. GLCC, Minutes of the Court of Directors, book 1 (MCD-b1), 18, 1812/08/05.

42. *Times*, 1804/12/11.

43. Winsor to B&W, 1804/08/20, BWA-MS-3147/3/539 #38.

44. Winsor 1806, viii.

45. Winsor to B&W, 1804/08/20, BWA-MS-3147/3/539 #38.

46. Lord Stanhope. See *Times*, 1810/06/02.

47. *Times*, 1804/09/20.

48. *Times*, 1804/10/10.

49. Golinski 1992; Morus 1998, 1–5, 70–98, 155ff.; Morus 1996b.

50. R. S. Kirby, *Kirby's Wonderful and Eccentric Museum* (R. S. Kirby, 1805), vol. 3, 38–40. See also "Mr. Winsor's Thermolampen-Ofen in London," *London und Paris* 15 (1805): 109–12.

51. Harris 1994.

52. Harris 2000, 110, 135.

53. Harris 2000; Mathias 1983, 145. On usury and the Bubble Act, see Cottrell 1980. On special interests, see Harris 2000, 135. The Bubble Act was repealed in 1825.

54. *Times*, 1804/12/11. See also "Useful improvements and discoveries," *The Literary Magazine, and American Register* 3 (1805): 296–7.

55. *Times*, 1804/12/11; Winsor 1806, viii.

56. Winsor 1806, viii–ix.

57. *Times*, 1804/12/19.

58. Winsor 1806, vii–x; GLCC MCD-b1, 18, 1812/08/05.

59. Grant and Accum 1808, 14.

60. *Hampshire Telegraph and Sussex Chronicle*, 1807/02/09.

61. Harris 2000, 117; Millward 2005, 59.

62. Kindleberger 1993, 193; "Joint stock companies," *The Spirit of the Public Journals* 11 (1808): 324–9.

63. Harris 2000, 216–7. For a list of companies seeking acts of incoporation at this time, see Thomas Tooke and William Newmarch, *A History of Prices, and of the State of the Circulation, from 1793 to 1837*, vol. 1 (Longmans, 1838), 277–285.

64. GLCC MCD-b1, 19, 91, and 96–7, 1812/08/07, 1813/03/09 and 1813/03/19.

65. Grant and Accum 1808, appendix.

66. Winsor 1806, xi.

67. The first mention of 97 is in *Morning Chronicle*, 1806/07/28.

68. *Times*, 1804/08/18, 1804/09/24 and 1804/09/28.

69. GLCC MCD-b1, 97, 1813/03/19.

70. Winsor 1806; *Morning Chronicle*, 1806/07/28 and 1806/08/07.

71. Winsor 1806, 7.

72. *Times*, 1806/11/22; *Morning Chronicle*, 1806/11/26, 1806/12/04.

73. *Morning Post*, 1806/11/01.

74. *Morning Chronicle*, 1806/10/25.

75. There are advertisements in the *Times* on November 22, 28, December 2, 4, 10, 11, 12, 15, 16, 18, 27, 29, January 1, 9, 16, 29, February 10, 27, 28, March 24, April 4, 6, May 8, 14, 15, June 4, 16, 18, 20, 23, July 13, 15, 16. This is not a complete list, and Winsor also advertised in other papers, including the *Morning Post* and the *Morning Chronicle*, with a similar frequency.

76. See, for example, *La Belle Assemblée,* Nov. 1807, 33; *Times*, 1807/10/15.

77. *Morning Chronicle*, 1807/10/15; *Times*, 1807/04/15.

78. *Times*, 1807/01/29.

79. Winsor 1807a.

80. *Times*, 1807/01/21; *Morning Post* 1807/01/21.

81. *Letters from an Irish Student in England to His Father in Ireland,* vol. 2 (Cradock & Joy, 1809), 230–3.

82. James Sayers, *Heroic Epistle to Mr. Winsor* (R. Spencer, 1808). Many minor satirical references can be found; see, e.g., "Horace in London," *The Poetical Register, and Repository of Fugitive Poetry for 1808–1809* 7: 246; *Woburn-Abbey Georgics*, second edition (C. Chapple, 1813), 35.

83. William Nicholson, "Remarks on Mr. Winsor's projected Heat and Light Company," *A Journal of Natural Philosophy, Chemistry and the Arts* 16 (1807): 73–5.

84. Webster 1807.

85. Winsor 1807b.

86. William Nicholson, "A few remarks on a pamphlet entitled 'Mr. W. Nicholson's attack, in his Philosophical Journal, on Mr. Winsor and the National Light and Heat Company: with Mr. Winsor's defence,'" *A Journal of Natural Philosophy, Chemistry and the Arts* 16 (1807): 308–10.

87. "Review of Parkes' *Chemical Catechism*," *The Anti-Jacobin Review and Magazine* 27 (1807): 26.

88. *Times*, 1807/03/24.

89. *Times*, 1807/05/14.

90. *Times*, 1807/01/16; *Morning Chronicle*, 1807/01/10.

91. *Times*, 1807/01/09.

92. *Times*, 1807/02/27; *Morning Chronicle,* 1807/02/27 and 1807/03/04.

93. *The Parliamentary register; or, an impartial report of the debates that have occurred in the two Houses of Parliament, in the course of the First (and only) session of the Third Parliament of the United Kingdom of Great Britain and Ireland,* vol. 2, (Stockdale, 1807), 474.

94. *Times*, 1807/04/06.

95. *Morning Chronicle*, 1807/03/05.

96. *Times*, 1807/04/04.

97. *Morning Chronicle*, 1807/06/01.

98. For a fairly detailed description of the display, see "Account of the first experiment of the public use of gas lights," *The Monthly Magazine* 23 (1807): 520.

99. *Times*, 1807/05/08.

100. "Account of the first experiment of the public use of gas lights," *The Monthly Magazine* 23 (1807): 520.

101. *Times*, 1807/06/16; *Morning Chronicle*, 1807/06/17.

102. *Times*, 1807/07/18; *Morning Chronicle*, 1807/07/20.

103. *Considerations on the nature and objects of the intended Light and Heat Company* 1808, 93; Grant and Accum 1808, 15–6.

104. William James, *The Naval History of Great Britain: from the Declaration of War by France in 1793, to the Accession of George IV*, second edition, vol. 3 (Richard Bentley 1837), 151.

105. *Times*, 1807/09/24; Grant and Accum 1808.

106. *Morning Chronicle*, 1807/08/18.

107. Joseph Haydn, *The Book of Dignities* (Longmans, 1851), 569.

108. Willem G. J. Kuiters, "Paxton, Sir William (1743/4–1824)," in *DNB*; Thorne 1986, vol. 1, 323.

109. Beatson 1807, 75. A William Devaynes, perhaps the current one's father, was also the chairman or deputy chairman of the East India Company sporadically in the years 1774–1794; Joseph Haydn, *The Book of Dignities* (Longmans, 1851), 273. See also Thorne 1986.

110. "Murray, Lord George (1761–1803).", in the *DNB*.

111. *Times*, 1807/06/20; Winsor 1807d.

112. *Times*, 1807/07/16.

113. *Morning Chronicle*, 1806/12/27 and 1806/12/27. For a report on his patent, see "Mr. Edward Heard's discovery," *The Monthly Magazine* 23 (1807): 67.

114. "Some account of the experiment made at Golden-Lane to illuminate streets by coal-gas lights," *The Athenaeum* 2 (1807): 187–8; "Lighting of the Golden Lane Brewery," *The Monthly Magazine* 24 (1807): 193.

115. *Times*, 1807/07/16.

116. *Times*, 1807/07/15.

117. *Times*, 1807/09/15 and 1807/08/20.

118. *Times*, 1807/08/20.

119. *Times*, 1807/09/15.

120. *Times*, 1807/09/22.

121. *Times*, 1807/10/15 and 1807/10/24.

122. *Times*, 1807/09/30.

123. Leveson-Gower and Ponsonby 1916, 281–2.

124. Winsor 1807c; Winsor 1799; Winsor 1798.

125. *Times*, 1807/11/19.

126. *Morning Chronicle*, 1807/11/20, 1807/11/21, 1807/11/24.

127. *Morning Chronicle*, 1807/12/01.

128. Grant and Accum 1808, 3.

129. *Times*, 1807/08/20; Grant and Accum 1808, 4.

130. *Times*, 1807/10/22.

131. Grant and Accum 1808, 5.

132. "Pall Mall, South Side, Past Buildings: Nos 93–95 Pall Mall: F.A. Winsor and the development of gas lighting" 1960.

133. Grant and Accum 1808, appendix.

134. "Pall-Mall illuminated by coal gas lights," *The Athenaeum* 3 (1808): 74; "Pall-Mall lighted by patent gas-lights," *The Literary Panorama* 3 (1808): 1083.

135. Grant and Accum 1808, 5.

136. "Lighting of the Pall Mall," *The Monthly Magazine* 24 (1808): 581.

137. "Pall-Mall illuminated by coal gas lights," *The Athenaeum* 3 (1808): 74.

138. *Morning Chronicle*, 1807/12/07; Grant and Accum 1808, appendix.

139. *The Gas Light and Coke Company: an account of the progress of the Company from its incorporation by Royal Charter in the year 1812 to the present time, 1812–1912* 1912, 15–6; *Morning Chronicle*, 1808/03/15.

140. *Times*, 1807/12/23, 1807/12/24, and 1807/12/26; *Morning Chronicle*, 1807/12/26.

141. *Times*, 1808/01/02; Grant and Accum 1808, 6.

142. *Times*, 1808/01/07.

143. *The Gas Light and Coke Company: an account of the progress of the Company from its incorporation by Royal Charter in the year 1812 to the present time, 1812–1912* 1912, 15–6; *Morning Chronicle*, 1808/03/23.

144. Accum and Winsor 1816, 22; Grant and Accum 1808, 31–7.

145. Browne 1925; Golinski 1992, 246ff.; Accum 1800, 1803, 1804b, 1805, 1807.

146. Grant and Accum 1808, 31–2.

147. *Times*, 1807/11/28.

148. Grant and Accum 1808, appendix; Hall 1809, 37.

149. *Considerations on the nature and objects of the intended Light and Heat Company* 1808, 4–5, 6–7, 18.

150. Winsor to B&W, 1804/08/20, BWA-MS-3147/3/539 #38.

151. William Balston to B&W, 1807/01/23, BWA-MS-3147/3/42 #24.

152. JWjr to GAL, 1807/05/04, BWA-JWP-6/62, 148.

153. Ambrose Weston to JWjr, 1807/04/26, BWA-MIV/W11.

154. JWjr to GAL, 1807/05/04, BWA-JWP-6/62 148.

155. Berman 1978, 8–20; Knight 2004; quote from terms Rumford attached to the medal, as cited in Bektas and Crosland 1992, 48.

156. Miller 2000; Miller 2004.

157. Miller 2000, 2, 8, 17.

158. Miller 2000, 6.

159. Miller 1999. For a later period, see Morus 1996a, 417ff.

160. Murdoch 1808, 124.

161. MacLeod 2007, 74–5, 80–4, 97–9; Werrett 2007.

162. JWjr to JW, 1807/06/04, BWA-MII/13/1.

163. JW to JWjr, 1807/06/09, BWA-MIV Box 16/15.

164. For summary of what Wilson found, see Griffiths 1992, 224–249. Wilson to B&W, 1808/01/27 and 1808/01/29, BWA-MS-3147/3/363 #10 and 11.

165. Draft of RS paper, 1808/02/22, BWA-MS-3147/3/480 #24.

166. Murdoch 1808.

167. JWjr to JW, 1808/02/26, BWA-MII/13/2.

168. Griffiths 1992, 261; MRB to Joseph Banks 1808/12/26, BWA-Lunar Society #149; JWjr to Joseph Banks[?], 1809/01/18, BWA-JWP-6/65, 288.

169. *The Annual review and history of literature, for 1808*, vol. 7, 703; *The Annual Register, or, A View of the History, Politics, and Literature for the Year 1808*, 131; *The Athenaeum*, vol. 3, no. 16, April 1808, 372 and vol. 4, no. 20, August 1808, 153–55; *The Belfast Monthly Magazine* December 1808, vol. 1, no. 4, 280–1; *The British Critic: A New Review*, March 1809, vol. 33, 258; *The Critical Review, or, Annals of Literature*, January 1809, vol. 16, no. 1, 33; *The Eclectic Review*, vol. 5, part 1, May 1809, 443; *A Journal of Natural Philosophy, Chemistry and the Arts*, Oct. 1808, vol.

21, no. 92, 94; *The Literary Panorama* Sept. 1808, vol. 4, 1157–1160. *Monthly Magazine*, Jan 1, 1809, vol. 26, 546; *The New Annual Register, or General Repository . . . for the year 1808* (1809), 250; *The Philosophical Magazine*, December 1808, vol. 32, no. 127, 113; *The Repertory of Patent Inventions*, September 1808, second series, vol. 13, no. 76, 262; *Retrospect of philosophical, mechanical, chemical, and agricultural discoveries*, 1809, vol. 4, 198; *The Universal Magazine*, 1808, vol. 10, 58; *The Scots Magazine and Edinburgh Literary Miscellany*, 1808, vol. 70, pt 2, 819–823.

170. "An account of the application of Gas from coal to economical purposes, by Mr. Wm. Murdoch," *The Athenaeum* 4 (1808): 153–5.

171. JW to JWjr, 1808/02/22, BWA-JWP-LB/4 202; JWjr to JW, 1808/02/26, BWA-MII/13/2.

172. Grant and Accum 1808, 9.

173. *Times*, 1808/01/20; *Morning Chronicle*, 1808/01/06 and 1808/01/25.

174. Grant and Accum 1808, 9.

175. Ibid., 10.

176. *Times*, 1808/07/22.

177. Grant and Accum 1808, appendix.

178. *Patents for inventions. Abridgments of specifications relating to the production and applications of gas* 1860, 11. According to Chandler and Lacey (1949, 32), it was due to restrictions the Patent Office had placed on his forming a company based on the patent.

179. Grant and Accum 1808, appendix; *Considerations on the nature and objects of the intended Light and Heat Company* 1808, 27–8.

180. *Times*, 1808/04/12.

181. Grant and Accum 1808, 14.

182. *Times*, 1808/06/29; *Morning Chronicle*, 1808/07/23.

183. *Times*, 1808/06/29; *Morning Chronicle*, 1808/06/29.

184. *Times*, 1808/07/22; *Morning Chronicle*, 1808/07/23.

185. *Times*, 1808/06/04.

186. GLCC MCD-b1, 97, 1813/03/19.

187. *Patents for inventions. Abridgments of specifications relating to the production and applications of gas* 1860, 11.

188. *Morning Chronicle*, 1808/07/23.

189. Clifford 1885, vol. 1, 205; *Journal of the House of Commons* 64, 24 February 1809, 96.

190. James Weston to JWjr, 1809/03/09, BWA-MS-3147/3/480 #1.

191. JWjr to MRB, 1809/03/10, BWA-Lunar Society #115.

192. MRB to JWjr, 1809/03/13, BWA-MIV/B6.

193. *Times*, 1809/03/28; *Journal of the House of Commons* 64, 1809/03/24 and 1809/03/27, 175, 183.

194. James Weston to JWjr, 1809/03/30, BWA-MS-3147/3/480 #2.

195. MRB to JWjr, 1809/03/31, BWA-MS-3147/3/43 #5.

196. JW to Ambrose Weston, 1809/04/03, BWA-JWP-LB/4 257.

197. Ambrose Weston to JW, 1809/04/03, BWA-JWP-4/86.

198. JW to JWjr, 1809/04/05, BWA-MIV Box 16/16.

199. JWjr to MRB, 1809/04/05, BWA-Lunar Society #116.

200. Woodward to JW, 1807/05/26, BWA-MII/13/1.

201. MRB to JWjr, 1809/04/13, BWA-Lunar Society #117 and BWA-MS-3147/3/43 #6.

202. JWjr to MRB, 1809/04/20, BWA-Lunar Society #118.

203. *Times*, 1809/04/21; *Journal of the House of Commons* 64, 1809/04/20 and 1809/04/24, 226, 235.

204. *Remarks upon the bill for incorporating the Gas light and Coke company* (George Sidney, 1809) 4, 8, 16.

205. MRB to JWjr, 1809/04/22, BWA-MS-3147/3/43 #8.

206. JWjr to MRB, 1809/04/27, BWA-Lunar Society #121; JWjr to Lee, 1809/04/27, BWA-JWP-6/63, 232.

207. Lee to JWjr, 1809/04/29, BWA-MS-3147/3/480 #4.

208. JWjr to MRB, 1809/04/28, BWA-Lunar Society #121; Experiment notes, 1809/04/29, BWA-MS-3147/3/480 #5.

209. Murdoch 1809. On this exchange, see Werrett 2007.

210. Murdoch 1809, 3, 5, 7, 8, 10, 12–3.

211. David Macpherson, *An account of the South Sea scheme, and a number of other bubbles, which were encouraged by public infatuation, in the year 1720* (J. Cawthorne, 1806).

212. Harris 2000, 205ff, 235ff.

213. Clifford 1885, vol. 1, 205ff.

214. *Journal of the House of Commons* 64, 1809/05/05, 278; John Southern to JWjr, 1809/05/04 and 1809/05/05, BWA-MS-3147/3/332 #8 and 9.

215. Ambrose Weston to JW, 1809/05/10, BWA-JWP-4/86.

216. JWjr to John Southern, 1809/05/12, BWA-MS-3147/3/60 #16.

217. Ambrose Weston to JW, 1809/05/10, BWA-JWP-4/86; Browne to JW, 1809/04/24, BWA-JWP-4/86; JW to Browne, 1809/04/26, BWA-JWP-LB/4 266; JWjr to MRB, 1809/04/27, BWA-Lunar Society #120; Ambrose Weston to JW, 1809/05/10, BWA-JWP-4/86.

218. Hall 1809, 8–9, 12–25. Some of these claims were believed. See William Playfair and Joshua Jepson Oddy, *A Sketch for the Improvement of the Political, Commercial, and Local Interests of Britain* (J. J. Stockdale, 1810), 24–5.

219. Hall 1809, 29, 38.

220. John Southern to JWjr, 1809/05/09, BWA-MS-3147/3/332 #11; JW to Ambrose Weston, 1809/05/09, BWA-JWP-LB/4 266.

221. John Southern to JWjr, 1809/05/08, 1809/05/09, and 1809/05/10, BWA-MS-3147/3/332 #10, 11, 12, and 13. See also JW to Ambrose Weston, 1809/05/09, BWA-JWP-LB/4 266; JW to Weston, 1809/05/09, BWA-MIV Box 16/16.

222. 1809/05/09, BWA-MS-3147/3/479 #1; JWjr to John Southern 1809/05/12, BWA-MS-3147/3/60 #16. William Henry also contributed some evidence regarding ammonia, but it came too late to make any difference. GAL to JWjr, 1809/05/30, BWA-MS-3147/3/480 #19; William Henry to JWjr, 1809/05/29, BWA-MS-3147/3/480 #18.

223. JWjr to John Southern, 1809/05/12, BWA-MS-3147/3/60 #16.

224. Hall 1809, 38–45, 47–8.

225. MRB to JWjr, 1809/04/22, BWA-MS-3147/3/43 #8; Hall 1809, 45–6, 50. Davy considered Accum "a cheat and a Quack." Golinski 1992, 246.

226. Hall 1809, 49.

227. Hall 1809, 51–62.

228. *Times* 1809/05/18; *Journal of the House of Commons* 64, 1809/05/08, 1809/05/12, and 1809/05/17, 291, 308, 316.

229. Invitation, 1809/05/15, BWA-MS-3147/3/481 #26.

230. Clifford 1885, 207; *Times* 1809/05/18; *Journal of the House of Commons* 64, 1809/05/17, 316.

231. JWjr to JW, 1809/05/18, BWA-MII/13/2.

232. *Times* and *Morning Chronicle*, 1809/05/20.

233. Brougham to JWjr, 1809/05/17, BWA-MS-3147/3/480 #8; Rickman to Humphry Davy, 1809/05/17, BWA-MS-3147/3/480 #7; *In Parliament: Gas-light bill*, 1809.

234. *In Parliament: Gas-light bill*, 1809, 10–1.

235. JWjr to Thomas Plummer, 1809/05/20, BWA-MS-3147/3/480 #9; JW to JWjr, 1809/05/22, BWA-MIV Box 16/17; Duckworth to JWjr, 1809/05/23, BWA-MS-3147/3/480 #10; P. Wilson to JWjr, 1809/05/27, BWA-MS-3147/3/480 #15.

236. JWjr to JW, 1809/05/27, BWA-MII/13/2.

237. *Remarks on the gas light and coke bill*, 1809.

238. Clauses to be inserted, from Pedder, 1809/05/28, BWA-MS-3147/3/480 #33.

239. JWjr to JW, 1809/05/27, BWA-MII/13/2.

240. Brougham to JWjr, 1809/05/26, BWA-MS-3147/3/480 #12 and 13; James Weston to JWjr, 1809/05/26, BWA-MS-3147/3/480 #14; Ambrose Weston to JWjr, 1809/05/28, BWA-MS-3147/3/480 #17; JW to JWjr, 1809/05/22, BWA-MIV Box 16/17.

241. Henry Brougham, *Speeches of Henry Brougham, Esq delivered before a committee of the Honorable House of Commons, in opposition to a bill, for incorporating certain persons by the name of the Gas Light and Coke Company* (Strahan and Preston, 1809); Warren and Harrison 1809.

242. *Gas light bill third reading, Friday, June 2, 1809* (G. Sidney, 1809).

243. Further remarks in opposition to the Gas Light Bill, 1809/06/01, BWA-MS-3147/3/480 #31.

244. *Journal of the House of Commons* 64, 1809/06/02, 380.

245. *The Parliamentary debates from the year 1803 to the present time*, vol. 14 (1812), 860–1; Clifford 1885, 207–8; *Times*, 1809/06/03; Wilberforce to JWjr, 1809/05/27, BWA-MS-3147/3/480 #16.

246. JWjr to JW, 1809/06/06, BWA-MII/13/2.

247. Ewart to JWjr, 1809/06/05, BWA-MS–3147/3/249 #82; JW to JWjr, 1809/06/10, BWA-MIV Box 16/17; Lee to JWjr, 1809/06/05, BWA-MS–3147/3/480 #23.

248. John Southern to JWjr, 1809/06/05, BWA-MS-3147/3/332 #17.

249. JWjr to JW, 1809/06/06, BWA-MII/13/2.

250. JWjr to JW, 1809/06/06, BWA-MII/13/2; *Times* 1809/07/06; Box 1817, 9.

251. Hamilton to JWjr, 1809/07/02, BWA-MIV/H1.

252. Box 1817, 8.

253. Van Voorst 1809, iii.

254. Box 1817, 8–11; *Times*, 1809/07/03.

255. *Times*, 1809/07/05.

256. Box 1817, 11.

257. Van Voorst 1809; *Times* 1809/07/06 and 1809/07/09.

258. Van Voorst 1809, iii, vii–viii.

259. *Times*, 1809/07/08.

260. *Times*, 1809/07/12.

261. *Times*, 1809/08/08; Grant 1809.

262. Grant 1809, 10.

263. *Times*, 1809/08/30.

264. *Times*, 1809/09/18, 1809/09/29, 1809/10/09.

265. *An act [50 Geo.III cap. clxiii] for granting certain Powers and Authorities to a Company*, 1810, Articles I, II, XXVIII.

266. *Times*, 1810/01/21; *Journal of the House of Commons* 65, 1810/02/20, 107.

267. Brougham to JWjr, 1810/02/08, BWA-MIV/B7; Ewart to JWjr, 1810/02/18, BWA-MIV/ E2.

268. Copy of bill and note, 1810/03/15, BWA-MS-3147/3/481 #26 and 27.

269. *Times*, 1810/03/07, 1810/03/15; *Journal of the House of Commons* 65, 1810/03/06 (p. 150), 1810/03/14 (p. 178), 1810/03/30 (p. 233), 1810/04/04 (p. 246), 1810/04/05 (p. 251), and 1810/04/09 (p. 259).

270. *Times*, 1810/04/10 and 1810/04/11; *Caledonian Mercury*, 1810/04/14.

271. *Minutes of Evidence taken before the Lords Committees to whom was referred the Bill, intitled, "An act for enabling His Majesty to incorporate by Charter, to be called 'The Gas Light and Coke Company' for making Inflammable Air for the lighting of the Streets of the Metropolis, and for procuring Coke, Oil, Tar, Pitch, Asphaltum, Ammoniacal Liquor, and essential Oil, from Coal, and for other Purposes relating thereto, Journal of the House of Lords* 47 (1810), 1810/04/10, 581.

272. *Morning Chronicle*, 1810/05/23; *Times*, 1810/05/12, 1810/05/16, 1810/05/23, 1810/06/10; *Journal of the House of Lords* 47, 1810/05/15, 1810/05/18, 1810/05/22, 1810/05/28; 1810/06/01, 1810/06/06; *Journal of the House of Commons* 65, 1810/06/02, 1810/06/05.

273. *Journal of the House of Lords* 47, 1810/05/28, 706; *The Parliamentary debates from the year 1803 to the present time*, vol. 36 (1817), 561.

274. *An act [50 Geo.III cap. clxiii] for granting certain Powers and Authorities to a Company* 1810, Article III. *Recital of the Act for granting certain powers and authorities to a Company to be incorporated, called "The Gas Light and Coke Company"*(1812/04/30).

275. Although the joint-stock form was the standard, the expansion of the scale of business in finance and transportation was also creating strains with these legal models that had been largely set around 1720. Harris 2000, 9–10, 99.

276. Flichy 2007, 87–9.

NOTES TO CHAPTER 5

1. Henderson and Escher 1968, 54–5.

2. Hughes 1983, 18ff. See also Israel 1998, 167–8.

3. Platt 1991.

4. Joerges 1988, 38; van der Vleuten 2006, 300; Caron 1988, 72ff.

5. Hughes 1983, chapter 9.

6. Salsbury 1988, 38. See also Caron, "The Birth of a Network Technology: The First French Railway System," in Berg and Bruland 1998; Wilson 1995, 37–9; Gourvish 1973; Gourvish 1972, 167–82; Kirby 1994; Schmitz 1995, 11ff. Radkau 1994, 52.

7. Chandler 1977, 96–98; Salsbury 1988, 41.

8. Tarr and Gabriel 1988, xiii–xvii; Ball and Sunderland 2001, 31, 264; Melosi 2000.

9. Grundmann 1999, 251.

10. Hall 1809, 26–7, 37–8. The plan for St. James parish presented to Parliament envisaged six stations supplying 800 lamps. The stations were to be in rented houses, each manned by two or three employees.

11. For a summary of approaches to users and technology, see Oudshoorn and Pinch 2003. On users and telephone see Fischer 1992. On electricity see Nye 1990.

12. On contracts see Mackay et al. 2000, 744, 752.

13. On the expansion of technical systems see Hughes 1987, 66.

14. On standardization see Joerges 1988, 30; Chatzis 1999, 82; Mokyr 1999, 64; Marsden and Smith 2005, 9–10.

15. White 2007, 452.

16. Ibid., 418.

17. Ibid., 68.

18. Ibid., 22–3, 109, 163–5, 175ff, 187–9.

19. Ibid., 131–40.

20. Owen and MacLeod 1982, 23ff.; Webb and Webb 1906–08, vol. 1, 228ff., 239, vol. 2, 606–10, 690–2, vol. 3, 212–31, vol. 5, 235ff, 276ff.; White 2007, 449.

21. Millward 2005, 54.

22. White 2007, 255–7, 334, 337, 384–8, 390.

23. Ibid., 387ff.

24. Estcourt 1828, 128, 201; *Times*, 1819/05/18.

25. White 2007, 22ff.

26. *Recital of the Act for granting certain powers and authorities to a Company to be incorporated, called "The Gas Light and Coke Company"* (1812/04/30).

27. *Morning Chronicle*, 1812/05/15.

28. MCP, 1, 1812/06/24.

29. *Recital of the Act* gives a very few details about the directors.

30. MCD-b1, 9, 1812/07/10.

31. MCD-b1, 9 and 14, 1812/07/17 and 1812/07/31.

32. MCD-b1, 16, 1812/08/05; MCD-b1, 17, 1812/08/07.

33. MCD-b1, 29, 1812/09/04.

34. MCD-b1, 27, 1812/08/12; MCD-b1, 29, 1812/09/04; MCD-b1, 12, 1812/07/24.

35. MCD-b1, 31, 1812/09/11.

36. For Hargreaves see MCD-b1, 36, 75, 77, 102. For Barlow see MCD-b1, 25, 33, 36, 40, 49.

37. MCD-b1, 28, 1812/09/04. For some earlier discussion of the matter, see MCD-b1, 24 and 27, 1812/08/14 and 1812/08/28; MCD-b1, 37 and 40, 1812/10/10 and 1812/11/07.

38. MCD-b1, 45, 1812/09/16; Chemistry Minutes, 5, 1812/09/16.

39. MCD-b1, 54–5, 1812/12/11–12; Chemistry Minutes, 13, 1812/12/04.

40. MCD-b1, 47 and 50, 1812/11/20 and 1812/12/01; MCD-b1, 48, 1812/11/28; MCD-b1, 51, 1812/12/04.

41. "The Manor and Liberty of Norton Folgate," 1957.

42. MCD-b1, 53, 1812/12/10; MCD-b1, 40, 1812/11/07; MCD-b1, 225, 1813/12/01.

43. MCD-b1, 80, 1813/02/16.

44. MCD-b1, 40, 1812/11/07.

45. MCD-b1, 55, 1812/12/11.

46. MCD-b1, 125, 1813/05/07.

47. MCD-b1, 56, 1812/12/11.

48. Samuel Clegg, "Improved Apparatus for extracting Carbonated Hydrogen Gas from Pit Coal," *Transactions of the Society, Instituted at London, for the Encouragement of Arts, Manufactures, and Commerce* 26 (1808): 202–6. See also Awty 1974.

49. "Materials for a memoir of Mr. Samuel Clegg," *Mechanics' Magazine* 22 (1835): 470–2; Clegg 1841, 13–16; Matthews 1827a, 52; Clegg 1820, 3–5; Chandler and Lacey 1949; Elton 1958.

50. MCD-b1, 77, 1813/02/09.

51. MCD-b1, 15, 1812/08/05.

52. MCP, 10, 1813/02/25; MCD-b1, 96–8, 1813/03/19.

53. MCD-b1, 43–4, 1812/11/13.

54. MCD-b1, 47, 1812/11/20; MCD-b1, 55, 1812/12/12; Chemistry minutes, 12, 1812/12/11; MCD-b1, 61, 1812/12/22.

55. MCP, 2, 1813/01/09.

56. *Times*, 1813/02/05.

57. MCD-b1, 77, 1812/02/09.

58. MCP, 9, 1813/02/08.

59. He was also the son of George III's saddle maker. *Jackson's Oxford Journal* of 1807/12/19 refers to his wedding. The *Morning Chronicle* of 1810/08/25 mentions his chairmanship of a canal company meeting. The *Morning Chronicle* of 1811/11/18 mentions his chairmanship of a proposed canal company. Pollock published a book in 1813: *Tables, exhibiting the various Particulars requisite to be attended to, in pursuance of the standing Orders of the two Houses of Parliament, in soliciting such private Bills as usually commence in the House of Commons, corrected to August 1813. By David Pollock, Esq. of the Middle Temple, Barrister at Law.* See *British Critic*, 42 (1813), 206. For more details, see Everard 1949, 115 footnote; Hutchinson 1985, 260ff.; Ingpen 1912, 292.

60. MCP, 10, 1813/02/25.

61. MCP, 16–7, 1813/05/17.

62. MCD-b1, 131, 1813/05/18.

63. MCP, 18–20, 1813/07/06.

64. MCD-b1, 162, 1813/07/22.

65. MCP, 25, 1813/11/01; MCD-b1, 162, 1813/07/22; MCD-b1, 209, 1813/11/02.

66. MCD-b1, 81, 1813/02/16; MCD-b1, 124, 1813/05/06.

67. MCD-b1, 123, 1813/05/04.

68. MCD-b1, 82, 1813/02/19; MCD-b1, 90, 1813/03/04.

69. MCD-b1, 90, 1813/03/04.

70. MCD-b1, 14, 1812/07/31; MCD-b1, 99, 1813/03/24; MCD-b1, 117, 1813/04/20. *Minutes of Evidence taken before the Lords Committees to whom was referred the Bill, intitled, "An act to alter and enlarge the Powers of Two Acts of His present Majesty for granting certain Powers to the Gas Light and Coke Company*, 1816, 29.

71. MCD-b1, 116, 1813/04/15; MCD-b1, 100, 1813/03/25.

72. MCD-b1, 90 and 94, 1813/03/05 and 1813/03/16; MCD-b1, 100, 1813/03/25; MCD-b1, 133, 1813/05/25. The Secretary of the Treasury had to give permission: MCD-b1, 110 and 138, 1813/04/06 and 1813/06/04.

73. MCD-b1, 171, 1813/08/10. For a very brief description of the Peter Street site in June 1813, see "Gas light company," *The Monthly Magazine* 35 (1813): 533–4.

74. MCD-b1, 173 and 176, 1813/08/13 and 1813/08/20; Everard 1949, 35.

75. MCD-b1, 198–9, 1813/10/12 and 1813/10/14; MCD-b1, 220 and 222, 1813/11/19 and 1813/11/23.

76. MCD-b1, 221–3, 1813/11/19, 1813/11/23 and 1813/11/26.

77. MCD-b1, 92, 1813/03/13; MCD-b1, 108, 1813/04/01; MCD-b1, 125, 1813/05/07.

78. MCD-b1, 56–7, 1812/12/12; MCD-b1, 115, 1813/04/13; MCD-b1, 135 and 139, 1813/05/28 and 1813/06/08; MCD-b1, 155, 1813/07/15. It was located at what is now Hearn and Worship Streets on Curtain Road. At that time Worship was a street that changed names as it crossed Curtain Road, with the name Providence Worship Street to the west and Hog Lane to the east of Curtain Road.

79. MCD-b1, 165, 1813/07/26; MCD-b1, 196 and 197, 1813/10/07 and 1813/10/08.

80. MCD-b1, 200, 1813/10/14.

81. MCD-b1, 181, 1813/09/03; MCD-b1, 201, 1813/10/15; MCD-b1, 217, 1813/11/13.

82. MCD-b1, 191, 1813/09/23.

83. MCD-b1, 162, 1813/07/22.

84. MCD-b1, 180, 1813/09/02.

85. MCD-b1, 191, 1813/09/23.

86. MCD-b1, 166, 1813/07/29.

87. MCD-b1, 115–6, 1813/04/13 and 1813/04/15; MCD-b1, 128, 1813/05/14.

88. MCD-b1, 127, 1813/05/11.

89. MCD-b1, 149, 1813/07/01.

90. MCD-b1, 155 and 157, 1813/07/15 and 1813/07/17.

91. MCD-b1, 155 and 157, 1813/07/15 and 1813/07/17; MCD-b1, 160, 1813/07/17.

92. Banks 1823, 4; Matthews 1827a, 67–68; MCD-b1, 207 and 212–3, 1813/11/02 and 09; MCP, 24, 1813/11/01; *Minutes of Evidence taken before the Lords Committees to whom was referred the Bill, intitled, "An act to alter and enlarge the Powers of Two Acts of His present Majesty for granting certain Powers to the Gas Light and Coke Company* 1816, 8–9.

93. MCD-b1, 203, 1813/10/27.

94. MCD-b1, 205, 1813/10/29.

95. MCP, 26, 1813/11/01.

96. MCD-b1, 211–2, 1813/11/05.

97. MCD-b1, 209, 1813/11/02.

98. MCD-b1, 231, 1813/12/10; MCP, 35, 1814/01/15; *Morning Chronicle*, 1814/01/25.

99. Joseph Nightingale, *London and Middlesex* (W. Wilson, 1815), vol. 3, 748.

100. Amelia M. Murray, *Recollections from 1803 to 1837: with a conclusion in 1868* (Longmans, 1868), 45–7. It may have been the house mentioned in this sale: *Morning Chronicle*, 1819/06/16.

101. MCP, 35, 1814/01/15.

102. MCP, 13, 1813/03/29.

103. MCP, 28–9, 1813/11/30.

104. MCD-b1, 234 and 237, 1813/12/17 and 1813/12/21; MCD-b1, 246, 1814/01/04.

105. MCD-b1, 211 and 221–3, 1813/11/05, 1813/11/19, 1813/11/23, and 1813/11/26.

106. MCD-b1, 227, 231 and 238, 1813/12/03, 1813/12/10 and 1813/12/21.

107. MCD-b1, 210–1, 1813/11/04 and 1813/11/05; MCD-b1, 231, 237 and 241, 1813/12/10, 1813/12/21 and 1813/12/24.

108. MCD-b1, 249, 1814/01/07; MCD-b1, 274, 1814/02/22.

109. *Morning Chronicle*, 1813/12/20; MCD-b1, 232, 1813/12/14; MCP, 32, 1814/12/07.

110. MCP, 32, 1814/01/07.

111. MCP, 32, 1814/01/07.

112. MCD-b1, 245, 1814/01/04.

113. MCD-b1, 307–10, 1814/04/26.

114. MCD-b1, 286, 1814/03/22.

115. MCP, 39, 1814/05/25; Accum 1819, 55; Peckston 1819, 146–7.

116. MCD-b1, 248, 1814/01/07; MCD-b1, 278 and 287, 1814/03/01 and 1814/03/25; MCD-b1, 266, 1814/02/08.

117. MCD-b1, 257–8, 1814/01/28.

118. MCP, 40 and 42, 1814/05/25.

119. MCP, 39, 1814/05/25; MCD-b1, 318, 1814/05/06.

120. MCD-b1, 242, 1813/12/28; MCD-b1, 232, 1813/12/10; MCD-b1, 240, 1813/12/24; MCD-b1, 245 and 249, 1814/01/04 and 1814/01/07; MCD-b1, 235, 1813/12/17; MCD-b1, 237, 1813/12/21.

121. MCP, 40, 1814/05/25.

122. MCD-b1, 250 and 258–9, 1814/01/18 and 1814/01/28; MCD-b1, 268, 1814/02/15; MCD-b1, 278, 279 and 284–5, 1814/03/01, 1814/03/04 and 1814/03/18; MCD-b1, 485, 1814/11/18.

123. MCD-b1, 262–3, 1814/02/04.

124. MCD-b1, 270, 1814/02/18.

125. MCD-b1, 366, 1814/07/05.

126. MCD-b1, 306–7 and 311, 1814/04/26 and 1814/04/29.

127. MCD-b1, 336, 1814/05/27.

128. MCD-b1, 314, 1814/04/29. MCD-b1, 371, 1814/07/08. The mains went from Folgate to Bishopsgate to Leadenhall.

129. MCD-b1, 385 and 387, 1814/07/29; MCD-b1, 388, 1814/08/03.

130. MCD-b1, 383, 1814/07/26; MCD-b1, 397, 1814/08/12; MCD-b1, 399 and 403, 1814/08/19; MCD-b1, 411, 1814/08/30; MCD-b1, 415, 1814/09/02; MCD-b1, 419, 1814/09/06; MCD-b1, 460, 1814/10/21; MCD-b1, 495, 1814/11/25.

131. MCD-b1, 291, 1814/04/01; MCD-b1, 302, 1814/04/15; MCD-b1, 332, 1814/05/24; MCD-b1, 288, 1814/03/25; MCD-b1, 296 and 303, 1814/04/07 and 1814/04/15.

132. MCD-b1, 294, 1814/04/01. The cost of fitting up a lamp in 1818 was £4. Bennet 1818, 227.

133. MCD-b1, 324, 1814/05/10; MCD-b1, 336, 1814/05/27; MCD-b1, 333, 1814/05/25; MCD-b1, 341 and 342, 1814/06/03.

134. MCD-b1, 315, 1814/05/03; MCD-b1, 386–7, 1814/07/31.

135. MCD-b1, 330 and 332, 1814/05/20 and 1814/05/24; MCD-b1, 341, 1814/06/03.

136. MCD-b2, 838 and 840, 1816/03/12; MCW-b87, 30, 1815/11/16; MCW-b87, 35, 1815/11/25.

137. MCD-b1, 343–347, 1814/06/03 and 1814/06/10; MCD-b1, 356, 1814/06/21.

138. MCD-b1, 386, 1814/07/31.

139. *Caledonian Mercury*, 1814/08/06.

140. MCD-b1, 252, 254, and 256, 1814/01/21, 1814/01/25 and 1814/01/28.

141. MCD-b1, 261–264, 1814/02/04 and 1814/02/08.

142. MCD-b1, 274 and 276, 1814/02/22 and 1814/02/25; MCD-b1, 282, 1814/03/11.

143. Banks 1823.

144. MCD-b1, 306, 1814/04/26; MCD-b1, 322, 5, 7, and 8, 1814/05/10, 1814/05/13, 1814/05/17, and 1814/05/20.

145. MCD-b1, 276 and 306, 1814/02/24 and 1814/04/26.

146. *An act [54 Geo. III cap. cxvi] for enlarging the powers of an act of His Present Majesty for granting certain powers and authorities to the Gas Light and Coke Company*, 1814.

147. MCP, 45, 1814/07/04.

148. MCP, 44, 1814/07/04.

149. MCP, 54, 1814/08/05.

150. *Morning Post*, 1814/08/16.

151. MCD-b2, 555, 558, 562, 567–8 and 571, 1815/02/07, 1815/02/7, 1815/02/14, 1815/02/21 and 1815/02/28.

152. Pollard 1965, 51.

153. Wilson 1995, 23; Wilson and Thomson 2006, 52.

154. Wilson 1995, 23.

155. Pollard 1965, 61–103.

156. Pollard 1965, 102; Wilson 1995, 36; Kirby 1994, 126–8.

157. Wilson and Thomson 2006, 11–2, 52–5; Kirby 1994, 113–8.

158. Wilson 1995, 11–4; Wilson and Thomson 2006, 21; Chandler and Takashi 1990.

159. Wilson and Thomson 2006, 42–5; Kirby 1994, 128ff.; Pollard 1965, 87–9.

160. Kirby 1994, 118–26.

161. Schmitz 1995, 10.

162. Wilson 1995, 37–41.

163. Chandler 1977; Salsbury 1988, 38, 41. See also Caron, "The Birth of a Network Technology: the First French Railway System," in Berg and Bruland 1998; Wilson 1995, 37–9; Gourvish 1973; Gourvish 1972, 167–82; Kirby 1994; Schmitz 1995, 11ff.

164. Kirby 1994, 128.

165. Kirby 1994, 129–34.; Wilson and Thomson 2006, 37–41, 56–7; Wilson 1995, 37–41; Gourvish 1972.

166. MCD-b2 555, 558, 562, 567–8 and 571, 1815/02/03, 1815/02/7, 1815/02/14, 1815/02/21 and 1815/02/28.

167. The directors required full details and plans for all technology: MCD-b3, 5, 1815/06/11.

168. The allowable capital was originally £200,000, half of which was raised by 1812. It was expanded by £200,000 in 1816 and again in 1819. See the acts 56 Geo.III cap. lxxxvii 1816, Clause II; 59 Geo. III cap. xx 1819, Clause II. 4 Geo.IV cap. cxix 1823, Clause I, added another £300,000. On dividends see MCP, 99, 1817/07/07.

169. Hutchinson 1985, 262.

170. Everard 1949, 114–5.

171. Williot 1999, 28.

172. MCP, 146–7, 1820/11/07 and 1820/11/09.

173. Peckston worked for the GLCC but was fired by the Court in 1819. MCW-b88, 214, 1819/08/16; MCD-b4, 341, 1819/06/29.

174. Accum 1819, 51, 60.

175. Peckston 1819, xii.

176. Accum 1819, 57–8.

177. Creighton 1824, 451.

178. MCD-b2, 670, 1815/06/25.

179. Accum 1819, 55.

180. MCD-b1, 287, 1814/03/25; MCD-b1, 343, 1814/06/03; MCD-b1, 345, 1814/06/07; MCD-b1, 375, 1814/07/12.

181. Accum 1819, 54–7.

182. Accum 1819, 60–1.

183. MCW-b87, 5 and 117, 1815/02/07 and 1816/04/18.

184. MCD-b3, 30, 37 and 40, 1816/07/16, 1816/07/23 and 1816/07/25; Accum 1819, 62. These described extensive tests of 2, 3, 4 retorts/fire.

185. MCD-b3, 14 and 26, 1816/06/21 and 1816/07/12.

186. Accum 1819, 63–4.

187. Ibid., 65.

188. MCD-b2, 833, 5, 1816/05/16.

189. MCW-b87, 181, 1816/08/26; MCW-b87, 194, 1816/09/12; MCW-b87, 232, 1816/12/12.

190. MCW-b87, 294, 1817/06/12. For a description of these rotary retorts, see Accum 1819. For the patent granted for the retort design, see "Specification of the Patent granted to Samuel Clegg" 1816.

191. MCW-b87, 175, 1816/08/15; MCD-b3, 169, 1817/01/21.

192. MCD-b3, 190, 1817/02/18; MCD-b3, 302 and 306, 1817/07/18 and 1817/07/26.

193. MCD-b3, 134, 1816/12/13; MCD-b3, 149, 1816/12/24.

194. Creighton has suggested just such a setting in 1808. See Henry Creighton to B&W, 1808/01/25, BWA-MS-3147/3/247 #43.

195. MCW-b88, 11 and 18, 1818/02/05 and 1818/03/02.

196. MCW-b87, 267, 270 and 272, 1817/04/10 and 1817/04/14.

197. MCD-b3, 229, 1817/04/15; MCD-b3, 239, 1817/04/29.

198. MCW-b87, 278, 1817/05/01; MCW-b87, 274, 1817/04/21; MCW-b87, 276, 1817/04/24.

199. MCW-b87, 330, 1817/10/30.

200. MCW-b87, 346, 1817/12/22.

201. MCW-b88, 3, 1818/01/15.

202. MCW-b88, 9, 1818/02/02 and 1818/02/05.

203. Accum 1819, 67–8.

204. MCW-b88, 32–33, 34 and 36, 1818/04/06. The carbonization ratio decreased from 47% to 16%, a huge drop.

205. MCW-b88, 165ff, 1819/03/29; Accum 1819, 60–1.

206. Accum 1819, 67; Thomas S. Peckston, *Cursory observations on different processes adopted for the distillation of coal, drawn from actual experiment* (T. Baker, 1819), 6. The company was experimenting with 7 to 10 retorts per oven in 1820. See MCD-b5, 156 and 158, 1820/08/11 and 1820/08/18.

207. Clegg 1841, 69; Peter Nicholson, *Practical masonry, bricklaying and plastering, both plain and ornamental* (T. Kelly, 1838), 138.

208. MCW-b88, 14, 1818/02/09.

209. MCW-b88, 195, 1819/06/21; MCD-b4, 334 and 340, 1819/06/22 and 1819/06/29. See, among many others, MCD-b4, 275, 1819/03/26

210. Accum: MCD-b4, 165, 1818/10/09; MCD-b4, 168, 1818/10/13; MCD-b4, 343, 1819/07/02 and 1819/07/16; William Congreve, and J. Whiting, *A short account of a patent lately taken out by Sir William Congreve, Bart* (T. Egerton, 1819). Outhett: MCD-b4, 425 and 440, 1819/11/16; *Patents for inventions. Abridgments of specifications relating to the production and applications of gas* (G. E. Eyre and W. Spottiswoode, 1860, 28–9.

211. Accum 1819, 52.

212. Peckston 1819, 186.

213. MCD-b3, 214, 1817/03/21.

214. MCW-b87, 261, 1817/03/24; MCW-b87, 276, 1817/04/24

215. Matthews 1827b, 22–3.

216. Ibid.

217. MCW-b88, 8, 1818/02/02; MCW-b88, 11, 1818/02/05.

218. Peckston 1819, 128.

219. MCD-b4, 360, 373 and 376, 1819/07/30.

220. MCW-b88, 274, 1819/12/20; MCD-b4, 458, 1819/12/31; Matthews 1827b, 21–4.

221. MCW-b88, 164, 1819/03/27.

222. "Specification of the Patent granted to Edward Heard, of London, Chemist," *The repertory of arts, manufactures, and agriculture*, second series, 10 (1807): 31–2.

223. Matthews 1827b, 25; Henry 1805, 68–9; Henry 1808, 302–3. Clegg junior bitterly accused Henry of stealing his father's idea, but they were both preceded by Heard, and the basic principle was hardly new.

224. Clegg 1820, 3–5.

225. Matthews 1827a, 52.

226. Accum 1819, 141–3.

227. MCD-b2, 632, 1815/05/12; MCD-b2, 643, 1815/05/26; MCD-b2, 687, 1815/07/15.

228. MCD-b2, 758, 1815/11/14; MCW-b87, 28, 1815/11/16; MCW-b87, 146, 1816/06/03.

229. "Specification of the Patent granted to Samuel Clegg," 1816; Accum 1819, 149–157; Chandler and Lacey 1949, 62; MCW-b87, 165, 1816/07/25.

230. MCW-b87, 197, 1816/09/16; MCW-b87, 298, 1817/06/19.

231. MCD-b3, 101, 1816/10/29.

232. MCD-b3, 116, 1816/11/22.

233. MCD-b3, 209, 1817/03/14; MCD-b3, 212, 1817/03/18.

234. MCD-b3, 147, 1816/12/24.

235. MCW-b87, 268, 1817/04/10; MCD-b3, 274 and 277, 1817/06/13 and 1817/06/17.

236. MCW-b87, 293, 1817/06/09; MCW-b87, 298, 1817/06/19.

237. MCW-b87, 307, 1817/08/07; MCD-b3, 365, 1817/10/31; MCD-b4, 70, 1818/05/12.

238. MCW-b88, 54, 1818/06/04.

239. MCW-b88, 56, 1818/06/11; MCW-b88, 57, 58, 61 and 62, 1818/06/15, 1818/06/25 and 1818/06/29.

240. Matthews 1827b, 24–31.

241. MCW-b88, 323, 1820/04/10.

242. "Specification of the Patent granted to Daniel Wilson," *The Repertory of Arts, Manufactures, and Agriculture*, second series 32 (1817): 11–6.

243. MCD-b2, 456, 1814/10/18; MCD-b3, 145, 1816/12/24. See also MCD-b3, 124, 1816/11/29. It was located on Brick Lane (now Central Street) at Old Street.

244. MCD-b3, 212, 1817/03/18; MCD-b3, 320, 1817/08/22.

245. "Specification of the Patent granted to Rueben [*sic*] Phillips," *The Repertory of Arts, Manufactures, and Agriculture* second series 33 (1818): 67–70.

246. MCD-b3, 350, 1817/10/07.

247. MCD-b3, 362, 1817/10/28.

248. MCD-b3, 403, 1817/12/23; MCD-b3, 448–9, 1818/03/07; MCD-b4, 61, 1818/06/02.

249. Matthews 1827b, 29.

250. "Specification of the Patent granted to George Holworthy Palmer," *The Repertory of Arts, Manufactures, and Agriculture*, second series, 34 (1819): 196–8. 1818/04/16, MCW-b88, 38. 1819/03/26, MCD-b4, 274. Charcoal and muriate of lime were also used to purify, suggested by Grafton 1819/10/29, MCD-b4, 410 413.

251. MCW-b88, 14, 1818/02/05; MCW-b88, 18, 1818/02/26.

252. MCW-b88, 17, 1818/02/23; MCD-b5, 157, 1820/08/18; MCD-b5, 167–8, 1820/09/01.

253. Bowditch 1867, 16–7, 19–21.

254. MCD-b4, 331, 1819/06/18.

255. Clegg 1841, 166.

256. MCD-b1, 348, 1814/06/10. Syphons were definitely in use a few months later. See MCD-b2, 506, 1814/12/09.

257. MCW-b87, 13, 1815/05/19.

258. What was called a syphon in the gas industry was not what is normally called a siphon. For a rant against the gas industry's naming conventions, see Peckston 1819, 279–280.

259. The GLCC decided to use portable pumps to empty syphons. See MCW-b87, 219, 1816/11/12; MCD-b3, 114, 1816/11/15; MCD-b3, 195, 1817/02/25; MCD-b3, 416, 1818/01/20.

260. MCD-b3, 278, 1817/06/17; MCD-b3, 281, 1817/06/20; MCW-b87, 287, 1817/05/15.

261. Peckston 1819, 279–282.

262. MCD-b1, 415, 1814/09/02; MCD-b1, 420, 1814/09/06.

263. MCD-b2, 482–3, 1814/11/15; MCD-b2, 500, 1814/12/02.

264. MCW-b87, 13, 1815/05/19.

265. MCD-b2, 739, 1815/10/15; MCD-b2, 743, 1815/10/20.

266. MCD-b2, 748, 1815/10/27.

267. MCD-b3, 270, 1817/06/10.

268. MCD-b1, 424, 1814/09/09.

269. MCD-b2, 500, 1814/12/02.

270. MCW-b87, 292, 1817/06/05.

271. MCD-b3, 304, 1817/07/22; MCW-b87, 324, 1817/10/06.

272. MCW-b87, 265, 1817/04/03.

273. MCW-b87, 286, 1817/05/15.

274. Creighton 1824, 454–6; Matthews 1827a, 79ff.

275. Clegg 1841, 133–4.

276. "Specification of the Patent granted to Samuel Clegg." The date stated in this article is 1816, but the actual date was 1815. See *Patents for inventions. Abridgments of specifications relating to the production and applications of gas*, 1860, 16.

277. See Peckston 1819; Accum 1819; Creighton 1824; Matthews 1827b. Even Clegg junior still mentions them (Clegg 1841, 133–4). "Review: The Theory and Practice of Gas-lighting," *Philosophical Magazine and Journal* 53 (1819): 455–60, takes Peckston to task for discussing counterweights as if they were still relevant.

278. MCW-b87, 123, 1816/04/25.

279. MCD-b1, 437, 1814/09/23; MCD-b1, 441, 1814/09/30; MCD-b2, 453, 1814/10/14.

280. MCW-b87, 13, 1815/05/19.

281. MCD-b2, 643, 1815/05/26.

282. MCD-b2, 646, 1815/05/31; MCD-b2, 671, 1815/06/27.

283. MCD-b2, 741, 1815/10/17; MCD-b2, 745, 1815/10/24.

284. MCD-b2, 595, 1815/03/23; MCD-b2, 652, 1815/06/09; MCD-b2, 672, 1815/06/30.

285. MCD-b2, 753, 1815/11/03.

286. MCW-b87, 26, 1815/11/10.

287. MCD-b2, 772, 1815/12/01.

288. MCD-b2, 786, 1815/12/22.

289. MCD-b2, 879, 1816/05/03.

290. MCD-b2, 884, 1816/05/14.

291. MCW-b87, 138, 1816/05/16.

292. MCD-b2, 894, 1816/05/31; MCW-b87, 146, 1816/06/03; MCD-b2, 575, 1815/03/03.

293. MCW-b87, 147, 1816/06/06.

294. MCD-b3 p, 56, 1816/08/20.

295. MCD-b3, 87, 1816/10/08; MCD-b2, 538, 1815/01/13. It began using pipes instead in 1815: MCD-b2, 551, 1815/01/27; MCD-b3, 104, 1816/11/01.

296. *Observations on such parts of a report lately submitted to the House of Commons, on gas light establishments: as relate to the dangers of explosion* (Rivingtons and Cochran, 1823),13.

297. MCW-b88, 64, 1818/07/02.

298. MCW-b88, 237, 239, 1819/10/04 and 1819/10/07.

299. MCW-b88, 237, 239, 244, and 246, 1819/10/04, 07, 1819/10/14 and 1819/10/18.

300. MCW-b88, 246–7, 1819/10/18; MCD-b4, 397 and 401, 1819/10/12 and 1819/10/19.

301. MCW-b88, 250–1, 1819/10/21 and 1819/10/25.

302. MCW-b87, 245, 1817/01/27.

303. MCW-b88, 257–8, 1819/11/08 and 1819/11/11.

304. MCD-b1, 423–4, 1814/09/09. Clegg suggested basic rules of thumb for main size: four times the current load if there are no lamps attached, eight times the current load if there are lamps attached.

305. MCD-b2, 448, 1814/10/07.

306. MCD-b2, 588, 1815/03/17; MCD-b2, 617, 1815/04/28; MCD-b2, 684, 1815/07/11. On the company's mains in early 1815, see "Interesting particulars respecting gas lights," *The Entertaining Magazine* 3 (1815): 76–7.

307. MCD-b2, 770–1, 1815/11/20.

308. MCD-b2, 738, 1815/10/13; MCW-b87, 325, 1817/10/09; A. J. B. Defauconpret, *Six mois à Londres en 1816* (Alexis Eymery et Delaunay, 1817), 38.

309. MCW-b87, 265, 1817/04/03; MCW-b87, 302, 1817/07/14.

310. MCW-b88, 254, 1819/11/04; MCD-b4, 129, 1818/08/28; MCW-b88, 308, 1820/03/09. Clegg 1841, 169.

311. MCW-b87, 331, 1817/11/03; MCW-b88, 83, 1818/09/03.

312. MCW-b88, 79–81, 1818/08/31.

313. MCD-b4, 132, 1818/09/01.

314. MCW-b88, 94–96, 1818/10/05.

315. MCW-b88, 106 and 109, 1818/10/22 and 1818/10/26; MCW-b88, 258, 1819/11/11.

316. MCD-b3, 226, 1817/04/11.

317. MCW-b88, 100–1, 1818/10/12.

318. MCD-b3, 340, and 344, 1817/09/26, 28, and 1817/09/30.

319. MCD-b4, 395, 1819/10/08; MCW-b88, 249, 1819/10/21.

320. Henderson and Escher 1968, 162.

321. Dubergier, *Le glaneur à Londres, ou l'observateur français* (Paris, 1820), 69.

322. M. d'Avot, *Lettres sur l'Angleterre, ou Deux années à Londres* (C. Painparré, 1821), 270.

323. *The Life and Correspondence of Robert Southey*, ed. C. C. Southey (Longmans, 1849–50), vol. 5, 116.

324. Thomas Pennant, and John Wallis, *Wallis's Guide to London*, fourth edition (Sherwood, 1814), 123. For other references to the brilliance of gaslights, see *The Annual Register; or, A view of the history, politics, and literature for the year 1815* (Baldwin, Cradock and Joy, 1824), 301; Fisher 1866, vol. 1, 145–6.

325. *Trewman's Exeter Flying Post or Plymouth and Cornish Advertiser*, 1815/06/01.

326. MCD-b2, 705, 1815/08/25.

327. MCD-b3, 296, 1817/07/11.

328. For an example of a dispute between a paving commission and the GLCC leading to a lawsuit, see *Morning Chronicle*, 1817/12/05.

329. MCD-b2, 631, 1815/05/09.

330. MCD-b2, 683, 1815/07/11.

331. MCD-b2, 740, 1815/10/17.

332. MCD-b2, 753, 1815/11/03.

333. *Morning Chronicle*, 1815/11/03.

334. MCD-b2, 474, 1814/11/08.

335. MCD-b2, 615–6, 1815/04/25 and 28.

336. MCD-b2, 720–1, 1815/09/19 and 22.

337. "Specification of the Patent granted to Samuel Clegg," 1816, 5–7; MCD-b2, 686, 1815/07/14; MCD-b2, 722, 1815/09/22.

338. *Minutes of Evidence taken before the Lords Committees to whom was referred the Bill, intitled, "An act to alter and enlarge the Powers of Two Acts of His present Majesty for granting certain Powers to the Gas Light and Coke Company"* 1816, 37. Meters were tried in 1818 but failed again: MCD-b4, 116 and 128, 1818/08/11 and 1818/08/25; MCD-b4, 158, 1818/09/29; MCD-b4, 176, 1818/10/23.

339. *Morning Chronicle*, 1815/12/27.

340. MCD-b2, 793 and 795, 1816/01/02 and 1816/01/05. Although is not clear whether the Harris the GLCC was dealing with was Thomas Harris (d. 1820) or his son Henry, Thomas handed over ownership in 1809. Henry Saxe Wyndham, *The Annals of Covent Garden Theater from 1732 to 1897* (Chatto and Windus, 1906), vol. 2, 293.

341. MCD-b2, 839, 841, 1816/03/15.

342. MCD-b2, 894, 1816/05/31; MCD-b3, 44, 1816/07/30.

343. MCW-b87, 183, 1816/08/29.

344. MCD-b2, 774, 1815/12/01.

345. MCW-b87, 168, 1816/08/01.

346. MCW-b87, 186, 1816/08/29.

347. MCW-b87, 186, 1816/08/29; MCD-b3, 69, 1816/09/10.

348. MCD-b3, 72, 1816/09/13; MCD-b3, 77, 1816/09/24.

349. MCD-b3, 112, 1816/11/12.

350. MCW-b87, 192, 1816/09/09.

351. MCD-b3, 222, 1817/04/01.

352. MCD-b3, 281, 1817/06/20.

353. MCD-b3, 303, 1817/07/22.

354. MCD-b3, 328 and 344, 1817/09/05 and 1817/09/30.

355. MCW-b88, 92, 1818/10/02.

356. MCW-b88, 116, 1818/11/16; MCD-b4, 194, 1818/11/17.

357. MCD-b4, 198, 1818/11/24. For a description of the theater lamps at this time, see *Morning Chronicle*, 1818/09/05.

358. MCW-b88, 145, 1819/02/01.

359. MCW-b88, 175, 1819/04/22.

360. MCD-b4, 332, 1819/06/18; MCD-b4, 337, 1819/06/25.

361. MCD-b4, 344, 1819/07/02; MCD-b4, 349, 1819/07/09; MCD-b4, 353, 1819/07/16.

362. MCD-b4, 382, 1819/09/17.

363. MCW-b88, 224, 1819/09/27.

364. MCD-b4, 388, 1819/10/01.

365. MCD-b4, 432, 1819/11/23.

366. MCD-b4, 436, 1819/11/30; *Times*, 1819/12/11.

367. MCD-b4, 443, 1819/12/07.

368. MCD-b5, 137, 1820/07/14.

369. MCD-b5, 143, 1820/07/21; MCD-b5, 170, 1820/09/08.

370. MCD-b5, 170, 1820/09/08.

371. MCD-b5, 173, 1820/09/14.

372. MCD-b5, 181, 1820/09/22; MCD-b5, 202, 1820/10/27.

373. Schivelbusch 1988, 28–9.

374. MCD-b1, 294, 1814/04/05.

375. MCD-b1, 356 and 359, 1814/06/21 and 1814/06/24; MCD-b1, 407, 1814/08/26.

376. MCD-b2, 446, 1814/10/04.

377. MCD-b2, 592, 1815/03/21; MCD-b2, 602, 1815/03/31. Later service pipes would be made of wrought iron or pewter. Clegg 1841, 186.

378. MCW-b87, 9, 1815/04/11; MCW-b87, 21, 1815/10/06; MCD-b3, 267, 1817/06/03.

379. MCD-b2, 633, 1815/05/12.

380. MCW-b87, 13, 1815/05/19; MCD-b2, 676, 1815/07/03.

381. MCD-b2, 748, 1815/10/27.

382. MCW-b87, 33, 1815/11/22; MCD-b2, 772, 1815/12/01.

383. MCD-b2, 735, 1815/10/10.

384. MCD-b2, 615, 1815/04/25.

385. MCD-b2, 700, 1815/08/18.

386. MCD-b2, 772, 1815/12/01.

387. MCD-b3, 86, 1816/10/08.

388. MCW-b87, 171, 1816/08/05.

389. MCD-b2, 737, 1815/10/10.

390. MCD-b2, 785, 1815/12/19; MCD-b2, 872, 1816/04/23.

391. MCW-b87, 151, 1816/06/13; MCW-b87, 161, 1816/07/04.

392. MCD-b3, 181 and 184, 1817/02/07 and 1817/02/11.

393. "Specification of the Patent granted to Samuel Clegg," 1816, 5–7.

394. MCD-b2, 884, 1816/05/14; MCW-b87, 153, 1816/06/13; MCD-b3, 10, 1816/06/21; MCW-b87, 159, 1816/07/01; *Minutes of Evidence taken before the Lords Committees to whom was referred the Bill, intitled, "An act to alter and enlarge the Powers of Two Acts of His present Majesty for granting certain Powers to the Gas Light and Coke Company*, 1816, 33, 37; Harvey 1823, 82–4.

395. "Specification of the Patent granted to Samuel Clegg," 1816; "John Malam's Improved Gas-Meter," *Transactions of the Society, Instituted at London, for the Encouragement of Arts, Manufactures, and Commerce* 37 (1820): 167–71; "Crossley v. Beverley," in *Law reports of patent cases [1602–1842]*, ed. W. Carpmael (A. Macintosh, 1843), 480–8. "Crossley v. Beverley," in *Decisions on the law of patents for inventions rendered by English Courts since the beginning of the seventeenth century*, ed. B. V. Abbott, vol. 1 (Charles R. Brodix, 1887), 409–16.

396. Clegg 1841, 127.

397. Matthews 1827b, 44–6; Matthews 1827a, 77; Matthews 1983, 37–42; Matthews 1986, 253.

398. MCD-b3, 201, 1817/03/04.

399. MCD-b2, 446, 1814/10/04; MCD-b2, 598, 1815/03/28.

400. MCD-b2, 649, 1815/06/02.

401. MCD-b2, 530, 1814/12/30.

402. MCD-b2, 538, 1815/01/13.

403. MCD-b2, 575, 1815/03/03.

404. MCD-b2, 601, 1815/03/31.

405. MCD-b2, 652 and 672, 1815/06/09 and 1815/06/30.

406. MCD-b2, 702 and 728, 1815/08/18.

407. MCD-b2, 753, 1815/11/03.

408. MCW-b87, 33, 1815/11/22; MCW-b87, 161, 1816/07/04; MCW-b87, 219, 1816/12/05.

409. MCW-b87, 279, 1817/05/01.

410. MCD-b2, 747, 1815/10/27.

411. MCD-b2, 681, 1815/07/07.

412. MCD-b3, 59 and 62, 1816/08/23 and 1816/08/30.

413. MCD-b3, 79, 1816/09/24; MCD-b3, 116, 1816/11/19.

414. MCD-b2, 742, 1815/10/20.

415. MCD-b2, 742, 1815/10/20; MCD-b2, 830, 1816/03/01.

416. MCD-b3, 117, 1816/11/22.

417. MCD-b3, 74, 1816/09/20; MCD-b3, 171, 1817/01/24.

418. MCD-b3, 251, 1817/05/13.

419. MCD-b3, 93, 1816/10/18; MCD-b3, 167, 1817/01/17.

420. MCD-b4, 148, 1818/09/18.

421. MCW-b88, 130–1, 1818/12/21.

422. MCD-b4, 419, 1819/11/12.

423. Johann Valentin Hecke, *Reise durch die Vereinigten Staaten von Nord-Amerika in den Jahren 1818 und 1819, und Rückreise durch England* (H. Ph. Petri., 1820), vol. 2, 286.

424. *Minutes of Evidence taken before the Lords Committees to whom was referred the Bill, intitled, "An act to alter and enlarge the Powers of Two Acts of His present Majesty for granting certain Powers to the Gas Light and Coke Company,"* 1816, 5, 6, 10. The statements are inconsistent, but it appears that there were 700 public lamps, 7,000 lamps in 1,700 homes, and 1,000 shops. The list of public customers on pp. 31 and 32 includes 586 lamps and 21 contracts.

425. MCP, 94, 1817/01/06.

426. MCD-b1, 366, 1814/07/05; Gas Light and Coke Company, "Extracts of the Proceeding of the Half-yearly meeting of the Court of Proprietors," 6 May 1820, 2. In 1822 the GLCC supplied 31,1334 lamps and had 122 miles of mains. Banks 1823, 27–32.

427. Peckston 1819, 283–4.

428. J. Schweigger, "Ueber Gasbeleuchtung in London," *Journal für Chemie und Physik* 17 (1816): 376–83; J. Prechtl, "Aus mehreren Briefen," *Annalen der Physik* 58 (1818): 111.

429. Williot 1999, 34, 50, 55, 62–3; "Specification of the Patent granted to John Grafton," *The Repertory of Arts, Manufactures, and Agriculture*, second series, 38 (1821): 272–5.

430. Petersen 1990, 37.

431. Körting 1963; Hill 1950; Hutchinson 1985, 256–8; Treue 1967, 1006; Kooij 2006, 5–6.

BIBLIOGRAPHY

ARCHIVAL SOURCES

The records of the Gas Light and Coke Company are held in the London Metropolitan Archives. These consist of a series of minute books:

Minutes of the Court of Proprietors document B/GLCC/64/1. (MCP)

Minutes of the Court of Directors document B/GLCC/1/1–5. (MCD)

Minutes of the Committee of Works document B/GLCC/1/87–88. (MCW)

Minutes of the Committee of Chemistry B/GLCC/146.

The Boulton & Watt archives are in the Birmingham Central Library, and some items are available on microfilm from Adam Matthew Publishers.

PARLIAMENTARY ACTS AND GOVERNMENT PAPERS

An act [4 Geo. IV cap. cxix] to enlarge the powers of the Gas Light and Coke Company.

An act [50 Geo. III cap. clxiii] for granting certain Powers and Authorities to a Company to be incorporated by Charter, to be called The Gas Light and Coke Company for making Inflammable Air for the lighting of the Streets of the Metropolis, and for procuring Coke, Oil, Tar, Pitch, Asphaltum, Ammoniacal Liquor, and essential Oil, from Coal, and for other Purposes relating thereto. 9 June 1810.

An act [54 Geo. III cap. cxvi] for enlarging the powers of an act of His Present Majesty for granting certain powers and authorities to the Gas Light and Coke Company. 17 June 1814.

An act [56 Geo. III cap. lxxxvii] to alter and enlarge the Powers of Two Acts of His Present Majesty, for granting certain Powers to the Gas Light and Coke Company. 2 July 1816.

An Act [59 Geo. III cap. xx] to alter and enlarge the powers of The Gas Light and Coke Company and to amend three acts of His present Majesty, relating to the said company. 7 April 1819.

Banks, Joseph. 1823. Report of Royal Society on gas-lights; Report of inspectors on gas-light establishments in the Metropolis. In House of Commons papers: accounts and papers. V.303.

Bennet, Henry Gray. 1818. Third report from the Committee on the State of the Police of the Metropolis: with minutes of the evidence taken before the committee; and an appendix. In House of Commons papers. Reports of committees. VIII.1 (423).

Estcourt, Thomas Grimston Bucknall. 1828. Report from the Select Committee on the Police of the Metropolis. In House of Commons papers. Reports of committees. VI.1. (533).

Hall, James. 1809. Select committee on Gas-Light and Coke Company's bill to incorporate persons for procuring coke, oil, tar, pitch, ammoniacal liquor and inflammable air from coal: minutes of evidence. In House of Commons papers: reports of committees. III.315.

Harvey, Eliab. 1823. Select Committee on Report of Royal Society on Gas-Lights, and Reports on Gas-Light Establishments. Report, Minutes of Evidence, Appendix. In House of Commons papers: reports of committees. V.215. (529).

Minutes of Evidence taken before the Lords Committees to whom was referred the Bill, intitled, "An act for enabling His Majesty to incorporate by Charter, to be called 'The Gas Light and Coke Company' for making Inflammable Air for the lighting of the Streets of the Metropolis, and for procuring Coke, Oil, Tar, Pitch, Asphaltum, Ammoniacal Liquor, and essential Oil, from Coal, and for other Purposes relating thereto." 1810. In House of Lords papers: reports of committees. XXXI.265.

Minutes of Evidence taken before the Lords Committees to whom was referred the Bill, intitled, "An act to alter and enlarge the Powers of Two Acts of His present Majesty for granting certain Powers to the Gas Light and Coke Company. 1816. In House of Lords papers: reports of committees. LXXXVII.38.

PRIMARY SOURCES

Accum, Frederick Christian. 1800. *Chemical Recreations: Comprehending a variety of entertaining and striking experiments, which may be performed with ease and facility.* n.p.

Accum, Frederick Christian. 1803. *A System of Theoretical and Practical Chemistry.* G. Kearsley.

Accum, Frederick Christian. 1804a. "Description of an improved Gazometer." *The Repertory of Arts, Manufactures, and Agriculture*, second series, 5: 179–182.

Accum, Frederick Christian. 1804b. *A Practical Essay on the Analysis of Minerals.* G. Kearsley, J. Johnson and J. Callow.

Accum, Frederick Christian. 1805. *Catalogue of Chemical Preparations for Philosophical Chemistry, and Apparatus and Instruments.* Hayden.

Accum, Frederick Christian. 1807. *System of Theoretical and Practical Chemistry*, second edition. G. Kearsley.

Accum, Frederick Christian. 1815. *A Practical Treatise on Gas-Light; exhibiting a summary description of the apparatus and machinery best calculated for illuminating streets, houses, and manufactories, with carburetted hydrogen, or coal-gas.* G. Hayden.

Accum, Frederick Christian. 1819. *Description of the Process of Manufacturing Coal Gas, for the Lighting of Streets, Houses, and Public Buildings, with elevations, sections, and plans of the most improved sorts of apparatus now employed at the gas works in London.* T. Boys.

Accum, Friedrich, and Wilhelm August Lampadius. 1816. *Praktische Abhandlung über das Gaslicht: eine vollständige Beschreibung des Apparats und der Maschinerie, um Straßen, Häuser und Manufacturen damit zu beleuchten, enthaltend.* Landes-Industrie-Comptoir.

Accum, Frederick Christian, and Frederick Albert Winsor. 1816. *Traité pratique de l'éclairage par le gaz inflammable contenant une description sommaire de l'appareil et du mécanisme employés pour l'illumination des rues, des maisons, et des manufactures à l'aide du gaz hydrogène carburé, tiré du charbon de terre.* Nepveu.

Aiton, William. 1811. *General View of the Agriculture of the County of Ayr; with observations on the means of its improvement; drawn up for the consideration of the Board of agriculture, and internal improvements.* A. Napier.

Bailey, John. 1810. *General View of the Agriculture of the County of Durham with observations on the means of its improvement; drawn up for the consideration of the Board of Agriculture and Internal Improvement.* Richard Phillips.

Beatson, Robert. 1807. *A Chronological Register of Both Houses of the British Parliament, from the union in 1708, to the third Parliament of the United Kingdom of Great Britain and Ireland, in 1807.* J. Chalmers for Longman, Hurst, Rees, and Orme.

Becher, Johann Joachim. 1682. *Närrische Weissheit und weise Narrheit.* Johann Peter Zubrods.

Beddoes, Thomas, and James Watt. 1795. *Considerations on the Medicinal Use, and on the Production of Factitious Airs.* J. Johnson.

Berzelius, Jöns Jacob. 1841. *Lehrbuch der Chemie*, third edition. Arnold.

Blagdon, Francis William. 1803. *Paris As It Was and As It Is; or, a sketch of the French capital*, volume 1. C. and R. Baldwin.

Bowditch, W. R. 1867. *The Analysis, Technical Valuation, Purification and Use of Coal Gas.* E. and F. N. Spon.

Box, James. 1817. *An Account of the Society, Denominated the Associated Proprietors of the Gas Light and Coke Company containing particulars of the rise, the progress, the principles, the object, and the usefulness of that society.* Printed and sold for the author.

Boyle, Robert. 1665. *New Experiments and Observations Touching Cold, or an experimental history of cold, begun.* John Crook.

Boyle, Robert. 1672. *Tracts Written by the Honourable Robert Boyle containing new experiments, touching the relation betwixt flame and air, and about explosions, an hydrostatical discourse occasion'd by some objections of Dr. Henry More against some explications of new experiments made by the author of these tracts.* Printed for Richard Davis.

Boyle, Robert. 1685. *An Essay of the Great Effects of Even Languid and Unheeded Motion. Whereunto is annexed An experimental discourse of some little observed causes of the insalubrity and salubrity of the air and its effects.* M. Flesher.

Brisson, Mathurin-Jacques. 1797. *Traité élémentaire ou principes de physique; fondés sur les connoissances les plus certaines, tant anciennes que modernes, & confirmés par l'expérience,* second edition, volume 2. Moutard.

Clegg, Samuel. 1820. *Description of an Apparatus, by Which Twenty-Five Thousand Cubic Feet of Gas, Are Obtained from Each Chaldron of Coal, Without Producing Either Tar or Ammoniacal Liquor.* R. Ackerman.

Clegg, Samuel, Jr. 1841. *A Practical Treatise on the Manufacture and Distribution of Coal-Gas: its introduction and progressive improvement, illustrated by engravings from working drawings, with general estimates.* J. Weale.

Clifford, Frederick. 1885. *A History of Private Bill Legislation.* Butterworth.

Considerations on the Nature and Objects of the Intended Light and Heat Company: published by authority of the committee. T. Davison for James Ridgway, 1808.

Considerations on the Nature and Objects of the Intended Light and Heat Company: published by authority of the committee, with improvements and additional experiments, second edition. T. Davison for James Ridgway, 1808.

Cooper, Thomas. 1816. *Some Information Concerning Gas Lights.* John Conrad.

Creighton, Henry. 1824. Gas-lights. In *Encyclopaedia Britannica, or A dictionary of arts, sciences and miscellaneous literature. Supplement to the Fourth, Fifth, and Sixth Editions.* A. Constable.

Ehrmann, Friedrich Ludwig. 1780. *Déscription et usage de quelques lampes à air inflammable.* J. H. Heitz.

Fisher, George Park. 1866. *Life of Benjamin Silliman, M.D., LL. D., Late Professor of Chemistry, Mineralogy, and Geology in Yale college. Chiefly from his manuscript reminiscences, diaries, and correspondence.* Scribner.

Genssane, Antoine de. 1770. *Traité de la fonte des mines par le feu du charbon de terre.* Vallat-la-Chapelle.

Granger, Joseph, and William Appleby. 1794. *General View of the Agriculture of the County of Durham, particularly that part of it extending from the Tyne to the Tees; with observations on the means of its improvement*. Colin Macrae.

Grant, James Ludovic. 1809. *The Second Report of James Ludovic Grant, Esq. Chairman, and the Other Acting Trustees of the Fund for Assisting Mr. Winsor in His Experiments to the Subscribers of That Fund, at a meeting convened at a meeting convened by public advertisements, at the Crown and Anchor Tavern in the Strand, on the 29th of August 1809*. G. Woodfall.

Grant, James Ludovic, and Frederick Christian Accum. 1808. *The Report of James Ludovic Grant, Esq. Chairman, and the Other Acting Trustees of the Fund for Assisting Mr. Winsor in His Experiments to the Subscribers of That Fund, at a meeting convened . . . on the 26th of May, 1808*. G. Sidney.

Henry, William. 1805. "Experiments on the gases obtained by the destructive distillation of pit coal, &c., with a view to the theory of their combustion when employed as sources of artificial light." *A Journal of Natural Philosophy, Chemistry and the Arts* 11: 65–74.

Henry, William. 1808. "Description of an apparatus for the analysis of the compound inflammable gases by slow combustion; with experiments on the gas from coal, explaining its application." *Philosophical Transactions of the Royal Society of London* 98: 282–303.

Hollunder, Christian. 1824. *Tagebuch einer metallurgisch-technologischen Reise, durch Mähren-Böhmen, einen Theil von Deutschland und der Niederlande*. J. L. Schrag.

"Hugo Franz Altgraf zu Salm-Reifferscheid-Krautheim, Herr der Mährische herrschaften Raitz, Jedownitz und des Ollmützer Lehens Blansko." *Taschenbuch für die vaterländische Geschichte* 29, new series (1840), no. 11: 523–596.

Ingenhousz, Jan. 1782. *Vermischte Schriften phisisch-medizinischen Inhalts*. Johann Paul Krauss.

Knapp, Friedrich Ludwig. 1847. *Lehrbuch der chemischen Technologie: zum Unterricht und Selbststudium*. Vieweg.

Krünitz, Johann Georg. 1793. Lampe. In *Oekonomische-Technologische Encyclopädie*, volume 59, ed. J. G. Krünitz. J. Pauli.

Lampadius, W. A. 1816. *Neue Erfahrungen im Gebiete der Chemie und Hüttenkunde*. Landes-Industrie-Comptoir.

Lebon, Philippe. 1799. Moyens nouveaux d'employer les combustibles plus utilement & à la chaleur & à la lumière, & d'en receuillir leur sous-produits. In René Masse, *Le Gaz*, Annexe I, 1914.

Lebon, Philippe. 1801a. Additions. In René Masse, *Le Gaz*, Annexe I, 1914.

Lebon, Philippe. 1801b. "Thermolampes." *Journal général de la littérature de France: ou répertoire méthodique des livres nouveaux, cartes géographiques, estampes et oeuvres de musique* 4: 253.

Lebon, Philippe. 1801c. *Thermolampes, ou poêles qui chauffent, éclairent avec economie, et offrent, avec plusiers produits precieux, une force motrice applicable à toute espèce de machines*. Pougens.

"Lebons Thermolampen." *London und Paris* 8 (1801): 206–213.

Liebig, Justus. 1849. *Handwörterbuch der Reinen und Angewandten Chemie*, volume 4. Vieweg.

Marchais, Jean Baptiste Charles. 1801. "Note sur la 1ere expérience du thermolampe." *Mémoires des sociétés savantes et littéraires de la République française* 1: 384–347.

Matthews, William. 1827a. *An Historical Sketch of the Origin, Progress, & Present State of Gas-Lighting*. R. Hunter.

Matthews, William. 1827b. *Compendium of Gas-Lighting, adapted for the use of those who are unacquainted with chemistry; containing an account of some new apparatus lately introduced*. R. Hunter.

Minckelers, Jean Pierre. 1784. *Mémoire sur l'air inflammable tiré de différentes substances*. n.p.

Murdoch, William. 1808. "An account of the application of the gas from coal to economical purposes." *Philosophical Transactions of the Royal Society of London* 98: 124–132.

Murdoch, William. 1809. *A Letter to a Member of Parliament, from Mr. William Murdock in Vindication of His Character and Claims, in reply to a recent publication by the committee for conducting through Parliament a bill for incorporating a Gas-Light & Coke Company*. Galabin and Marchant.

Patents for Inventions. Abridgments of specifications relating to the production and applications of gas. G. E. Eyre and W. Spottiswoode, 1860.

Peckston, Thomas S. 1819. *The Theory and Practice of Gas-Lighting: in which is exhibited an historical sketch of the rise and progress of the science, and the theories of light, combustion, and formation of coal, with descriptions of the most approved apparatus for generating, collecting, and distributing, coal-gas for illuminating purposes*. Thomas and George Underwood.

Poppe, Johann Heinrich Moritz von. 1811. *Geschichte der Technologie seit der Wiederherstellung der Wissenschaften bis an das ende des achtzehnten Jahrhunderts*, volume 3. J. F. Röwer.

Salm, Hugo Franz Altgraf. 1836. In *Oesterreichische National-Encyklopädie, oder, Alphabetische Darlegung der wissenswürdigsten Eigenthümlichkeiten des österreichischen Kaiserthumes*, ed. J. Czikann and F. Gräffer. Beck'schen Universitäts-Buchhandlung.

"Specification of the patent granted to Frederick Albert Winsor for an improved oven, stove, or apparatus, for the purpose of extracting air, oil, pitch, tar, and acids, from, and reducing into, coke and charcoal, all kinds of fuel." *The Repertory of Arts, Manufactures, and Agriculture*, second series, 5 (1804): 172–174.

"Specification of the Patent granted to Samuel Clegg." *The Repertory of Arts, Manufactures, and Agriculture*, second series, 30 (1816): 1–10, 65–75, 129–137.

"Thermolampe." *Englische Miscellen* 19 (1805): 68–74.

The Parliamentary Debates from the Year 1803 to the Present Time. T. C. Hansard, 1812–1820.

Van Voorst, John. 1809. *An Address to the Proprietors of the Intended Gas Light and Coke Company to which is annexed, an epitome of the evidence taken before the committee of the House of Commons*. Printed for the author.

Volta, Alessandro. 1777. *Lettere del Signor Don Alessandro Volta . . . sull'aria infiammabile nativa delle paludi*. G. Marelli.

Volta, Alessandro. 1778. "Due lettere a Jean Senebier sopra esperienze e considerazioni sull'aria infiammabile." In *Le opere di Alessandro Volta*, volume. 6.. Hoepli [1918].

Volta, Alessandro. 1783. "Seconds Partie du memoire de M. de Volta sur les Isolements Imparfaits." *Observations sur la physique, sur l'histoire naturelle et sur les arts* 23: 3–16.

Volta, Alessandro. 1784a. "Memoria sopra i fuochi de terreni e delle fontane ardenti in generale, e sopra quelli di Pietra-Mala in particolare." *Opuscoli scelti sulle scienze e sulle arti: tratti dagli atti delle accademie, e dalle altre collezione filosofiche, e letterarie, dalle opere più recenti inglesi, tedesche, francesi, latine, e italiane, e da manoscritti originali, e inediti* 7: 321–333.

Volta, Alessandro. 1784b. "Memoria sopra i fuochi de' terreni e delle fontane ardenti in generale, e spora quelli di Pietra-Mala in particolare." *Memorie di matematica e fisica della Società italiana* 2: 662.

Volta, Alessandro. 1918. *Le opere di Alessandro Volta*. Hoepli.

Volta, Alessandro. 1949. *Epistolario di Alessandro Volta*. N. Zanichelli.

Wagner, Johan. 1802. "Versuche über Lebons Thermolampen, und deren Beschreibung." *Annalen der Physik* 10: 491–499.

Wankel, Heinrich. 1882. *Bilder aus der mährischen Schweiz und ihrer Vergangenheit*. Adolph Holzhauser.

Warren, Charles, and William Harrison. 1809. *Speeches of Charles Warren, Esq. and William Harrison, Esq. delivered before a committee of the Honourable House of Commons in support of a bill, to authorise His Majesty to grant a charter of incorporation to certain persons, by the name of the Gas Light and Coke Company*. G. Woodfall.

Watt, James Muirhead James Patrick. 1854. *The Origin and Progress of the Mechanical Inventions of James Watt*. J. Murray.

Webster, John. 1807. "Account of the discovery of the means of illumination by gas from coal." *A Journal of Natural Philosophy, Chemistry and the Arts* 16: 83.

Winsor, Frederick Albert. 1798. *Addresse aux souverains de l'Europe etc. etc en deux parties*. G. G. & J. Robinson, Wm. Richardson.

Winsor, Frederick Albert. 1799. *Prosperity of England Midst the Clamors of Ruin*. Printed for the author.

Winsor, Frederick Albert. 1802. *Description of the Thermolamp Invented by Lebon, of Paris, Published with Remarks by F. A. W.* Vieweg.

Winsor, Frederick Albert. 1804a. *Account of the Most Ingenious and Important National Discovery for Some Ages. British Imperial Patent Light Ovens and Stoves, respectfully dedicated to both Houses of Parliament, and all Patriotic Societies*. T. Maiden.

Winsor, Frederick Albert. 1804b. *The Superiority of the New Patent Coke over the Use of Coals.* Richardson.

Winsor, Frederick Albert. 1806. *To Be Sanctioned by Act of Parliament. A National Light and Heat Company for providing our streets and houses with light and heat, on similar principles, as they are now supplied with water, demonstrated with the patentee's authority and instructions, by Professor Hardie, at the Theatre of Sciences, No. 98, Pall Mall*, first edition. Watts & Bridgewater for the patentee.

Winsor, Frederick Albert. 1807a. *Analogy between Animal and Vegetable Life. Demonstrating the beneficial application of the patent light stoves to all green and hot houses, etc.* Richardson.

Winsor, Frederick Albert. 1807b. *Mr. W. Nicholson's Attack, in His Philosophical Journal, on Mr. Winsor and the National Light and Heat Company: with Mr. Winsor's defence.* London.

Winsor, Frederick Albert. 1807c. *National Deposit Bank, or, The bulwark of British security, credit and commerce, in all times of difficulty, changes, and revolutions with the mercantile policy of separating the discounts of merchants and manufacturers from the law of national interest, as practised in the first commercial cities on the continent.* G. Sidney.

Winsor, Frederick Albert. 1807d. *To Be Sanctioned by Act of Parliament. A National Light and Heat Company, for providing our streets and houses with hydrocarbonic gas-lights, on similar principles, as they are now supplied with water, four tables of calculation, founded on official experiments, prove the immense national profits and increase of revenue by the adoption of this plan, which is to be had at the National Light & Heat Company's office, no.97, Pall Mall*, new edition. Watts & Bridgewater.

Winsor, Frederick Albert. 1816. *Notice historique sur l'utilisation du gaz hydrogène pour l'éclairage, avec un extrait du procès-verbal d'enquête faite par le parlement d'Angleterre sur cet éclairage, et l'application aux arts et métiers des produits tirés de la distillation du charbon de terre.* Nouzou.

Winzler, Zachäus Andr. 1803a. *Berichtigungs-Magazin der Einwürfe, Zweifel und Bedenklichkeiten gegen den neu erfundenen Koch- Heiz- Leucht- und Sparofen, oder die deutsche Thermolampe.* k. k. priv. Buchdruckerey.

Winzler, Zachäus Andr. 1803b. *Die Thermolampe in Deutschland: oder, vollständige, sowohl theoretisch- als praktische Anleitung, den ursprünglich in Frankreich erfundenen, nun aber auch in Deutschland entdekten Universal- Leucht- Heiz- Koch- Sud- Destillir- und Sparoven zu errichten. Mit vier Kupfertafeln.* Franz Karl Siedler.

"Winzler, Zachäus Andr." 1835. In *Oesterreichische National-Encyklopädie, oder, Alphabetische Darlegung der wissenswürdigsten Eigenthümlichkeiten des österreichischen Kaiserthumes*, ed. J. Czikann and F. Gräffer. Beck'schen Universitäts-Buchhandlung.

zu Salm-Reiferscheid, Hugo. 1808. "Holzverkohlung im Grossen vermittelst der Thermolampe." *Annalen der Physik* 30: 402–403.

SECONDARY SOURCES

Adry, Eg. 1925. *Un siècle d'éclairage, 1824–1924*. Ratinckx frères.

Albert, William, Michael J. Freeman, and Derek Howard Aldcroft, eds. 1983. *Transport in the Industrial Revolution*. Manchester University Press.

Allen, Edward A. 1984. "Deforestation and fuel crisis in pre-revolutionary Languedoc, 1720–1789." *French Historical Studies* 13: 455–473.

Allen, Robert C. 2009. *The British Industrial Revolution in Global Perspective*. Cambridge University Press.

Ashworth, William J. 2008. "The ghost of Rostow: Science, culture and the British industrial revolution." *History of Science* 46: 249–274.

Austerfield, Peter J. 1981. The Development of Large-Scale Production and Utilisation of Lighter-than-Air Gases in France, Britain and the Low Countries from 1783 to 1821 with Reference to Aeronautics and the Coal-Gas Industry. Ph.D. dissertation, University of London.

Awty, Brian G. 1974. "The introduction of gas-lighting to Preston." *Transactions of the Historic Society of Lancashire and Cheshire* 125: 82–118.

Ball, Michael, and David Sunderland. 2001. *An Economic History of London, 1800–1914*. Routledge.

Beaver, S. H. 1951. "Coke Manufacture in Great Britain: A Study in Industrial Geography." *Transactions and Papers (Institute of British Geographers)* 17: 133–148.

Bektas, M. Yakup, and Maurice Crosland. 1992. "The Copley Medal: The Establishment of a Reward System in the Royal Society, 1731–1839." *Notes and Records of the Royal Society of London* 46: 43–76.

Benoit, Serge. 1990. "La consommation de combustible végétal et l'évolution des systèmes techniques." In *Forges et forêts: Recherches sur la consommation et proto-industrie de bois*, ed. D. Woronoff. École des Hautes Études en Sciences Sociales.

Bensaude-Vincent, Bernadette, and Christine Blondel, eds. 2008. *Science and Spectacle in the European Enlightenment*, Ashgate.

Berg, Maxine. 1994. *The Age of Manufactures, 1700–1820: Industry, Innovation, and Work in Britain*, second edition. Routledge.

Berg, Maxine. 2007. "The genesis of 'useful knowledge'." *History of Science* 45: 123–133.

Berg, Maxine, and Kristine Bruland, eds. 1998. *Technological Revolutions in Europe: Historical Perspectives*. Elgar.

Berg, Maxine, and Pat Hudson. 1992. "Rehabilitating the Industrial Revolution." *Economic History Review* 45: 24–50.

Berman, Morris. 1978. *Social Change and Scientific Organization: The Royal Institution, 1799–1844.* Cornell University Press.

Brenni, Paolo. 2003. "Volta's electric lighter and its improvements: The birth, life and death of a peculiar scientific apparatus which became the first electric household appliance." In *Musa Musaei: Studies on Scientific Instruments and Collections in Honour of Mara Miniati,* ed. M. Beretta, P. Galluzzi, and C. Triarico. L. S. Olschki.

Brereton, Austin. 1903. *The Lyceum and Henry Irving.* Lawrence & Bullen.

Bret, Patrice. 2004. "Un bateleur de la science: Le «machiniste-physicien» François Bienvenu et la diffusion de Franklin et Lavoisier." *Annales historiques de la Révolution française* no. 338: 95–127.

Brose, Eric Dorn. 1985. "Competitiveness and obsolescence in the German charcoal iron industry." *Technology and Culture* 26: 532–559.

Browne, Charles Albert. 1925. "The life and chemical services of Frederick Accum." *Journal of Chemical Education* 2: 829–851, 1008–1034, 1140–1149.

Bruland, Kristine, and David C. Mowery. 2005. "Innovation through time." In *The Oxford Handbook of Innovation,* ed. J. Fagerberg, D. Mowery, and R. Nelson. Oxford University Press.

Buchanan, B. J. 1986. "The evolution of the English turnpike trusts: Lessons from a case study." *Economic History Review* 39: 223–243.

Cardwell, Donald. 1972. *Turning Points in Western Technology: A Study of Technology, Science and History.* Science History Publications.

Caron, François. 1988. "The evolution of the technical system of railways in France from 1832 to 1937." In *The Development of Large Technical Systems,* ed. R. Mayntz and T. Hughes. Campus.

Cartwright, F. F. 1967. "The association of Thomas Beddoes, M.D. with James Watt, F.R.S." *Notes and Records of the Royal Society of London* 22: 131–143.

Chandler, Alfred D. 1977. *The Visible Hand: The Managerial Revolution in American Business.* Belknap.

Chandler, Alfred D., and Hikino Takashi. 1990. *Scale and Scope: The Dynamics of Industrial Capitalism.* Belknap.

Chandler, Dean, and A. Douglas Lacey. 1949. *The Rise of the Gas Industry in Britain.* British Gas Council.

Chatzis, Konstantinos. 1999. "Designing and operating storm water drain systems: empirical findings and conceptual developments." In *The Governance of Large Technical Systems,* ed. O. Coutard. Routledge.

Clow, Archibald, and Nan L. Clow. 1952. *The Chemical Revolution: A Contribution to Social Technology.* Books for Libraries Press.

Cohen, H. Floris. 2004. "Inside Newcomen's fire engine, or: The scientific revolution and the rise of the modern world." *History of Technology* 25: 111–132.

Cottrell, P. L. 1980. *Industrial Finance, 1830–1914: The Finance and Organization of English Manufacturing Industry.* Methuen.

Crafts, N. F. R. 1994. "The Industrial Revolution." In *The Economic history of Britain since 1700*, volume 1: *1700–1860*, ed. R. Floud and D. McCloskey. Cambridge University Press.

Crafts, N. F. R. 1995. "Exogenous or endogenous growth? The Industrial Revolution reconsidered." *Journal of Economic History* 55: 745–772.

De Clercq, Peter. 1988. "Science at court: The eighteenth-century cabinet of scientific instruments and models of the Dutch Stadholders." *Annals of Science* 45: 113–152.

Deane, Phyllis. 1979. *The First Industrial Revolution*, second edition. Cambridge University Press.

Dickinson, H. W. 1936. *James Watt, Craftsman and Engineer.* Cambridge University Press.

Dickinson, H. W. 1937. *Matthew Boulton.* Cambridge University Press.

Divall, Colin, and Sean Johnston. 2000. *Scaling Up: The Institution of Chemical Engineers and the Rise of a New Profession.* Kluwer.

Duchesne, Ricardo. 2005. "Peer Vries, the Great Divergence, and the California School: Who's in and who's out?" *World History Connected* 2.

Duchesne, Ricardo. 2006. "Max Weber is the measure of the West: A further argument on Vries and Goldstone." *World History Connected* 4.

Edwards, John Richard. 1989. *A History of Financial Accounting.* Routledge.

Elton, Arthur. 1958. Gas for light and heat. In *A History of Technology*, volume IV: *The Industrial Revolution c. 1750 to c. 1850*, ed. C. Singer, E. Holmyard, A. Hall, and T. Williams. Oxford University Press.

Everard, Stirling. 1949. *The History of the Gas Light and Coke Company, 1812–1949.* Benn.

Falkus, Malcolm. 1977. "The development of municipal trading in the nineteenth century." *Business History* 19: 134–161.

Falkus, Malcolm E. 1982. "The early development of the British gas industry, 1790–1815." *Economic History Review* 35: 217–234.

Farrar, W. V. 2008. "Reichenbach, Karl (or Carl) Ludwig." In *Complete Dictionary of Scientific Biography.* Scribner.

Fester, Gustav Anselm. 1923 [1969]. *Die Entwicklung der chemischen Technik bis zu den Anfängen der Grossindustrie; ein technologisch-historischer Versuch.* M. Sändig.

Fischer, Claude S. 1992. *America Calling: A Social History of the Telephone to 1940.* University of California Press.

Flichy, Patrice. 2007. *Understanding Technological Innovation: A Socio-Technical Approach.* Elgar.

Fox, Celina. 2009. *The Arts of Industry in the Age of Enlightenment.* Yale University Press.

Franklin, Benjamin. 1904. *The Works of Benjamin Franklin, including the private as well as the official and scientific correspondence together with the unmutilated and correct version of the Autobiography,* ed. J. Bigelow. Putnam and Knickerbocker.

Gericke, Hans Otto. 1998. "Von der Holzkohle zum Koks. Die Auswirkungen der "Holzkrise" auf die Mansfelder Kupferhütten." *Vierteljahrschrift für Sozial- und Wirtschaftsgeschichte* 85: 156–195.

Gillispie, Charles C. 1957a. "The discovery of the Leblanc process." *Isis* 48: 152–170.

Gillispie, Charles C. 1957b. "The natural history of industry." *Isis* 48: 398–407.

Gli Strumenti di Alessandro Volta: Il Gabinetto di fisica dell'Università di Pavia. Hoepli, 2002.

Golinski, Jan. 1992. *Science as Public Culture: Chemistry and Enlightenment in Britain, 1760–1820.* Cambridge University Press.

Golinski, Jan. 2008. "Joseph Priestley and the chemical sublime in British public science." In *Science and Spectacle in the European Enlightenment,* ed. B. Bensaude-Vincent and C. Blondel. Ashgate.

Gourvish, Terence R. 1972. *Mark Huish and the London & North Western Railway: A Study of Management.* Leicester University Press.

Gourvish, Terence R. 1973. "A British business elite: The chief executive managers of the railway industry, 1850–1922." *Business History Review* 47: 289–316.

Griffiths, John Charles. 1992. *The Third Man: The Life and Times of William Murdoch, 1754–1839, the Inventor of Gas Lighting.* A. Deutsch.

Grundmann, Reiner. 1999. "On control and shifting boundaries: Modern society in the web of systems and networks." In *The Governance of Large Technical Systems,* ed. O. Coutard. Routledge.

Guerlac, Henry. 1957a. "Joseph Black and fixed air: A bicentenary retrospective, with some new or little known material." *Isis* 48: 124–151.

Guerlac, Henry. 1957b. "Joseph Black and fixed air: Part II." *Isis* 48: 433–456.

Hall, A. Rupert. 1974. "What did the Industrial Revolution in Britain owe to science?" In *Historical Perspectives: Studies in English Thought and Society, in Honour of J. H. Plumb,* ed. N. McKendrick. Europa.

HalvaDenk, Helma. 1991. Bedeutende Südmährer. Offizielle Homepage des Südmährischen Landschaftsrats (www.suedmaehren.eu).

Harris, J. R. 1966. "Copper and shipping in the eighteenth century." *Economic History Review* 19: 550–568.

Harris, J. R. 1976. "Skills, coal and British industry in the eighteenth century." *History* 61: 167–182.

Harris, J. R. 1998. *Industrial Espionage and Technology Transfer: Britain and France in the Eighteenth Century*. Ashgate.

Harris, Ron. 1994. "The Bubble Act: its passage and its effects on business organization." *Journal of Economic History* 54: 610–627.

Harris, Ron. 2000. *Industrializing English Law: Entrepreneurship and Business Organization, 1720–1844*. Cambridge University Press.

Hatwagner, Gabriele. 2008. Die Lust an der Illusion—über den Reiz der „Scheinkunstsammlung" des Grafen Deym, der sich Müller nannte. Dissertation, Universität Wien.

Henderson, W. O., and Hans Caspar Escher. 1968. *Industrial Britain under the Regency: The Diaries of Escher, Bodmer, May, and de Gallois, 1814–18*. A. M. Kelley.

Hill, N. K. 1950. The History of the Imperial Continental Gas Association 1824–1900: A Study in British Economic Enterprise on the Continent of Europe in the 19th Century. Ph.D. dissertation, University of London.

Hills, R. L., and A. J. Pacey. 1972. "The measurement of power in early steam-driven textile mills." *Technology and Culture* 13: 25–43.

Hills, Richard Leslie. 2002–2006. *James Watt*. Landmark.

Hochadel, Oliver. 2003. *Öffentliche Wissenschaft: Elektrizität in der deutschen Aufklärung*. Wallstein.

Holmes, Frederic Lawrence. 2000. "Phlogiston in the air." In *Nuova Voltiana: Studies on Volta and His Times*, volume 2, ed. F. Bevilacqua and L. Fregonese. Hoepli.

Home, Roderick W. 2000. "Volta's English connections." In *Nuova Voltiana: Studies on Volta and His Times*, volume 1, ed. F. Bevilacqua and L. Fregonese. Hoepli.

Hudson, Pat. 1986. *The Genesis of Industrial Capital: A Study of the West Riding Wool Textile Industry, c. 1750–1850*. Cambridge University Press.

Hughes, Thomas Parke. 1983. *Networks of Power: Electrification in Western Society, 1880–1930*. Johns Hopkins University Press.

Hughes, Thomas Parke. 1987. "The evolution of large technological systems." In *The Social Construction of Technological Systems: New Directions in the Sociology and History of Technology*, ed. W. Bijker, T. Hughes, and T. Pinch. MIT Press.

Hunter, Michael. 2008. "Boyle, Robert." In *Complete Dictionary of Scientific Biography*. Scribner.

Hutchinson, Kenneth. 1985. "The Royal Society and the foundation of the British gas industry." *Notes and Records of the Royal Society of London* 39: 245–270.

Ingpen, Arthur Robert. 1912. *The Middle Temple bench book: being a register of benchers of the Middle Temple from the earliest records to the present time, with historical introduction.* Chiswick.

Inkster, Ian. 1991. *Science and Technology in History: An Approach to Industrial Development.* Rutgers University Press.

Inkster, Ian. 2003. "Technology in history: Case studies and concepts, circa 1700–2000." In *The Dynamics of Technology: Creation and Diffusion of Skills and Knowledge,* ed. R. Narasimha, J. Srinivasan, and S. K. Biswas. Sage.

Israel, Paul. 1998. *Edison: A Life of Invention.* Wiley.

Jacob, Margaret C. 1988. *The Cultural Meaning of the Scientific Revolution,* first edition. Temple University Press.

Jacob, Margaret C. 1997. *Scientific Culture and the Making of the Industrial West.* Oxford University Press.

Jacob, Margaret C. 2007. "Mechanical Science on the factory floor: The early Industrial Revolution in Leeds." *History of Science* 45: 197–221.

Jacob, Margaret C., and David Reid. 2001. "Technical knowledge and the mental universe of Manchester's early cotton manufacturers." *Canadian Journal of History* 36: 283–304.

Jacob, Margaret C., and Larry Stewart. 2004. *Practical Matter: Newton's Science in the Service of Industry and Empire, 1687–1851.* Harvard University Press.

Jaspers, P. A. Th. M. 1983. *J. P. Minckelers, 1748–1824.* Stichting Historische Reeks.

Jaspers, P. A. Th. M., and J. Roegiers. 1983. "Le mémoire sur l'air inflammable de J. P. Minckelers (1748–1824)." *Lias* 10: 217–252.

Jay, Mike. 2009. *The Atmosphere of Heaven: The Unnatural Experiments of Dr. Beddoes and His Sons of Genius.* Yale University Press.

Jobert, Philippe. 1991. "Jean-Baptiste Mollerat. Un pionnier de la chimie française." *Histoire, économie et société* 10: 245–268.

Joerges, Bernward. 1988. "Large technical systems: Concepts and issues." In *The Development of Large Technical Systems,* ed. R. Mayntz and T. Hughes. Campus.

Jones, Peter M. 1999. "Living the Enlightenment and the French Revolution: James Watt, Matthew Boulton, and their sons." *Historical Journal* 42: 157–182.

Jones, Peter M. 2008a. "Industrial Enlightenment in practice: Visitors to the Soho Manufactory, 1765–1820." *Midland History* 33: 68–96.

Jones, Peter M. 2008b. *Industrial Enlightenment: Science, Technology and Culture in Birmingham and the West Midlands, 1760–1820.* Manchester University Press.

Kennelly, Arthur E. 1928. *Vestiges of Pre-Metric Weights and Measures Persisting in Metric-System Europe, 1926–1927.* Macmillan.

Kim, Mi Gyung. 2001. "The analytic ideal of chemical elements: Robert Boyle and the French didactic tradition of chemistry." *Science in Context* 14: 361–395.

Kim, Mi Gyung. 2006. "'Public' Science: Hydrogen balloons and Lavoisier's decomposition of water." *Annals of Science* 63: 291–318.

Kindleberger, Charles Poor. 1993. *A Financial History of Western Europe.* Oxford University Press.

Kirby, Maurice W. 1994. "Big business before 1900." In *Business Enterprise in Modern Britain: From the Eighteenth to the Twentieth Century,* ed. M. Kirby and M. Rose. Routledge.

Klein, Ursula. 1994. *Verbindung und Affinität: die Grundlegung der neuzeitlichen Chemie an der Wende vom 17. zum 18. Jahrhundert.* Birkhäuser.

Klein, Ursula. 2005. "Technoscience avant la lettre." *Perspectives on science* 13: 226–266.

Klein, Ursula, and Wolfgang Lefèvre. 2007. *Materials in Eighteenth-Century Science: A Historical Ontology.* MIT Press.

Klein, Ursula, and E. C. Spary, eds. 2010. *Materials and Expertise in Early Modern Europe: Between Market and Laboratory.* University of Chicago Press.

Knight, David M. 1992. *Humphry Davy: Science and Power.* Blackwell.

Knight, David M. 2004. "Thompson, Sir Benjamin, Count Rumford in the nobility of the Holy Roman empire (1753–1814)." In *Oxford Dictionary of National Biography,* ed. H. Matthew, B. Harrison, and L. Goldman. Oxford University Press.

Kooij, Pim. 2006. "'Where the action is': The introduction and acceptance of infrastructural innovations in Dutch cities 1850–1950." Presented at World Economic History Conference, Helsinki.

Körting, Johannes. 1963. *Geschichte der deutschen Gasindustrie mit Vorgeschichte und bestimmenden Einflüssen des Auslandes.* Vulkan.

Landes, David S. 2003. *The Unbound Prometheus: Technological Change and Industrial Development in Western Europe from 1750 to the Present,* second edition. Cambridge University Press.

Langins, Janis. 1983. "Hydrogen production for ballooning during the French Revolution: An early example of chemical process development." *Annals of Science* 40: 531–558.

Levere, Trevor Harvey. 2005a. "Lavoisier's gasometer and others: Research, control, and dissemination." In *Lavoisier in Perspective,* ed. M. Beretta. Deutsches Museum.

Levere, Trevor Harvey. 2005b. "The role of instruments in the dissemination of the Chemical Revolution." *Éndoxa: series filosóficas* 19: 227–242.

Levere, Trevor Harvey. 2007. "Dr. Thomas Beddoes (1760–1808): chemistry, medicine, and books in the French and Chemical Revolutions." In *New Narratives in Eighteenth-Century Chemistry: Contributions from the First Francis Bacon Workshop, 21–23 April 2005,* ed. L. Principe. Springer.

Leveson-Gower, Granville (1st Earl Granville), and Henrietta-Frances (Countess of Bessborough) Ponsonby. 1916. *Lord Granville Leveson Gower (first earl Granville): Private Correspondence, 1781 to 1821*, volume 2, ed. C. Granville. J. Murray.

Lindqvist, Svante. 1983. "Natural resources and technology: The debate about energy technology in eighteenth-century Sweden." *Scandinavian Journal of History* 8: 83–107.

Lindqvist, Svante. 1984. *Technology on Trial: The Introduction of Steam Power Technology into Sweden, 1715–1736*. Almqvist & Wiksell.

Lindqvist, Svante. 1990. "Labs in the woods: The quantification of technology during the Enlightenment." In *The Quantifying Spirit in the Eighteenth Century*, ed. T. Frangsmyr, J. Heilbron, and R. Rider. University of California Press.

Luxbacher, G. 1991. "Die Entwicklung der industriellen Holzdestillation in Österreich bis zur Methanolsynthese." *Blätter für Technikgeschichte* 53/54: 241–260.

Lynn, Michael. 2006. *Popular Science and Public Opinion in Eighteenth-Century France*. Manchester University Press.

Macfarlan, J. 1925. "George Dixon: Discoverer of gas light from coal." *Transactions of the Newcomen Society* 5: 53–55.

Mackay, Hugh, Chris Carne, Paul Beynon-Davies, and Doug Tudhope. 2000. "Reconfiguring the user: Using rapid application development." *Social Studies of Science* 30: 737–757.

MacLeod, Christine. 1988. *Inventing the Industrial Revolution: The English Patent System, 1660–1800*. Cambridge University Press.

MacLeod, Christine. 2004. "The European origins of British technological predominance." In *Exceptionalism and Industrialisation: Britain and its European Rivals, 1688–1815*, ed. L. Prados de la Escosura. Cambridge University Press.

MacLeod, Christine. 2007. *Heroes of Invention: Technology, Liberalism and British Identity, 1750–1914*. Cambridge University Press.

Marsden, Ben, and Crosbie Smith. 2005. *Engineering Empires: A Cultural History of Technology in Nineteenth-Century Britain*. Palgrave Macmillan.

Mason, Shena. 2009. *Matthew Boulton: Selling What All the World Desires*. Yale University Press.

Mather, A. S. 1992. "The forest transition." *Area* 24: 367–379.

Mather, A. S., J. Fairbairn, and C. L. Needle. 1999. "The course and drivers of the forest transition: The case of France." *Journal of Rural Studies* 15: 65–90.

Mathias, Peter. 1972. "Who unbound Prometheus? Science and technical change, 1600–1800." In *Science, Technology, and Economic Growth in the Eighteenth Century*, ed. A. Musson. Methuen.

Mathias, Peter. 1983. *The First Industrial Nation: The Economic History of Britain, 1700–1914*, second edition. Methuen.

Matthews, Derek. 1983. The London Gasworks: A Technical, Commercial and Labour History to 1914. Ph.D. dissertation, University of Hull.

Matthews, Derek. 1986. "Laissez-faire and the London gas industry in the nineteenth century: Another look." *Economic History Review* 39: 244–263.

Melosi, Martin V. 2000. *The Sanitary City: Urban Infrastructure in America from Colonial Times to the Present.* Johns Hopkins University Press.

Millburn, John R. 2000. *Adams of Fleet Street: Instrument Makers to King George III.* Ashgate.

Miller, David Philip. 1999. "The usefulness of natural philosophy: The Royal Society of London and the culture of practical utility in the later eighteenth century." *British Journal for the History Of Science* 32: 185–201.

Miller, David Philip. 2000. "'Puffing Jamie': The commercial and ideological importance of being a 'philosopher' in the case of the reputation of James Watt (1736–1819)." *History of Science* 38: 1–24.

Miller, David Philip. 2004. *Discovering Water: James Watt, Henry Cavendish, and the Nineteenth Century "Water Controversy."* Ashgate.

Miller, David Philip, and Trevor Harvey Levere. 2008. "Inhale it and see? The collaboration between Thomas Beddoes and James Watt in pneumatic medicine." *Ambix* 55: 5–28.

Millward, Robert. 1991. "Emergence of gas and water monopolies in nineteenth century Britain: Contested markets and public control." In *New Perspectives on the Late Victorian Economy: Essays in Quantitative Economic History 1860–1914*, ed. J. Foreman-Peck. Cambridge University Press.

Millward, Robert. 2005. *Private and Public Enterprise in Europe: Energy, Telecommunications and Transport, 1830–1990.* Cambridge University Press.

Mokyr, Joel. 1990. *The Lever of Riches: Technological Creativity and Economic Progress.* Oxford University Press.

Mokyr, Joel. 1994. "Technological change, 1700–1830." In *The Economic History of Britain Since 1700*, volume 1: *1700–1860*, ed. R. Floud and D. McCloskey. Cambridge University Press.

Mokyr, Joel. 1999. "Editor's introduction: The New Economic History and the Industrial Revolution." In *The British Industrial Revolution: An Economic Assessment*, ed. J. Mokyr. Westview Press.

Mokyr, Joel. 2002. *The Gifts of Athena: Historical Origins of the Knowledge Economy.* Princeton University Press.

Mokyr, Joel. 2005. "The intellectual origins of modern economic growth." *Journal of Economic History* 65: 285–351.

Mokyr, Joel. 2007. "Knowledge, enlightenment, and the Industrial Revolution: Reflections on the gifts of Athena." *History of Science* 45: 185–196.

Mokyr, Joel. 2009. *The Enlightened Economy: An Economic History of Britain, 1700–1850.* Yale University Press.

Mokyr, Joel. 2010. "Entrepreneurship and the Industrial Revolution in Britain." In *The Invention of Enterprise: Entrepreneurship from Ancient Mesopotamia to Modern Times*, ed. D. Landes, J. Mokyr and W. Baumol. Princeton University Press.

Morus, Iwan Rhys. 1996a. "Manufacturing nature: Science, technology and victorian consumer culture." *British Journal for the History of Science* 29: 403–434.

Morus, Iwan Rhys. 1996b. "The electric Ariel: Telegraphy and commercial culture in early Victorian England." *Victorian Studies* 39: 339–378.

Morus, Iwan Rhys. 1998. *Frankenstein's Children: Electricity, Exhibition, and Experiment in Early-Nineteenth-Century London.* Princeton University Press.

Mostyn, Dorothy A. 1980. *The Story of a House: Farnborough Hill*, second edition. Saint Michael's Abbey Press.

Musson, A. E., and Eric Robinson. 1969. *Science and Technology in the Industrial Revolution.* University of Toronto Press.

Neal, Larry. 1994. "The finance of business during the industrial revolution." In *The Economic history of Britain since 1700*, volume 1: *1700–1860*, ed. R. Floud and D. McCloskey. Cambridge University Press.

Nuvolari, Alessandro. 2004. "The emergence of science based technology. Comments on Floris Cohen's paper." *History of Technology* 25: 133–136.

Nye, David E. 1990. *Electrifying America: Social Meanings of a New Technology, 1880–1940.* MIT Press.

O'Brien, Patrick, Trevor Griffiths, and Philip Hunt. 1996. "Technological change during the first Industrial Revolution: The paradigm case of textiles, 1688–1851." In *Technological Change: Methods and Themes in the History of Technology*, ed. R. Fox. Harwood.

Oudshoorn, Nelly, and Trevor J. Pinch. 2003. *How Users Matter: The Co-Construction of Users and Technologies.* MIT Press.

Outland, Robert B. 2004. *Tapping the Pines: The Naval Stores Industry in the American South.* Louisiana State University Press.

Owen, David Edward, and Roy M. MacLeod. 1982. *The Government of Victorian London, 1855–1889: The Metropolitan Board of Works, the Vestries, and the City Corporation.* Belknap.

"Pall Mall, South Side, Past Buildings: Nos 93–95 Pall Mall: F.A. Winsor and the development of gas lighting." In *Survey of London*, volumes 29 and 30: *St James Westminster, Part 1*, ed. F. Sheppard. Athlone Press for London County Council, 1960.

Parascandola, John, and Aaron J. Ihde. 1969. "History of the pneumatic trough." *Isis* 60: 351–361.

Partington, J. R. 1961. *A History of Chemistry*. Macmillan and St. Martin's Press.

Pearson, Robin, and David Richardson. 2001. "Business networking in the Industrial Revolution." *Economic History Review* 54: 657–679.

Petersen, H. J. Styhr. 1990. "Diffusion of coal gas technology in Denmark, 1850–1920." *Technological Forecasting and Social Change* 38: 37–48.

Platt, Harold L. 1991. *The Electric City: Energy and the Growth of the Chicago area, 1880–1930*. University of Chicago Press.

Pollard, Sidney. 1964. "Fixed capital in the Industrial Revolution in Britain." *Journal of Economic History* 24: 299–314.

Pollard, Sidney. 1965. *The Genesis of Modern Management: A Study of the Industrial Revolution in Great Britain*. Harvard University Press.

Pomeranz, Kenneth. 2000. *The Great Divergence: China, Europe, and the Making of the Modern World Economy*. Princeton University Press.

Principe, Lawrence. 2007. "A revolution nobody noticed? Changes in early eighteenth-century chymistry." In *New Narratives in Eighteenth-Century Chemistry: Contributions from the First Francis Bacon Workshop*. Springer.

Quinn, Stephen. 2004. "Money, finance, and capital markets." In *The Cambridge Economic History of Modern Britain: Industrialisation 1700–1860*, ed. R. Floud and P. Johnson. Cambridge University Press.

Radkau, Joachim. 1983. "Holzverknappung und Krisenbewußtsein im 18. Jahrhundert." *Geschichte und Gesellschaft* 9: 513–543.

Radkau, Joachim. 1986. "Warum wurde die Gefährdung der Natur durch den Menschen nicht rechtzeitig erkannt? Naturkult und Angst vor Holznot um 1800." In *Ökologische Probleme im kulturellen Wandel*, ed. H. Lübbe and E. Ströker. Wilhelm Fink.

Rees, Gareth. 1971. "Copper sheathing: An example of technological diffusion in the English merchant fleet." *Journal of Transport History* 1: 85–94.

Robinson, Eric. 1954. "Training captains of industry: The education of Matthew Robinson Boulton [1770–1842] and the younger James Watt [1769–1848]." *Annals of Science* 10: 301–313.

Robinson, Eric H. 2004. "Watt, James (1769–1848)." In *Oxford Dictionary of National Biography*, ed. H. Matthew, B. Harrison, and L. Goldman. Oxford University Press.

Rosenberg, Nathan. 1982. *Inside the Black Box: Technology and Economics*. Cambridge University Press.

Rosenberg, Nathan, Ralph Landau, and David C. Mowery. 1992. *Technology and the Wealth of Nations*. Stanford University Press.

Salsbury, Stephen. 1988. "The emergence of an early large-scale technical system: The American railroad network." In *The Development of Large Technical Systems*, ed. R. Mayntz and T. Hughes. Campus.

Schivelbusch, Wolfgang. 1988. *Disenchanted Night: The Industrialization of Light in the Nineteenth Century*. University of California Press.

Schmitz, Christopher. 1995. *The Growth of Big Business in the United States and Western Europe, 1850–1939*. Cambridge University Press.

Schofield, Robert E. 1963. *The Lunar Society of Birmingham: A Social History of Provincial Science and Industry in Eighteenth-Century England*. Clarendon.

Shapin, Steven. 1996. *The Scientific Revolution*. University of Chicago Press.

Sieferle, Rolf Peter, and Michael Osman. 2001. *The Subterranean Forest: Energy Systems and the Industrial Revolution*. White Horse.

Siegfried, Robert. 1972. "Lavoisier's view of the gaseous state and its early application to pneumatic chemistry." *Isis* 63: 59–78.

Smith, John Graham. 1979. *The Origins and Early Development of the Heavy Chemical Industry in France*. Clarendon.

Stansfield, D. A., and R. G. Stansfield. 1986. "Dr Thomas Beddoes and James Watt: Preparatory work 1794–96 for the Bristol Pneumatic Institute." *Medical History* 30: 276.

Stewart, Larry. 1992. *The Rise of Public Science: Rhetoric, Technology, and Natural Philosophy in Newtonian Britain*. Cambridge University Press.

Stewart, Larry. 2007. "Experimental spaces and the knowledge economy." *History of Science* 45: 155–177.

Stewart, Larry. 2008. "Measure for measure: Projectors and the manufacture of enlightenment, 1770–1820." In *The Age of Projects*, ed. M. Novak. University of Toronto Press.

Storrs, F. C. 1966. "Lavoisier's technical reports: 1768–1794 Part I." *Annals of Science* 22: 251–275.

Strittmatter, Werner. 1986. "Wurde die Gefährdung der Natur durch den Menschen nicht rechtzeitig erkannt? Zur Diskussion der Thesen Radkaus." In *Ökologische Probleme im kulturellen Wandel*, ed. H. Lübbe and E. Ströker. Wilhelm Fink.

Szabadváry, Ferenc. 1966. *History of Analytical Chemistry*, first English edition. Pergamon.

Szostak, Rick. 1991. *The Role Of Transportation in the Industrial Revolution: A Comparison of England and France*. McGill–Queen's University Press.

Tann, Jennifer. 1978. "Marketing methods in the international steam engine market: The case of Boulton and Watt." *Journal of Economic History* 38: 363–391.

Tann, Jennifer. 2004a. "Boulton, Matthew (1728–1809)." In *Oxford Dictionary of National Biography*, ed. H. Matthew, B. Harrison, and L. Goldman. Oxford University Press.

Tann, Jennifer. 2004b. "Watt, James (1736–1819)." In *Oxford Dictionary of National Biography*, ed. H. Matthew, B. Harrison, and L. Goldman. Oxford University Press.

Tarr, Joel A., and Dupuy Gabriel. 1988. *Technology and the Rise of the Networked City in Europe and America.* Temple University Press.

Taylor, E. G. R. 1966. *The Mathematical Practitioners of Hanoverian England, 1714–1840.* Cambridge University Press for Institute of Navigation.

The Gas Light and Coke Company: An account of the progress of the Company from its incorporation by Royal Charter in the year 1812 to the present time, 1812–1912. G.L. & C.C., 1912.

"The manor and liberty of Norton Folgate." In *Survey of London*, volume 27: *Spitalfields and Mile End New Town*, ed. F. Sheppard. Athlone Press for London County Council, 1957.

Thorne, R. G. 1986. *The House of Commons, 1790–1820.* Secker & Warburg for History of Parliament Trust.

Tomory, Leslie. 2009. "Let it burn: Distinguishing inflammable airs 1766–1790." *Ambix* 56: 253–272.

Treue, W. 1967. "Die Entwicklung der chemischen Industrie von 1770 bis 1870." *Chemie Ingenieur Technik* 39: 1002–1008.

Trew, Alexander. 2010. "Infrastructure finance and industrial takeoff in the United Kingdom." *Journal of Money, Credit and Banking* 42: 985–1010.

Uglow, Jennifer S. 2002. *The Lunar Men: The Friends Who Made the Future, 1730–1810.* Faber and Faber.

Van Der Vleuten, Erik. 2006. "Understanding network societies: Two decades of large technical system studies." In *Networking Europe: Transnational Infrastructures and the Shaping of Europe, 1850–2000*, ed. E. van der Vleuten and A. Kaijser. Science History Publications.

Veillerette, François. 1987. *Philippe Lebon, ou, L'homme aux mains de lumière: la vie et l'oeuvre de l'illustre inventeur français du gaz d'éclairage et du chauffage au gaz.* N. Mourot.

Von Tunzelmann, G. N. 1993. "Technological and organizational change." In *The Industrial Revolution and British Society*, ed. P. O'Brien and R. Quinault. Cambridge University Press.

Vries, Peer. 2005. "Is California the measure of all things global? A rejoinder to Ricardo Duchesne, 'Peer Vries, the Great Divergence, and the California School: Who's in and who's out?'" *World History Connected* 2.

Ward, J. R. 1974. *The Finance of Canal Building in Eighteenth-Century England.* Oxford University Press.

Webb, Sidney, and Beatrice Webb. 1906–1908. *English Local Government from the Revolution to the Municipal Corporations Act: The Parish and the County*. Longmans, Green.

Wengenroth, Ulrich. 2003. "Science, technology, and industry." In *From Natural Philosophy to the Sciences: Writing the History of Nineteenth-Century Science*, ed. D. Cahan. University of Chicago Press.

Werrett, Simon. 2007. "From the grand whim to the gasworks: 'Philosophical fireworks' in Georgian England." In *The Mindful Hand: Inquiry and Invention from the Late Renaissance to Early Industrialisation*, ed. L. Roberts, S. Schaffer, and P. Dear. Bristol.

West, Ian. 2008. Light Satanic Mills—The Impact of Artificial Lighting in Early Factories. Dissertation, School of Archaeology and Ancient History, University of Leicester.

White, Jerry. 2007. *London in the Nineteenth Century: 'A Human Awful Wonder of God'*. Jonathan Cape.

Whited, Tamara L. 2005. *Northern Europe: An Environmental History*. ABC-Clio.

Williams, Justin. 1935. "English mercantilism and Carolina naval stores, 1705–1776." *Journal of Southern History* 1: 169–185.

Williot, Jean-Pierre. 1999. *Naissance d'un service public: le gaz à Paris*. Rive droite-Institut d'histoire de l'industrie.

Wilson, J. F. 1995. *British Business History, 1720–1994*. Manchester University Press.

Wilson, J. F., and A. W. J. Thomson. 2006. *The Making of Modern Management: British Management in Historical Perspective*. Oxford University Press.

Woronoff, Denis. 1984. *L'industrie sidérurgique en France pendant la Révolution et l'Empire*. École des hautes études en sciences sociales.

Woronoff, Denis. 1994. *Histoire de l'industrie en France: du XVIe siècle à nos jour*. Seuil.

Wright, Thomas. 1992. "Scale models, similitude and dimensions: Aspects of mid-nineteenth-century engineering science." *Annals of Science* 49: 233–254.

Wrigley, E. A. 1988. *Continuity, Chance and Change: The Character of the Industrial Revolution in England*. Cambridge University Press.

INDEX

Accum, Fredrick, 59, 143, 145, 153, 158, 159, 166, 177–179, 182, 184, 186, 187, 196, 198, 202

Adams, George, Jr., 33

Advertising. *See* Displays and demonstrations

Albion Flour Mill, 88

Ammonical liquor, 15, 17, 127, 159, 160, 197, 202, 203

Ballooning, 28–30, 34, 46, 50, 126

Banks, Joseph, 28, 116, 123, 124, 130, 136, 146, 148, 150, 192, 241

Barbier de Tinan, Jacques Théodore, 27

Barlow, James, 178, 179, 182

Beddoes, Thomas, 68, 73, 78, 80, 81, 88, 115, 203, 241

Beehive oven, 40, 41

Benjamin Gott & Company, 92, 112, 113

Berthollet, Claude-Louis, 2, 28, 31

Bienvenu, François, 32, 33

Birley & Co., 109, 113

Black, Joseph, 18, 22, 78

Bloxam, Matthew, 128, 140, 163, 164

Boulton, Matthew Robinson, 73

Boulton, Robinson, 75, 76, 85, 88, 113, 114, 154–158

Boulton & Watt, 7, 62, 63, 67–73, 110–116
  advantages, 5, 60, 61, 69, 70
  challenges, 96, 98, 108, 109
  competition, 30, 90, 91, 118
  customers, 87–92, 108, 109, 114, 115
  design and development work, 59–62, 75–83, 91–98, 203
  efficiency testing and results, 95–108
  expansion, 90–92, 108–112
  gas plants (*see* Gasometers; Gas plants)
  invention and experimentation, 71–75, 203
  management, 75, 76, 193 (*see also* Finance and accounting)
  marketing, 88–92 (*see also* Displays and demonstrations)
  Parliament, 154–163
  patents. *See* Patents
  prestige, 69–71, 124
  research and development, 50, 51, 83–95

Boyle, Robert, 17, 36

Brander, Georg Friedrich, 27

British Tar Company, 43

Brougham, Henry, 157, 158, 160–164

Browne, Isaac Hawkins, 158

Bubble Act of 1720, 131, 158

Burners. *See* Lamps

Byproducts, 38, 44, 46, 60, 98, 154, 161, 166, 208

Carbon dioxide, 15, 18, 22, 30, 35, 52, 202

Carbonic acid. *See* Carbon dioxide

Carbonic oxide. *See* Carbon monoxide

Carbon monoxide, 15, 78

Carburetted hydrogen. *See* Methane

Cardwell, Donald, 68

Carlton House lighting spectacle, 138–141, 191

Caron, François, 170

Cartwright, John, 33

Cavendish, Henry, 10, 17, 21

Champion, John, 38–42, 61, 70

Chandler, Alfred, 5, 171, 194

Chaptal, Jean-Antoine, 61

Chemical Catechism (Parkes), 137

City of London Gas Light Company, 187, 203

Clavering, George Nassau, 27

Clegg, Samuel, 65, 67, 71, 77, 86, 90, 113, 169, 180–187, 190, 196, 199, 203–216, 230–232, 236

Cochrane, Archibald, 38, 43, 45, 50, 56, 60, 63, 64, 75, 79, 127, 137, 160

Cockerell, Charles, 177

Coke, 2, 8, 17, 37, 40–44, 55, 79, 107, 122–124, 127, 130, 154, 158–160, 197. *See also* Beehive oven; Industrial distillation

Compagnie Anglaise, 237

Compagnie de l'Ouest, 237

Condensers, 15, 111, 152, 203, 204. *See also* Purification of gas

Congreve, William, 191, 202, 237

Continental gas lighting companies, 237

Counterweights. *See* Gasometers

Covent Garden Theater, 176, 224–227, 236

Creighton, Henry, 79, 89–92, 95–98, 101–109, 113, 149, 150

Davy, Humphry, 124, 130, 145, 160

Demonstrations. *See* Displays and demonstrations

Description of the Process of Manufacturing Coal Gas (Accum), 196

Diller, Charles, 9, 25, 30–36, 45, 54, 137, 156, 157, 241

Displays and demonstrations, 11, 14, 24, 67, 141, 144, 241, 242

  Boulton & Watt, 69, 77, 88–91, 116

  Champion, 42

  Diller, 25, 30–36, 54

GLCC, 177, 228, 241

  Lebon, 7, 47, 48, 117

  scientific, 30, 69

  Volta, 29

  Winsor, 121, 126, 129–134, 138–143, 152, 168, 178

  Winzler, 52–56, 58

Distillation. *See* Industrial distillation; Retorts

Dixon, George, II, 8, 38, 42–44, 50, 62, 63

Dual traditions leading to gas lighting, 7–11, 39, 59–62, 240, 241

Dumotiez brothers, 24, 32, 33

Electricity, 2–5, 13, 17, 27, 134, 170, 208, 234–236, 239

Edison, Thomas, 170, 234, 236

Ehrmann, Frédéric-Louis, 27

Enlightenment ideals, 3, 4, 10–15, 34, 63, 68, 69, 240, 241

Entrepreneurship, 8, 9, 34, 39, 50, 62, 67, 76, 121, 137, 167–170, 195, 241, 242

Ethylene, 9, 15

Eudiometer, 25, 29

Ewart, Peter, 84, 85, 162, 164

Experimental philosophy, 5, 10, 13, 28, 31, 239

Experimentation. *See also* Pneumatic chemistry

  Boulton & Watt, 8, 68, 74–84, 87, 93–96, 100–107, 115, 116, 119, 147–150, 159, 160

  Cochrane, 43

  Diller, 35

  GLCC, 177, 179, 184, 188, 190, 194–202, 208, 236

  Lanoix, 44

  Lebon, 47, 48

  scientific, 18–22, 25, 31, 53, 59, 78, 79, 115, 239, 240

  technological, 35, 42, 59

  Winsor, 118, 126, 130, 133, 136–141, 147, 152

  Winzler, 38, 53–55

Finance and accounting, 2, 123, 152, 173.
    See also Joint-stock companies
  Boulton & Watt, 76, 93, 103, 107, 118,
    160, 169, 193
  depreciation, 107
  GLCC, 154, 161, 164, 167, 168, 179, 180,
    187, 189, 193–195, 221, 236, 244
  infrastructure, 156, 171, 193, 240
  large-scale, 3, 118, 122, 123, 147, 164, 193,
    240
  stock market, 123, 133, 193
  Winsor, 122, 132–135, 140, 146, 147, 153
Fixed air. *See* Carbon dioxide
Fourcroy, Antoine-François, 31–35, 47, 127
French gas lighting companies, 237

Gas fitters, 172, 220, 230–236, 242
Gas leakage, 56, 97, 171, 172, 186, 195. *See
    also* Mains and pipes; Gasometers
Gas Light and Coke Company (GLCC), 70,
    125, 126, 166–168
  advertising. *See* Displays and demonstrations
  competition, 113, 114, 118, 119
  consumer relations, 220–230
  Court of Directors, 140, 145, 177–184,
    187–190, 194–198, 207, 215, 225, 231,
    233, 244
  customers (*see* Covent Garden theater;
    Street lighting; Users)
  early years, 118, 119, 176–186
  experimentation. *See* Experimentation
  gas plants, 170, 197–220, 230–234
  infrastructure development, 122, 195–197
  joint-stock financing, 122, 123
  location, 172–176
  management and policies, 177–195
  patent issues, 123, 124
  supply. *See* Gas supply; Mains and pipes
Gasometers
  Boulton & Watt, 70, 78–83, 86, 93–96, 109,
    113, 115, 212
  counterweights, 22, 24, 80–82, 96, 111,
    212, 213

  GLCC, 184, 186, 189–192, 203, 206, 208,
    213, 216, 219, 225
  Lavoisier, 18, 22–24, 35, 80, 83, 212, 241
  Philips & Lee, 82, 85, 86, 89, 95, 98, 101,
    111
  in plants, 15, 16
  smoothing gas flows and storage, 60, 70,
    80, 115, 208, 225
  Winsor, 159
  Winzler, 54
Gas plants, 8, 11, 18, 117, 221, 227
  Boulton & Watt, 68, 69, 76–78, 92–94, 98,
    103–106, 118, 170, 220, 243
  Clegg, 15–17, 180, 182, 189, 205
  GLCC, 178, 182–192, 197, 199, 200, 203,
    206, 207, 214, 218–220, 224–226, 233
  industrial-scale, 8, 65, 77–79, 105, 106, 119,
    168
  laboratory instruments, 21–24, 52, 68, 78,
    87, 115
  oil-gas plant, 227
  Philips & Lee, 79, 83–85, 88, 95, 99, 103,
    168
  urban, 118, 119, 122, 167, 169, 220, 221,
    243, 244
  Winsor, 143, 171
  Winzler, 52, 53
Gas supply, 159, 169, 189
  GLCC, 208–220
  large-scale, 35, 118, 135, 170, 171
  reliability, 34, 60, 70, 79, 135, 208, 212,
    215–220
  scheduling, 80, 95, 97, 188, 214, 221,
    229
Gasworks. *See* Gas plants
Gilbert (Giddy), Davies, 162, 166
Gilbert, Ludwig, 48–51
Grafton, John, 237
Grant, James Ludovic, 140, 146, 152, 153,
    156, 163, 177, 182, 187, 188, 198

Hales, Stephen, 18–22
Hall, James, 158

Hardie, Frank, 133, 134
Hargreaves, James, 140, 177–179, 182, 183,
  187, 198
Harris, Henry, 225–227
Harrison, William, 158, 161, 162
Heard, Edward, 128, 141, 205
Henry, William, 100, 101, 150, 188
Hutton, Eidington Smeaton, 85, 91, 92
Hyde incident, 144, 145
Hydraulic main, 15, 21, 201, 203
Hydrogen sulphide, 15, 18, 22, 35, 60, 113,
  202, 205. *See also* Purification of gas;
  Byproducts; Lime purifier; Lime
  water

Imperial Continental Gas Association,
  237
Industrial distillation, 5, 8–11, 17, 20, 22,
  34–46, 50–64, 71, 74, 118, 122, 159, 168,
  240. *See also* Retorts;
  Thermolamps
Industrial Enlightenment, 4, 5, 239. *See also*
  Industrial Revolution; Science
Industrial Revolution
  Britain v. Continental Europe, 39, 61, 62,
    67, 115, 240
  education and knowledge, 63, 64, 78
  entrepreneurship, 8, 9, 167, 241, 242
  financing, 2, 123, 240,
  infrastructure development, 3, 119, 120,
    242–244
  management structure, 2, 192–194
  new wave, 1–3, 123, 239, 240
  science, 2–5, 13–15, 35, 36, 63, 78, 239
  technical inventiveness, 3–5, 39, 40, 63,
    123, 124, 239
  technological and industrial patterns, 1–5,
    13, 39, 63, 78, 115, 239
  war, 61, 62
Inflammable air, 9, 17, 18, 25–36, 52, 126,
  127, 145
Inflammable air lighter, 9, 25–36
Ingenhousz, Jan, 27, 35

Jacob, Margaret, 4, 68
Joint-stock companies, 3, 119, 122, 123, 128,
  131, 133, 157, 158, 166, 167, 193, 194,
  236

Kennedy, James, 89, 92, 108–111
Kirby's Wonderful and Eccentric Museum, 130,
  131

Laboratories, 3, 27, 35, 87, 116
Laboratory apparatus, 9, 14, 22, 35, 52, 53,
  63, 111. *See also* Scientific instruments
Lamplighters, 139, 144, 173, 223
Lamps, 31, 42, 44, 77. *See also* Thermolamps
  Argand, 7, 94, 104–107
  cockspur, 94, 104–108
  gas, 7, 17, 47, 60, 75, 79, 81, 86–88, 94–98,
    101, 103, 109, 110, 118, 128, 139–144,
    153, 159, 169–172, 177–180, 184,
    188–191, 209, 213–225, 228–235, 243
  inflammable air, 25–33
  oil, 7, 122, 173, 175, 178, 180, 184, 189,
    190, 222, 223, 228
  street (*see* Street lighting)
  theatre, 224, 226
Lanoix, Jean-Baptiste, 8, 38, 44, 45, 50,
  61–63
Lavoisier, Antoine, 10, 18, 22–24, 28, 31, 35,
  80, 83, 127
Lawson, James, 157–160
Leadbetter, Richard, 196, 200, 202, 206, 233
Lebon, Philippe, 7–11, 30, 33, 34, 37–40,
  44–52, 55, 58–64, 69, 70, 76, 96, 117, 121,
  122, 126, 127, 167, 241, 242
Lee, George Augustus, 69, 79, 83–91,
  94–114, 147–150, 156–162, 168. *See also*
  Philips & Lee
Lighthouses, 42, 130
Lime purifier, 16, 17, 186, 203–208
Lime water, 15–18, 22, 35, 49, 60, 186, 205
Lindqvist, Svante, 68, 69
Livesey, Thomas, 186, 188, 195
Lunar Society, 73, 149

MacLeod, Christine, 4, 39, 69, 123, 149
Mains and pipes, 89, 101, 111, 141
  design, 208–212, 216–220, 225, 226
  GLCC, 172, 184–191, 195, 208, 209, 215,
    216, 221, 224–226, 233–236,
    244
  leaks and obstructions, 86, 98, 109, 169,
    202, 203, 209–217, 221, 231,
    233
  service pipes, 211, 216, 228, 229, 232, 235
  under streets, 120, 154, 170, 177, 184, 214,
    223
  urban network, 1, 117, 118, 122,
    159, 167–169, 180, 195, 203, 208, 209,
    212, 216–220, 235, 236, 243, 244
  Winsor, 139, 141
  within plants, 15–17, 68, 82–86, 89,
    93, 94, 97, 98, 109, 203
Malam, John, 196, 206, 207, 230
Manby, Aaron, 237
Marketing. See Displays and demonstrations
Marsland, Peter, 89, 108
McConnell & Kennedy, 89, 92, 108, 109
Mégnié, Pierre, 24
Mellish, William, 154, 162, 164
Methane, 9, 15, 35, 86, 216
Meusnier, Jean-Baptiste, 24
Miller, David, 69, 148
Mills. See Textile mills
Minckelers, Jan-Pieter, 9, 25, 28–30, 35, 36,
    39, 50, 240
Mokyr, Joel, 3, 4, 63, 68, 239
Mollerat, Jean-Baptiste, 59
Montgolfier brothers, 28
Murdoch, William, 7–9, 34, 38, 39, 45, 50,
    59, 62, 69, 70–79, 83–91, 96, 97, 101–108,
    113–118, 121–124, 137, 147–159, 167,
    168, 240, 241

Nairne, Edward, 27, 28
National Light and Heat Company
  competition, 30, 118, 146–151
  expositions, 132, 136–139, 164–166

incorporation efforts, 91, 121, 132–166
  investors, 135, 141, 142
  Parliament, 122, 124, 132–148, 151–166
Networks. See also Railways; Electricity;
  Water
  complexity and stabilization, 5, 6, 98, 99,
    169–172, 195–197, 203, 208
  construction and expansion, 65, 67, 70, 93,
    117–122, 169–172, 180, 195–197,
    211–221, 224, 230, 234–237, 240–244
  finance, 3, 119, 120, 166–168
  inspiration, 123, 127, 135, 159, 167,
    168
New Imperial Patent Company, 124, 128
Nicholson, William, 136–138
Northern, John, 84
Norton Folgate, 179, 184, 186, 189–191

O'Brien, Patrick, 63
Olefiant gas. See Ethylene
Origins of gas lighting, theories of, 13–15

Palmer, George, 208
Parkes, Samuel, 134, 137
Parliament, 179, 182–184, 188, 190, 236
Patents, 42, 43, 59, 160, 203, 205, 207, 236,
    237
  Boulton & Watt, 39, 73–76, 146, 148, 149,
    243
  Clegg, 200, 205, 214
  Lebon, 7, 44–50,
  Winsor, 122–137, 141, 142, 151–153, 156,
    168, 195
Paxton, William, 140, 177
Peckston, Thomas, 196, 198, 234, 235
Pedder, John, 146, 153, 154, 164, 177
Peel, Robert, 158, 159
Perks, John, 203
Philips, Reuben, 207, 208
Philips & Lee, 69, 71, 79, 82–111, 114–116,
    147, 149, 150, 157, 160. See also
  Gasometers; Gas Plants; Murdoch, William;
  Retorts

Philosophical fireworks, 9, 25, 30–33, 45, 144, 241. *See also* Diller, Charles

Pipes. *See* Mains and pipes

Pneumatic chemistry, 4–11, 14–25, 28–31, 34–40, 49–53, 62, 63, 78, 115, 240, 242

Pneumatic Institution, 88

Pneumatic medicine, 68, 78, 80–83

Pneumatic trough, 18–24, 27, 35

Pollock, David, 181, 187, 188, 195, 225, 234

Pollution, 170. *See also* Lime water

Practical Treatise on Gas-light (Accum), 196

Pressure regulators, 96, 208, 211–214, 217, 219, 235

Priestley, Joseph, 17, 18, 22, 73

Prince of Wales, 138, 139

Purification of gas. *See also* Condenser; Hydraulic main; Pneumatic trough; Lime purifier; Lime water; Woulfe apparatus
  Boulton & Watt, 70, 86, 93, 115
  Clegg, 113, 180
  difficulty, 35, 60–62, 115, 186, 195, 209, 241
  George Augustus Lee, 96, 98
  GLCC, 169, 170, 195, 197, 202–208

Rackhouse, Andrew, 196, 200, 202

Railways, 2, 3, 5, 119, 123, 170, 171, 180

Reichenbach, Karl Ludwig von, 55–61

Retorts, 205–208, 236
  Boulton & Watt, 60, 74, 75, 97, 101, 103, 105, 107, 115
  Clegg, 190
  cost, 107, 115, 169, 196–203
  design, 70, 74, 77–80, 86, 87, 93–99, 109–113
  GLCC, 184, 185, 189, 196–203
  industrial distillation, 44
  in gas plants, 15–17
  Minckelers, 29
  Philips & Lee, 84–89, 101
  recharging, 60, 79, 80, 86, 185
  scientific instruments, 17–21, 35, 36, 52, 53
  settings, 70, 93–96, 109, 115, 190, 196–203
  Winzler, 52, 53

Rouelle, Hilaire Martin, 22

Royal Navy, 41, 42, 140, 197

Royal Society, 157, 168, 195
  Boulton & Watt, 69–71, 103, 124, 146–151
  Murdoch's paper, 59, 106–108, 116, 122, 124, 148–151, 154, 158, 241
  presentations to, 27, 28, 32, 33
  report on gas lighting, 82, 192
  Winsor, 130

Rumford Medal, 148, 150, 158, 241

Ryss-Poncelet, Maxime, 49, 59, 61

Saltpeter, 51, 55, 63

Scale
  economies, 115, 116, 235
  industrial, 34, 38, 44, 67–69, 105, 241
  laboratory, 8, 35, 53, 70, 105, 139
  transformation, 9, 30, 34, 39, 58, 61, 62, 65–69, 79, 81, 87, 88, 105–107, 115, 116, 143, 147, 170, 179, 193, 197–202, 208–220, 235, 241

Scheele, Carl Wilhelm, 2, 18, 22

Science, 137, 148. *See also* Industrial Enlightenment; Royal Society; Scientific instruments
  cultural value, 123, 124
  displays, 30, 31, 130, 134
  experimentation, 5, 10, 13, 18, 20, 22, 25, 28, 31, 53, 59, 78, 79, 115, 239, 240
  men of, 3, 4, 27–29, 33, 51, 69, 124, 148, 239
  public, 13–15, 34, 36, 130, 134
  and technology, 2–4, 13–15, 34–36, 40, 68, 209, 239

Scientific instruments. *See also* Gasometers; Pneumatic trough; Retorts; Woulfe apparatus; Eudiometer
  Boulton & Watt, 68, 83, 115
  chemistry, 14, 52
  Diller, 30–33
  makers, 25–30, 33, 34, 68, 115, 145
  pneumatic chemistry, 3, 8–10, 14, 18, 25–31, 34–36, 52, 63, 68, 83, 239, 240, 243

Volta, 25–28, 31, 34
Scientific Revolution, 4, 239
Skills
  artisanal, 2, 68
  Boulton & Watt, 61, 67, 68, 73, 91, 115,
    121
  British, 5, 39, 64, 68
  engineering, 4, 91
  mechanical and iron working, 5,
    39, 61, 64, 67, 68, 115
  scientific instruments, 68
Soho Foundry and Manufactory, 74–76, 79,
    96, 97, 116, 147, 157
  experiments at, 101–105, 150, 159
  manufacturing gas apparatus, 76, 83–94,
    101, 113
  Murdoch, 72–75, 149
  steam engines, 76
Southern, John, 79, 90, 91, 94, 95, 157, 159,
    162, 163
South Sea Company Bubble, 131, 135, 142,
    157, 158, 162
Steam engines, 2, 13, 123, 127, 149, 239,
    241
  Boulton & Watt 7, 61, 68, 69, 73–76,
    83–85, 88, 91, 92, 96–99, 107, 114–118,
    146, 148, 168
  Lebon, 45
Stewart, Larry, 4, 68, 78
Stockholm Tar Company, 42
Stopcocks, 211, 215, 217, 229, 230, 232, 235.
    See also Valves
Street lighting, 1, 59, 117, 169, 222, 234
  GLCC, 169, 178, 179, 188–192, 221,
    230
  lamps, 179, 188, 211, 223
  Lebon, 47
  local authorities, 122, 174–178, 190, 192,
    221, 223
  Winsor, 134–144, 156, 159, 171
Sulphuretted hydrogen. See Hydrogen
    sulphide
Syphons, 209–211, 216, 217

Tar
  Boulton & Watt, 97, 98, 109, 111, 159
  byproduct, 8, 15, 17, 44, 98, 159, 160, 197,
    202
  Lebon, 49, 62, 70
  for maritime use, 10, 37–43, 61–64, 74,
    126, 159, 240
  purification, 17, 21, 98, 109, 111, 202, 203,
    209, 210
  wells, 209–211, 220
  Winsor, 126, 127, 159, 160
  Winzler and zu Salm, 56–60, 79
Textile mills, 67–73, 79, 83–109, 112–119,
    149, 160, 168–171, 180, 197, 205, 220,
    221, 243
Theory and Practice of Gas-Lighting (Peckston),
    196
Thermolampe in Deutschland (Winzler), 51
Thermolamps, 7, 37, 40, 44–62, 79, 117,
    121, 124, 126, 242
Thillaye-Platel, Antoine, 49
Thysbaert, François-Jean, 28, 29
Traité élémentaire de chimie (Lavoisier), 22–24,
    53, 80, 83
Trinity Corporation, 42, 130

Urban infrastructure, 1, 3, 65, 67, 70, 93,
    99, 118, 119, 169, 171, 195–197, 234
Users of gas lighting, 6, 93, 97, 99,
    170–172, 211, 214, 219–230, 235,
    242, 243

Valves, 25, 74, 96, 98, 110, 117, 203,
    208–217, 220, 223–226, 235. See also
    Stopcocks
Van Bochaute, Carolus, 29
Van Marum, Martinus, 24
Van Voorst, John, 163, 164
Volta, Alessandro, 9, 10, 17, 18, 25–36, 241

Warren, Charles, 158, 162
Warren, John, 182, 198, 225